Strategies and Mechanisms for Enhancing Stress Tolerance in Wheat

Strategies and Mechanisms for Enhancing Stress Tolerance in Wheat

Guest Editor

Xiao Wang

MDPI

Basel • Beijing • Wuhan • Barcelona • Belgrade • Novi Sad • Cluj • Manchester

Guest Editor
Xiao Wang
College of Agriculture
Nanjing Agricultural
University
Nanjing
China

Editorial Office
MDPI AG
Grosspeteranlage 5
4052 Basel, Switzerland

This is a reprint of the Special Issue, published open access by the journal *Plants* (ISSN 2223-7747), freely accessible at: https://www.mdpi.com/journal/plants/special_issues/RKHG4083L4.

For citation purposes, cite each article independently as indicated on the article page online and as indicated below:

Lastname, A.A.; Lastname, B.B. Article Title. *Journal Name* **Year**, *Volume Number*, Page Range.

ISBN 978-3-7258-7999-1 (Hbk)
ISBN 978-3-7258-8000-3 (PDF)
https://doi.org/10.3390/books978-3-7258-8000-3

Contents

plants

MDPI

Article

Effect of Exogenous Calcium on Tolerance of Winter Wheat to Cold Stress during Stem Elongation Stage

Maguje Masa Malko [1,2,†], Xinyue Peng [1,†], Xing Gao [1], Jian Cai [1], Qin Zhou [1], Xiao Wang [1,*] and Dong Jiang [1]

1 National Technique Innovation Center for Regional Wheat Production, Key Laboratory of Crop Ecophysiology, Ministry of Agriculture, Nanjing Agricultural University, Nanjing 210095, China; maguje700@gmail.com (M.M.M.); 2019101026@njau.edu.cn (X.P.); 2022101024@njau.edu.cn (X.G.); caijian@njau.edu.cn (J.C.); qinzhou@njau.edu.cn (Q.Z.); jiangd@njau.edu.cn (D.J.)

2 Department of Plant Science, College of Agriculture, Wolaita Sodo University, Wolaita Sodo P.O. Box. 138, Ethiopia

* Correspondence: xiaowang@njau.edu.cn

† These authors contributed equally to this work.

Abstract: Low-temperature stress during stem elongation is a major factor limiting wheat yield. While calcium (Ca^{2+}) is known to enhance stress tolerance, it's potential as an alternative to cold priming and the underlying mechanisms in wheat remains unclear. The current study assessed the effects of exogenous Ca^{2+} and calcium inhibitors on wheat growth and related physiology mechanisms under low-temperature stress. The results revealed that exogenous Ca^{2+} increased photosynthesis and antioxidant capacity, lowered cell membrane damage, and ultimately enhanced tolerance to low-temperature stress during the stem elongation stage, compared with the non-exogenous Ca^{2+} treatment. Moreover, exogenous Ca^{2+} induced endogenous Ca^{2+} content and triggered the upregulation of Ca^{2+} signaling and cold-responsive related genes. This study highlights the significance of exogenous Ca^{2+} in enhancing stress tolerance and contributing to wheat yield improvement under low-temperature stress.

Keywords: calcium; cold-responsive genes; low-temperature stress; priming; wheat

1. Introduction

Wheat (*Triticum aestivum* L.) is one of the world's three major grain crops and serves as a staple food for more than 60% of the world's population. Therefore, ensuring a stable and high wheat yield is essential for global food security [1]. However, with global climate change, extreme cold stress events, in terms of frequency, intensity, and duration, are increasing, resulting in disastrous impacts on agricultural production [2]. Although wheat is a low-temperature tolerant crop with a certain level of cold stress tolerance, it is highly sensitive to low-temperature stress during its stem elongation stage, which results in a decline in grain yield by 5–14% [3]. Low-temperature stress causes disruption of various physiological pathways, photosynthetic apparatus damage, cellular membrane peroxidation, lipid composition changes, and cytoskeleton arrangement changes, eventually affecting plant growth, development, and survival [4]. Thus, improving the low-temperature tolerance of wheat has been recognized as a crucial strategy to ensure stable and sustainable wheat production.

Calcium serves as the second messenger in eukaryotic signal transduction and is involved in signaling pathways that respond to various abiotic stresses including low-temperature stress [5]. In plants, there are a significant number of proteins that can bind to Ca^{2+}, referred to as Ca^{2+} receptors. These Ca^{2+} receptors include calcium-dependent protein kinases (CDPKs), calmodulin (CaM), calmodulin-like proteins (CMLs), and calcineurin B-like proteins (CBLs) [6]. These proteins decode the calcium signal, transmit it downstream, and regulate the amplification of cascades, ultimately leading to stimulus-specific physiological responses [7]. Previous studies have indicated that, after the plant perceives cold signals, the expression of genes of transcription factors, i.e., CBFs, is rapidly induced. The CBF protein then activates the cold-responsive gene by binding to the specific cis-elements of the downstream cold-responsive gene COR, thereby enhancing the plant's cold resistance capabilities [8,9]. The mitogen-activated protein kinase (MAPK) cascade also plays a critical role in eukaryotic cells' response to various external signals and regulates complex intracellular changes [10].

A previous study demonstrated that exogenous Ca^{2+} can alleviate leaf damage and growth inhibition under cold stress in peanuts [11]. In this regard, spraying Ca^{2+} on the peanut leaves could protect the photosystem from photoinhibition by promoting circulating electron flow and reducing the proton gradient across the thylakoid membrane [12]. Similarly, exogenous $CaCl_2$ treatment has been observed to increase photosynthetic rate, chlorophyll content, and protective leaf enzyme activity, thereby inducing cold resistance in tomato seedlings [13]. Exogenous Ca^{2+} treatment was also found to reduce chilling damage by maintaining higher biomass accumulation of roots and shoots in maize [14] and enhancing rice seedling cold tolerance [15]. However, the role of calcium signaling in wheat cold stress tolerance regulation and the mechanisms underlying the impact of exogenous Ca^{2+} on its cold stress tolerance remains unclear.

Earlier reports revealed that cold priming can enhance wheat's cold resistance by activating its antioxidant system and improving its photosynthesis, thus increasing the biomass of wheat plants [16]. Furthermore, cold priming significantly increased the photochemical efficiency of the photosystem, antioxidant capacity, and expression levels of cold-responsive genes in wheat [17]. However, cold priming faces challenges in its field application, and thus, alternative methods are needed to alleviate low-temperature stress occurring at the stem elongation stage. Given this context, we investigated whether Ca^{2+} shows similar functions in enhancing wheat's tolerance to low temperatures during its stem elongation stage. In this regard, exogenous Ca^{2+} was used to determine its effect on wheat's cold stress tolerance, and various parameters, including antioxidant capacity, endogenous Ca^{2+} content, gene expression related to Ca^{2+} signaling, and cold-responsive genes, were analyzed. These analyses aimed to explore the physiological mechanisms underlying the effect of exogenous Ca^{2+} on wheat's cold stress tolerance.

2. Results

2.1. Effects of Exogenous Calcium on Wheat Grain Yield under Low-Temperature Stress during the Stem Elongation Stage

Compared with the control treatment (CC), low-temperature stress significantly decreased the grain yield in CL and CaL treatments, whereas an increased grain yield was observed in the CaC treatment. Notably, CaL exhibited a higher grain yield than CL, primarily due to the higher grain weight observed in CaL. Importantly, the higher grain yield observed in CaC was attributed to the higher grain weight as compared to the CC treatment (Table 1).

Table 1. Effect of calcium on wheat yield and its components under low-temperature stress during stem elongation.

Treatment	Spikes	Kernels per Spike	Thousand Grain Weight (g)	Yield (g)
CC	13.00 ab	51.67 b	50.35 b	33.03 b
CaC	13.25 ab	57.13 a	54.60 a	39.98 a
CL	11.25 b	41.53 d	48.35 b	20.76 c
CaL	14.25 a	46.00 c	52.71 a	33.00 b

Note: CC: no $CaCl_2$ sprayed and no low-temperature stress; CaC: $CaCl_2$ sprayed without low-temperature stress; CL: no $CaCl_2$ sprayed with low-temperature stress; and CaL: $CaCl_2$ sprayed with low-temperature stress. Significant differences at $p < 0.05$ were computed, and means were separated according to Duncan's multiple range test. The means with the same small case letters are statistically non-significant.

2.2. Exogenous Calcium Effect on Leaf Lethal Temperature and Photosynthetic Capacity under Low-Temperature Stress during the Stem Elongation Stage

Compared with CL sprayed with Ca^{2+}, a significantly decreased LT50 was found in CaL primarily due to the fact that Ca^{2+} can alleviate cell membrane damage under low-temperature stress (Figure 1).

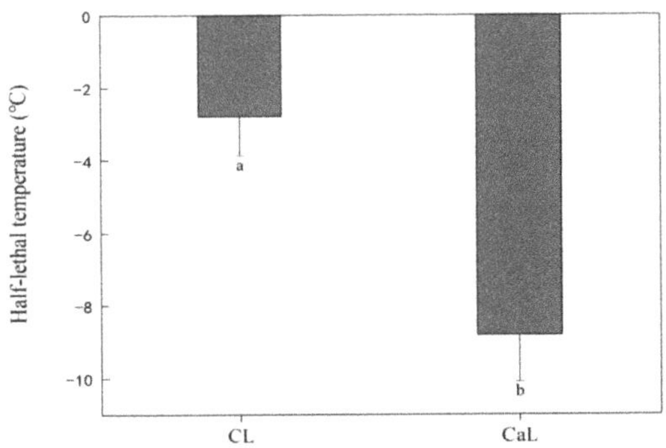

Figure 1. Effect of calcium on the half-lethal temperature of wheat under low-temperature stress at stem elongation stage. CL: no $CaCl_2$ sprayed with low-temperature stress; and CaL: $CaCl_2$ sprayed with low-temperature stress. Data are means ± SE across three biological replicates. Significant differences at $p < 0.05$ were computed, and means were separated according to Duncan's multiple range test. The means with the same small case letters are statistically non-significant.

Compared with CL, treatment with Ca^{2+} noticeably induced the net photosynthetic rate (Pn), quantum yield of photosystem II (Qy), and stomatal conductance (gs) in CaL. Except for photochemical quenching (qp), the other three parameters were increased with CaL and CaC compared to CL and CC, respectively. Due to the enhanced efficiency of the photochemical yield of PSII, the net photosynthesis increased in CaL and CaC (Figure 2).

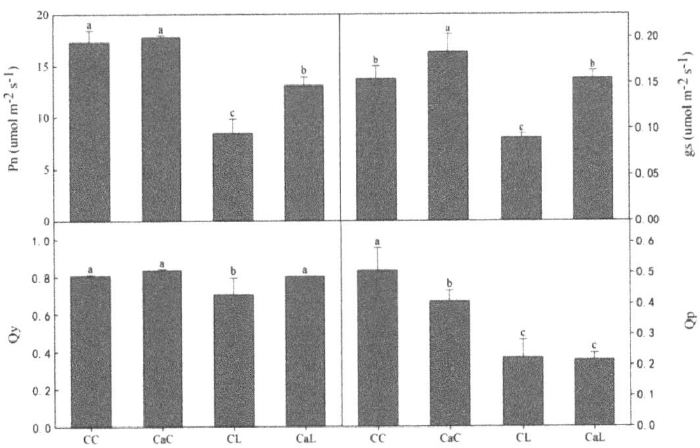

Figure 2. Effect of calcium on the gas exchange parameters of wheat under low-temperature stress at stem elongation stage. Pn: net photosynthetic rate; gs: stomatal conductance; Qy: quantum yield of photosystem II; qP: photochemical quenching; CC: no $CaCl_2$ sprayed and no low-temperature stress; CaC: $CaCl_2$ sprayed without low-temperature stress; CL: no $CaCl_2$ sprayed with low-temperature stress; and CaL: $CaCl_2$ sprayed with low-temperature stress. Data are means ± SE of three biological replicates. Significant differences at $p < 0.05$ were computed, and means were separated according to Duncan's multiple range test. The means with the same small case letters are statistically non-significant.

2.3. Effect of Exogenous Calcium on Osmotic Regulation under Low-Temperature Stress at the Stem Elongation Stage

Compared with CL, except for free amino acids, spraying with Ca^{2+} significantly increased the contents of total sugar, fructose, and sucrose in CaL. However, no significant difference was observed for all osmotic regulators among CaC and CC (Table 2).

Table 2. Effect of calcium on osmotic regulators under low-temperature stress during stem elongation.

Treatment	Total Sugar (mg g^{-1} DW)	Fructose (mg g^{-1} DW)	Sucrose (mg g^{-1} DW)	Free Amino Acids (mg g^{-1} DW)
CC	35.96 ^{c}	21.23 ^{c}	4.79 ^{c}	45.68 ^{a}
CaC	46.45 ^{c}	30.34 ^{c}	9.37 ^{c}	46.98 ^{a}
CL	108.53 ^{b}	83.58 ^{ab}	47.20 ^{b}	48.14 ^{a}
CaL	133.49 ^{a}	91.23 ^{a}	71.81 ^{a}	45.73 ^{a}

Note: CC refers to blank control, CaC refers to only $CaCl_2$ spraying, CL refers to direct low-temperature treatment, and CaL refers to low-temperature treatment after $CaCl_2$ spraying. Significant differences at $p < 0.05$ were computed, and means were separated according to Duncan's multiple range test. The means with the same small case letters are statistically non-significant.

2.4. Effect of Exogenous Calcium on Cell Membrane Peroxidation Rate and Antioxidant Enzyme Activity under Low-Temperature Stress at the Stem Elongation Stage

Compared with CC, the MDA content was highly increased in CL. However, the spraying with Ca^{2+} significantly reduced the MDA content in CaC and CaL, whereas there was no statistically significant difference observed between CC, CaC, and CaL (Figure 3).

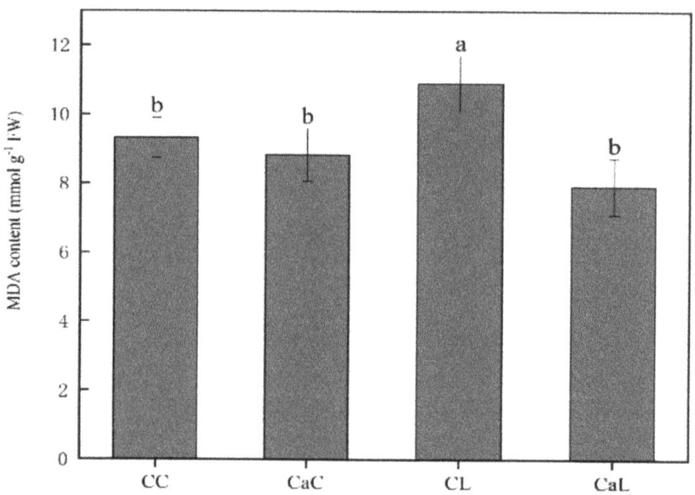

Figure 3. Ca^{2+} priming modifies the scavenging capacity of reactive oxygen species under the late-spring cold stress. CC: no $CaCl_2$ sprayed and no low-temperature stress; CaC: $CaCl_2$ sprayed without low-temperature stress; CL: no $CaCl_2$ sprayed with low-temperature stress; and CaL: $CaCl_2$ sprayed with low-temperature stress. Data are means ± SE of three biological replicates. Significant differences at $p < 0.05$ were computed, and means were separated according to Duncan's multiple range test. The means with the same small case letters are statistically non-significant.

Compared with CC, the treatment sprayed with Ca^{2+} enhanced the SOD, POD, APX, MDHAR, and GR activity in CaC and CaL, while no significant difference for CAT, DHAR, and GPX activity was observed in CaC and CaL (Table 3).

Table 3. Effect of calcium on the activity of antioxidant enzymes in wheat leaves under low-temperature stress at stem elongation.

Enzymes	**CC**	**CaC**	**CL**	**CaL**
SOD (U mg^{-1} protein)	4.01 ^{c}	10.34 ^{a}	6.56 ^{b}	11.37 ^{a}
POD (umol mg^{-1} min^{-1} protein)	2.21 ^{c}	3.06 ^{a}	2.59 ^{b}	3.26 ^{a}
CAT (umol mg^{-1} min^{-1} protein)	1.70 ^{ab}	1.79 ^{ab}	1.29 ^{b}	2.09 ^{a}
APX (umol mg^{-1} min^{-1} protein)	0.55 ^{b}	1.14 ^{a}	0.23 ^{b}	1.15 ^{a}
MDHAR (umol mg^{-1} min^{-1} protein)	1.16 ^{b}	1.64 ^{a}	0.79 ^{c}	1.69 ^{a}
DHAR (umol mg^{-1} min^{-1} protein)	7.10 ^{b}	10.73 ^{b}	8.02 ^{b}	17.91 ^{a}
GPX (umol mg^{-1} min^{-1} protein)	0.90 ^{bc}	1.46 ^{ab}	0.82 ^{c}	1.86 ^{a}
GR (umol mg^{-1} min^{-1} protein)	3.89 ^{b}	5.42 ^{a}	4.20 ^{b}	5.99 ^{a}

Note: CC: no $CaCl_2$ sprayed and no low-temperature stress; CaC: $CaCl_2$ sprayed without low-temperature stress; CL: no $CaCl_2$ sprayed with low-temperature stress; and CaL: $CaCl_2$ sprayed with low-temperature stress. Significant differences at $p < 0.05$ were computed, and means were separated according to Duncan's multiple range test. The means with the same small case letters are statistically non-significant.

2.5. *Effect of Exogenous Ca^{2+} and EGTA on Endogenous Calcium Content under Low-Temperature Stress at the Stem Elongation Stage*

Compared with CC, the Ca^{2+} and EGTA sprayed treatments remarkably induced endogenous calcium content in CaC, CaL, EC, and EL. Increased endogenous calcium content is associated with the exogenous Ca^{2+} treatment in CaC and CaL (Figure 4).

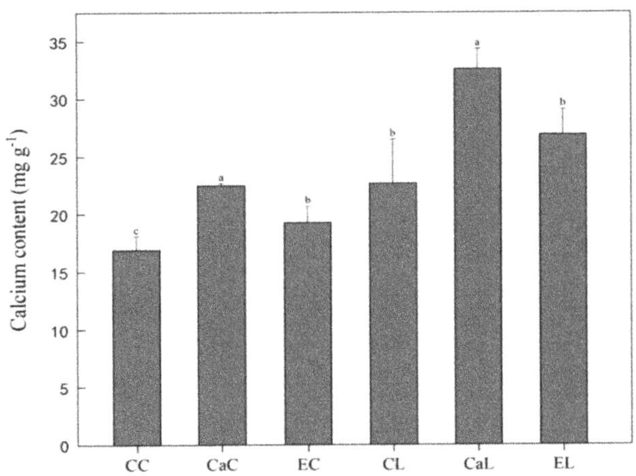

Figure 4. Effect of exogenous calcium and EGTA on endogenous Ca^{2+} content under low-temperature stress during stem elongation. CC: no $CaCl_2$ sprayed and no low-temperature stress; CaC: $CaCl_2$ sprayed without low-temperature stress; CL: no $CaCl_2$ sprayed with low-temperature stress; CaL: $CaCl_2$ sprayed with low-temperature stress; EC: EGTA sprayed with no low-temperature stress; and EL: EGTA sprayed with low-temperature stress. Data are means ± SE of three biological replicates. Significant differences at $p < 0.05$ were computed, and means were separated according to Duncan's multiple range test. The means with the same small case letters are statistically non-significant.

2.6. *Effect of Exogenous Calcium on Gene Expression Related to Calcium Signaling and Cold-Responsive Genes under Low-Temperature Stress during the Stem Elongation Stage*

Compared with CC, except CIPK31 gene, the relative expression levels of three calcium signaling genes such as CAMTA, CIPK2, and CBL6 were increased with Ca^{2+} application in CaC, whereas compared with CL, the calcium signaling genes were upregulated with Ca^{2+} spray in CaL (Figure 5).

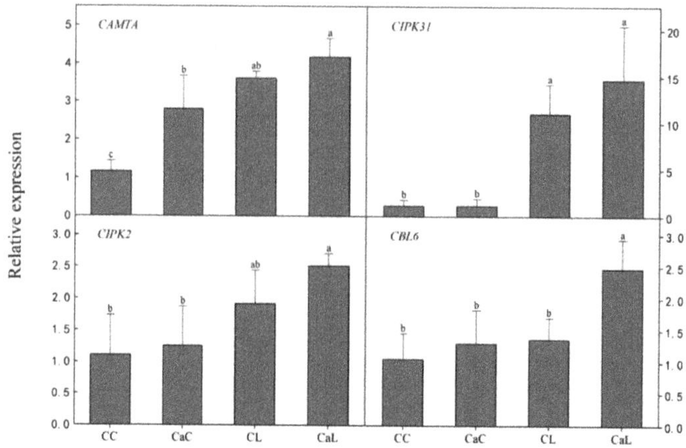

Figure 5. Calcium effect on gene expression of calcium signaling in wheat leaves under low-temperature stress at stem elongation. CC: no $CaCl_2$ sprayed and no low-temperature stress; CaC: $CaCl_2$ sprayed without low-temperature stress; CL: no $CaCl_2$ sprayed with low-temperature stress; and CaL: $CaCl_2$ sprayed with low-temperature stress. Data are means ± SE of three biological replicates. Significant differences at $p < 0.05$ were computed, and means were separated according to Duncan's multiple range test. The means with the same small case letters are statistically non-significant.

2.7. *Effect of Exogenous Calcium on Cold-Responsive Gene Expression under Low-Temperature Stress at Stem Elongation*

Compared with CC, the expression levels of all measured cold-responsive genes such as WCOR413, Wrab17, WCOR14, and WCOR410 were upregulated in CL and CaL. Moreover, Ca^{2+} spraying remarkably induced the expression level in CaL compared with CL, but Ca^{2+} spraying exhibited no significant difference between CaC and CC (Figure 6).

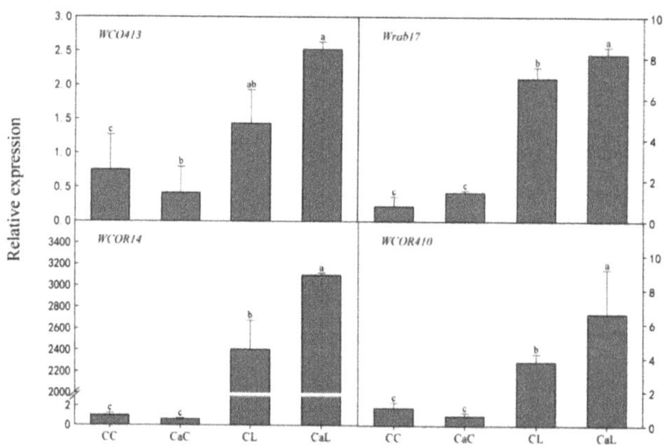

Figure 6. Effect of calcium on the expression of cold-responsive genes in wheat leaves under low-temperature stress at stem elongation. CC: no $CaCl_2$ sprayed and no low-temperature stress; CaC: $CaCl_2$ sprayed without low-temperature stress; CL: no $CaCl_2$ sprayed with low-temperature stress; and CaL: $CaCl_2$ sprayed with low-temperature stress. Data are means ± SE of three biological replicates. Significant differences at $p < 0.05$ were computed, and means were separated according to Duncan's multiple range test. The means with the same small case letters are statistically non-significant.

2.8. Correlation Analysis

The correlation estimation was conducted for 24 measured traits, and the correlation coefficient scores showed a significantly strong positive correlation for 17 traits and a negative correlation for 7 traits (Figure 7). It was indicated that the yield and photosynthetic traits were positively correlated with antioxidant enzymes like SOD, POD, APX, and CAT, while a negative correlation was observed with MDA and soluble sugars. However, except for free amino acids (FAAs), the other sugars showed a strong positive correlation with calcium signaling genes and cold-responsive genes. FAAs revealed a strong negative correlation with APX, SOD, and POD, while a positive correlation with MDA. The result showed that MDA negatively correlated with *CBL6* genes, Ca^{2+} content, APX, and POD and slightly correlated with *WCOR410* gene and *WCOR413* gene. Moreover, all measured calcium signaling genes were strongly correlated with cold-responsive genes and endogenous calcium content.

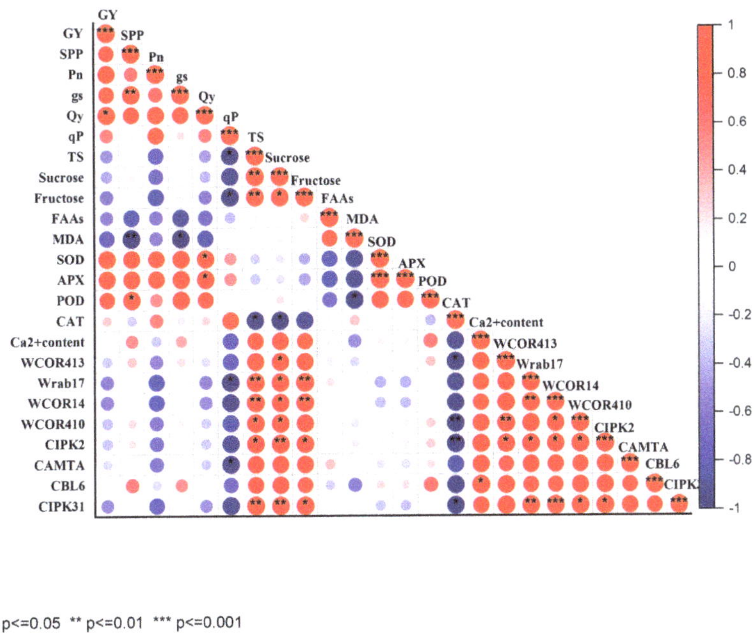

Figure 7. Correlation matrix of all measured indexes. GY: grain yield; SPP: spike per plant; TS: total sugar; and FAAs: free amino acids.

3. Discussion

3.1. Exogenous Calcium Improves Plant Tolerance to Low-Temperature Stress during Stem Elongation in Wheat

Plants are highly susceptible to temperature fluctuations, and membrane fluidity is a primary characteristic that is often targeted. Low temperature degrades the cell membrane and weakens the plant tolerance ability. The damage to the chloroplast membrane negatively affects the thylakoid structure, chlorophyll content, photosynthetic enzymes, and electron transport [18]. Photosynthetic capacity is closely associated with yield formation, including under low-temperature stress [19]. The LT_{50} is an indicator of the extent of membrane damage and plant tolerance to low temperatures [20]. Plants have diverse mechanisms for tolerating cold stress that work to improve their membrane stability and enable plants to avoid any low-temperature-induced damage [21]. The current study investigated the effects of Ca^{2+} application on plant tolerance to low-temperature stress during stem elongation in wheat. Our findings showed that plants sprayed with Ca^{2+} experienced reduced cell membrane damage than non-sprayed plants under low-temperature stress.

Photosynthesis is an exceptionally sensitive process that can be influenced by even slight changes in environmental conditions, particularly in response to extreme temperatures. Maintenance of the right balance between the light energy absorbed by photosystems and the overall light energy absorbed by the plant is essential [22]. Chlorophyll fluorescence is often used to reflect a plant's tolerance to abiotic stress [23]. Pigments play a critical role as antennae of the photosystem, trapping light energy to drive the charge separation in the reaction centers of PSII [24]. Previous studies have demonstrated that spraying Ca^{2+} can lead to enhanced maximum photochemical efficiency (Fv/Fm), enhanced PSII efficiency, and reduced relative electrolyte leakage that results in alleviating cold-induced membrane damage in rice [15]. Exogenous Ca^{2+} application was also shown to alleviate PSII photo-inhibition caused by low-temperature stress in tomatoes [13] and improve stomatal conductance, resulting in the decline of the photosynthetic rate in peanuts [25]. In line with previous studies, this study showed that wheat plants sprayed with Ca^{2+} maintained a relatively enhanced photosynthesis, likely a result of the higher stomatal conductance and reduced inhibition of photoelectron transport compared with non-sprayed plants under low-temperature stress. Our findings suggest that exogenous Ca^{2+} application may be particularly useful in improving plant tolerance to low-temperature stress, with a focus on maintaining efficient photosynthesis.

Inhibition of photosynthesis leads to an excess of electrons that can result in the formation of reactive oxygen species and membrane lipid peroxidation [26]. A direct product of membrane lipid peroxidation is malondialdehyde (MDA); thus, its concentration is an indicator of the extent of cell peroxidation [27]. In this study, low-temperature stress significantly induced the concentration of MDA when compared to the control. The lower concentration of MDA in CaL, when compared to CL, indicated lower cell membrane damage. Antioxidant systems are essential in scavenging ROS, and increasing the activity of antioxidant enzymes is crucial in the cellular defense strategy against oxidative stress [28]. Previous studies suggest that exogenous Ca^{2+} enhances ROS scavenging by increasing the activity of antioxidant enzymes and non-enzymatic antioxidants to maintain low ROS levels [29]. Consistent with this fact, our findings revealed that Ca^{2+} increased the content of all measured antioxidant enzymes under low-temperature stress when compared to non-sprayed plants. Taken together, the above results suggest that exogenous calcium application enhances plants' defense against oxidative stress caused by low-temperature stress by improving the cellular antioxidant system and reducing oxidative damage.

The accumulation of osmolytes is another critical element of plant adaptation to low-temperature stress [30]. Sugars have been confirmed to work as excellent osmotic agents, membrane stabilizers, and antioxidant defense molecules in response to low-temperature stress [31]. Total soluble sugar was positively correlated with low-temperature tolerance [32]. Previous reports have suggested that exogenous Ca^{2+} increases osmotic regulatory substances in maize under low-temperature stress. In this study, the application of Ca^{2+} maintained higher concentrations of sugar, fructose, and sucrose, which would have decreased under low-temperature stress, in comparison to non-sprayed plants. These findings suggest that exogenous Ca^{2+} application can enhance plants' ability to cope with low-temperature stress by improving osmolyte accumulation.

3.2. Exogenous Ca^{2+} Induced the Endogenous Ca^{2+} Content and Calcium Signaling Pathway

Calcium plays a crucial role as a universal messenger in various signal transduction pathways, including those responsive to abiotic stresses [33]. Previous reports suggested that Ca^{2+} pretreatment increased the endogenous Ca^{2+} concentration and upregulated *TaCaM* gene expression [34]. Furthermore, spraying Ca^{2+} on plants under cold stress significantly increased their expression level of all calcium-signaling genes [35]. It is known that the Ca^{2+} signals are triggered by low-temperature stress and the interacting protein kinases (CIPKs) receive the Ca^{2+} signal by interacting with the calcineurin B-like kinases (CBLs) that bind to Ca^{2+}, thereby activating the phosphorylation activity of CIPKs [36,37]. Previous studies have indicated that rice CIPK families are upregulated by low temperatures and play a crucial role in positively regulating cold tolerance [38]. In our study, the exogenous Ca^{2+} spray increased the endogenous Ca^{2+} content. The upregulation of Ca^{2+} signaling-related genes (*CAMTA, CBL6, CIPK31,* and *CIPK2*) showed a similar tendency with the change in endogenous Ca^{2+} content. The findings suggest that exogenous calcium application induced endogenous Ca^{2+} accumulation and the upregulation of genes associated with calcium signaling pathways. Previous studies have indicated that rice CIPK families are upregulated by low temperature and play a crucial role in positively regulating cold tolerance [39].

In response to low-temperature stress, several genes have been identified to play crucial roles in key mechanisms [40]. Among these genes, C-repeat/DREB binding factors (CBFs) are transcription factors that have been shown to play significant roles in low-temperature stress [41]. Numerous studies have reported the existence of the CBF/DREB pathway and its ability to increase the cold tolerance of rice plants [42]. In addition to CBFs, the *WRKY19* gene has been reported to regulate the antioxidant capacity under low temperatures in *Arabidopsis* [43]. The expression of interdependent plant metabolisms facilitates functioning to counteract the deleterious effects of low-temperature stress. A previous study revealed that, after perceiving the cold signal, the expression of genes of transcription factors, i.e., *CBFs*, can be rapidly induced, and the CBF proteins then activate the cold-responsive genes by binding to the specific cis-elements of the downstream cold-responsive genes CORs, thereby enhancing the plant's cold resistance [9,44].

COR410 is a dehydrin that is localized near the plasma membrane [45]. This protein is the terminal protein of the ICE1-CBF/DREB1-COR regulatory pathway in plant chilling injury [46]. The signal transductions and proteins that detect cold stress cause ICE genes to be activated, and as a result, ICE upregulates the expression of CBF genes. CBF proteins interact with CRT/DRE, the homeopathic component of the promoter of COR genes that activates the transduction [40]. During the cold signal transduction process, Ca^{2+} accumulation increases in the cytoplasm due to cold stress. As a result of the change in Ca^{2+} concentration, the expression of downstream genes is upregulated, indicating the positive regulation of calcium ions on the transcriptional expression of cold stress-related genes [47].

This study demonstrates that the exogenous Ca^{2+} can significantly increase the expression of cold-responsive genes (*WCOR413, WCOR410, WCOR14,* and *Wrab17*) during cold stress. This output suggests that Ca^{2+} plays a critical role in plant response to cold stress by inducing the expression of cold-responsive genes, which is consistent with previous studies [17]. Additionally, earlier studies have demonstrated that elevated expression of cold-responsive genes and antioxidant activity, cyanide-resistant respiration capacity, and molecular chaperone levels all contribute to the maintenance of cellular redox homeostasis and the safeguarding of the photosynthetic apparatus, thereby increasing wheat's ability to withstand low temperatures. Moreover, all the results including correlation analysis suggest that Ca^{2+} can induce the expression of cold-responsive genes, leading to enhanced cold resistance and improving the overall survivability of the plant under stress conditions.

4. Materials and Methods

4.1. Experimental Design

4.1.1. Experiment I

The greenhouse experiment was conducted at Nanjing Agricultural University using winter wheat, Yangmai 16, a predominant locally cultivated variety (obtained from Jiangsu Tomorrow Seed Technology Co., Ltd., Jiangsu, Nanjing, China), which was sown in plastic pots (22 cm in height and 25 cm in diameter). The plant population was reduced to seven plants per pot at the three-leaf stage. $CaCl_2$ (1.5 mM) was sprayed every afternoon (6:00 p.m.) for five days during the six-leaf stage. At the stem elongation stage (the eight-leaf stage), low-temperature stress was introduced by placing pots in a growth chamber with a temperature of 2 °C/0 °C (day/night), while the control chamber was maintained at a temperature of 16 °C/12 °C (day/night). A light intensity of 500 μmol m^{-2} s^{-1} was maintained for a photoperiod of 14 h. The low-temperature stress was applied for 3 days, after which all plants were shifted to the control chamber for a seven-day recovery. Finally, all plants were grown outside until maturity. The four treatments applied were as follows: no $CaCl_2$ sprayed with no low-temperature stress as a control (CC), $CaCl_2$ sprayed without low-temperature stress (CaC), no $CaCl_2$ sprayed with low-temperature stress (CL), and $CaCl_2$ sprayed with low-temperature stress (CaL).

4.1.2. Experiment II

The experiment employed the same winter wheat variety, Yangmai16, as in experiment I. Full and uniform wheat seeds were selected and disinfected with 2.5% NaClO for 10 min. The seeds were then placed evenly in wet plastic pots, on germination nets, and transferred to plastic pots (measuring 40 cm × 35 cm × 18 cm) containing Hoagland nutrient solution once they had sprouted. The seedlings were cultivated in an artificial climate chamber, with a temperature of 20 °C/16 °C (day/night, 12 h/12 h), a light intensity of approximately 500 μmol m^{-2} s^{-1}, and a relative humidity of 60%. At the four-leaf stage, the seedlings were divided into three groups: one group was sprayed with 1.5 mM $CaCl_2$, another with the inhibitor EGTA, and the last group was sprayed with water as a control (sprayed once every evening for five days starting at 17:00). At the six-leaf stage, half of the seedlings from the three groups were subjected to cold stress for three days in a climate chamber set at 2 °C/0 °C (day/night), while the other seedlings remained under the aforementioned growth conditions.

4.2. *Physiological Measurements*

4.2.1. Leaf Half-Lethal Temperature

The uppermost flag leaves were collected for biochemical and physiological examinations. The leaf half-lethal temperature (LT50), defined as the temperature at which 50% of ions in leaves leak out, was used to evaluate the cold tolerance of wheat leaves, with slight modifications made to the measurement method described by [48]. Fresh leaves were harvested after exposure to low-temperature cold stress and cut into 1.5 cm lengths. These leaf segments were then wrapped in a wet cotton cloth and incubated overnight at 4 °C. After that, the sample was exposed to various freezing temperatures (0 °C, −3 °C, −6 °C, −9 °C, −12 °C, and −15 °C) in a freezing bath (CDN-1007020F, Southeast Co., Ltd., Ningbo, China) before being thawed overnight at 4 °C. Thereafter, 15 mL of deionized water was added, and the leaves were extracted at room temperature for 12 h, following which, the electrolytic conductivity of the leachate was measured before and after boiling the leaves for 15 min with a conductivity meter at 20 °C (DDS-307A, LEX Instruments Co., Ltd., Shanghai, China).

4.2.2. Leaf Photosynthesis and Chlorophyll Fluorescence

Leaf photosynthetic rate (Pn) and stomatal conductance (gs) were measured using a Licor-6800 portable photosynthetic instrument (Li-COR Biosciences, Lincoln, NE, USA) between 9:00 and 11:00 a.m. The instrument was configured with a red and blue light source, with the CO_2 concentration maintained at 400 μmol mol^{-1} and photosynthetically active radiation set at 1000 μmol m^{-2} s^{-1}. Chlorophyll fluorescence characteristics were assessed using the Fluor Pen FP100 (Beijing Eketai Ecological Technology Co., LTD, Beijing, China) following the manufacturer's instructions [49]. The minimum fluorescence (Fo) and maximal fluorescence (Fm), which were dark-adapted, variable fluorescence (Fv) and light-adapted fluorescence, such as steady-state fluorescence (Fs), and maximal fluorescence (Fm′) were recorded. All the fluorescence parameters were calculated using the method reported in [50]. Accordingly, the maximum photochemical efficiency of PSII (Fv/Fm), Fv/Fm = (Fm − Fo)/Fm; the actual quantum yield of PSII (ΦPSII), (ΦPSII) = (Fm′ − Fo)/Fm; and the non-photochemical quenching (NPQ), NPQ = (Fm − Fm′)/Fm, were calculated.

4.2.3. Osmolyte Content

A total of 0.05 g of dry leaves were weighed and 14 mL of 80% ethanol was added to it three times in a 15 mL centrifuge tube, which was then shaken vigorously. The mixture was heated at 80 °C for 30 min in a water bath and centrifuged at 3000 rpm for 10 min. The supernatant obtained was combined, and the solution was diluted to a final volume of 15 mL. This extract was utilized for determining the total sugar content using the anthrone method [17] and the sucrose content using the resorcinol method according to [51].

4.2.4. Cell Membrane Damage and Antioxidant Enzyme Activities

The last fully expanded leaves were utilized for extracting cell membrane damage and antioxidant enzyme activities. Approximately 0.5 g of fresh leaves was homogenized in 5 mL of precooled buffer (50 mM HEPES buffer, pH 7.8) followed by centrifugation at 10,000 rpm for 20 min at 4 °C. The resulting supernatant was collected to determine antioxidant enzyme activity and MDA content, which was measured according to the method described by [52]. The reaction solution, containing trichloroacetic acid, thiobarbituric acid mixture, and 2 mL of the extract, was boiled in a water bath for 20 min. This solution was then centrifuged at 4000 rpm for 10 min, and the supernatant was measured at 450 nm, 532 nm, and 600 nm using a visible light spectrophotometer (UV-1780, Shimadzu (Suzhou) Instruments Manufacturing Co., Ltd., Suzhou, China). Enzyme activity, including that of guaiacol peroxidase (POD), catalase (CAT), superoxide dismutase (SOD), ascorbate peroxidase (APX), glutathione reductase (GR), dehydroascorbate reductase (DHAR), and monodehydroascorbate reductase (MDHAR), was determined according to the methods described by [21]. The total soluble sugar content was measured following the recommended methods outlined by [6].

4.2.5. Endogenous Calcium Content

The Ca^{2+} content in leaves was determined using a Calcium Colorimetric Assay Kit (Biyuntian, Nanjing). Fresh leaf samples (50 mg) were mixed with 500 μL of the cracking solution. Samples were fully cracked and then centrifuged at 10,000 rpm for 3–5 min at 4 °C, and the obtained supernatant was utilized for testing. Following the instructions provided with the kit, the OD value at 575 nm was measured using a Microplate Reader (BioTek Cytation 3, Nanjing, China) [53].

4.2.6. Gene Expression

In order to measure gene expressions, total RNA extraction, cDNA synthesis, and quantitative real-time PCR (qRT-PCR) were carried out as described by [53]. In Table 4, all primer sequences used in qRT-PCR are mentioned. Using the Actin gene as a reference gene, the relative expression levels of the various genes were determined using the $2^{-\Delta\Delta Ct}$ method.

Table 4. Primer used in the q-PCR analysis of genes expressed in wheat leaves.

Gene Name	Forward Primer Sequence	Reverse Primer Sequence
CAMTA	ATGGGAGTTGGGCGGTGATG	CATGTTCTCTTCGCCATCAC
CBL2	AGACGAGCAAGAAGGAGAGC	CTGAAGCATTTGGGTGAAAC
CBL4	TCAAGAAGAACCCGGCATCAC	CCGAATGCATCACAAAGCTCGG
CBL6	GACATACCAAATCGTCCCAAG	ATCACACCATCATCAACCACAG
CIPK2	TGCATTCCCCTAGTGATGTCTG	CCATACACAAACGCACTGTCC
CIPK29	ACGCGCAAGAAGGTCCACTT	ACACGAGCTGGCGGAAGTAA
CIPK31	CAGCCCACTGTGGAAGAGC	TTACAACAATCGGCCTTTCGC
MAPK2C	GATTGTAAGCTCAAAATATGTG	ACATAGTTCAGGTGCTCGGTA

Table 4. *Cont.*

Gene Name	Forward Primer Sequence	Reverse Primer Sequence
MAPKflrs	GCAAACTGTGACCTAAAA	ACAGAAGCTCTGGTGCC
ICE1	CAACAAGGTCGTAGGAGATG	CCAATCAGCATAAGAAAACG
ICE4	CAAGGGCAAGAAGAAGGG	GCGTCACCAAGGATTGAA
WCOR14	CGACCACCAGACCCAGACC	CGAGCGGCGAGGAAACAC
WCOR15	CTGGTTAGTCGTCCTCTGA	CCTTCTTCAACTCGTCGG
WCOR18	GTGGACGCACTGGGTGGTC	ACCTTGTTGCCGAGGCTGA
WCOR410	CCTCCTCGGCAACCTCCTC	TCTTGACCTCGGGCTCTTCC
WCOR413	GACAAGACGAACTGGAAGA	ACAACAACGAACGCAATC
Wrab17	GATGCCACCAAGGAGAAGTC	CTCACTTGTCTCCTCCCATC
Actin	ACAGTGTCTGGATCGGTGGC	GTGGACAATGCCGGGACCAG

4.3. Statistical Analysis

An analysis of variance (ANOVA) was conducted using the statistical software SPSS version 25.0 (SPSS Inc., Chicago, IL, USA). The means were separated using Duncan's multiple range testing at a significance level of $p < 0.05$ using SigmaPlot 12.5 (Systat Software Inc., San Jose, CA, USA). The correlations matrix was plotted according to Pearson's method by using Origin 2023 (OriginLab Inc., Northampton, MA, USA).

5. Conclusions

In conclusion, this study examined how exogenous Ca^{2+} affected the expression of genes, the physiological and biochemical responses, and the stress tolerance of wheat when exposed to low temperatures. The results show that Ca^{2+} induces the expression of Ca^{2+} signaling pathway genes and genes involved in cold stress responses of the plant. The upregulated expression of calcium-signaling genes (*CAMTA*, *CBL6*, *CIPK31*, and *CIPK2*) and cold-responsive genes (*WCOR413*, *WCOR410*, *WCOR14*, and *Wrab17*) maintained cellular redox homeostasis and a higher photosynthesis rate, and ultimately enhanced low-temperature stress tolerance in wheat. This study establishes the regulatory mechanisms of calcium for low-temperature stress tolerance in wheat, which provides a new avenue for future research on the adaptive mechanisms of plants to environmental stress.

Author Contributions: X.W. designed the experiment and wrote the manuscript. M.M.M. and X.P. performed the experiments and biochemistry analysis; X.G. analyzed the data; J.C., Q.Z. and D.J. were involved in the experiment setup; D.J. and X.W. were involved in the discussion and the provision of critical feedback on the article. All authors have read and agreed to the published version of the manuscript.

Funding: This study was supported by the Fundamental Research Funds for the Central Universities (XUEKEN2023013), the projects of the National Natural Science Foundation of China (32272213, 31771693, U1803235), the China Agriculture Research System (CARS-03), and Jiangsu Collaborative Innovation Center for Modern Crop Production (JCIC-MCP).

Data Availability Statement: Both the raw data and analyzed data sets are available from the corresponding author upon reasonable request.

Conflicts of Interest: The authors declare no conflict of interest.

References

1. Parry, M.A.; Reynolds, M.; Salvucci, M.E.; Raines, C.; Andralojc, P.J.; Zhu, X.-G.; Price, D.G.; Condon, A.G.; Furbank, R.T. Raising yield potential of wheat. II. Increasing photosynthetic capacity and efficiency. *J. Exp. Bot.* **2010**, *62*, 453–467. [CrossRef] [PubMed]
2. Augspurger, C.K. Reconstructing patterns of temperature, phenology, and frost damage over 124 years: Spring damage risk is increasing. *Ecology* **2013**, *94*, 41–50. [CrossRef] [PubMed]
3. Li, X.; Pu, H.; Liu, F.; Zhou, Q.; Cai, J.; Dai, T.; Cao, W.; Jiang, D. Winter wheat photosynthesis and grain yield responses to spring freeze. *Agron. J.* **2015**, *107*, 1002–1010. [CrossRef]
4. Thakur, P.; Kumar, S.; Malik, J.A.; Berger, J.D.; Nayyar, H. Cold stress effects on reproductive development in grain crops: An overview. *Environ. Exp. Bot.* **2010**, *67*, 429–443. [CrossRef]

5. Schulz, P.; Herde, M.; Romeis, T. Calcium-dependent protein kinases: Hubs in plant stress signaling and development. *Plant Physiol.* **2013**, *163*, 523–530. [CrossRef]
6. Xia, X.-J.; Wang, Y.-J.; Zhou, Y.-H.; Tao, Y.; Mao, W.-H.; Shi, K.; Asami, T.; Chen, Z.; Yu, J.-Q. Reactive oxygen species are involved in brassinosteroid-induced stress tolerance in cucumber. *Plant Physiol.* **2009**, *150*, 801–814. [CrossRef] [PubMed]
7. Zeng, H.; Xu, L.; Singh, A.; Wang, H.; Du, L.; Poovaiah, B.W. Involvement of calmodulin and calmodulin-like proteins in plant responses to abiotic stresses. *Front. Plant Sci.* **2015**, *6*, 600. [CrossRef] [PubMed]
8. Mori, K.; Renhu, N.; Naito, M.; Nakamura, A.; Shiba, H.; Yamamoto, T.; Suzaki, T.; Iida, H.; Miura, K. Ca^{2+}-permeable mechanosensitive channels MCA1 and MCA2 mediate cold-induced cytosolic Ca^{2+} increase and cold tolerance in Arabidopsis. *Sci. Rep.* **2018**, *8*, 550. [CrossRef] [PubMed]
9. Thomashow, M.F. Role of Cold-Responsive Genes in Plant Freezing Tolerance1. *Plant Physiol.* **1998**, *118*, 1–8. [CrossRef]
10. Kaur, N.; Gupta, A.K. Signal transduction pathways under abiotic stresses in plants. *Curr. Sci.* **2005**, *88*, 1771–1780.
11. Song, Q.; Liu, Y.; Pang, J.; Yong, J.W.H.; Chen, Y.; Bai, C.; Gille, C.; Shi, Q.; Wu, D.; Han, X.; et al. Supplementary calcium restores peanut (*Arachis hypogaea*) growth and photosynthetic capacity under low nocturnal temperatures. *Front. Plant Sci.* **2020**, *10*, 1637. [CrossRef]
12. Wu, D.; Liu, Y.; Pang, J.; Yong, J.W.H.; Chen, Y.; Bai, C.; Han, X.; Liu, X.; Sun, Z.; Zhang, S. Exogenous calcium alleviates nocturnal chilling-induced feedback inhibition of photosynthesis by improving sink demand in peanuts (*Arachis hypogaea*). *Front. Plant Sci.* **2020**, *11*, 607029. [CrossRef] [PubMed]
13. Zhang, G.; Liu, Y.; Ni, Y.; Meng, Z.; Lu, T.; Li, T. Exogenous calcium alleviates low night temperature stress on the photosynthetic apparatus of tomato leaves. *PLoS ONE* **2014**, *9*, e97322. [CrossRef] [PubMed]
14. Farooq, M.; Aziz, T.; Basra, S.M.A.; Wahid, A.; Khaliq, A.; Cheema, M.A. Exploring the role of calcium to improve chilling tolerance in hybrid maize. *J. Agron. Crop. Sci.* **2008**, *194*, 350–359. [CrossRef]
15. Jia, Y.; Zou, D.; Wang, J.; Sha, H.; Liu, H.; Inayat, M.A.; Sun, J.; Zheng, H.; Xia, N.; Zhao, H. Effects of γ-aminobutyric acid, glutamic acid, and calcium chloride on rice (*Oryza sativa* L.) under cold stress during the early vegetative stage. *J. Plant Growth Regul.* **2016**, *36*, 240–253. [CrossRef]
16. Li, X.; Cai, J.; Liu, F.; Dai, T.; Cao, W.; Jiang, D. Cold priming drives the sub-cellular antioxidant systems to protect photosynthetic electron transport against subsequent low-temperature stress in winter wheat. *Plant Physiol. Biochem.* **2014**, *82*, 34–43. [CrossRef]
17. Wang, W.; Wang, X.; Zhang, J.; Huang, M.; Cai, J.; Zhou, Q.; Dai, T.; Jiang, D. Salicylic acid and cold priming induce late-spring freezing tolerance by maintaining cellular redox homeostasis and protecting photosynthetic apparatus in wheat. *Plant Growth Regul.* **2019**, *90*, 109–121. [CrossRef]
18. Banerjee, A.; Roychoudhury, A. Regulation of photosynthesis under salinity and drought stress. *Environ. Photosynth. A Prospect* **2018**, 134–144.
19. Ji, H.; Xiao, L.; Xia, Y.; Song, H.; Liu, B.; Tang, L.; Cao, W.; Zhu, Y.; Liu, L. Effects of jointing and booting low-temperature stresses on grain yield and yield components in wheat. *Agric. For. Meteorol.* **2017**, *243*, 33–42. [CrossRef]
20. Su, L.; Dai, Z.; Li, S.; Xin, H. A novel system for evaluating drought–cold tolerance of grapevines using chlorophyll fluorescence. *BMC Plant Biol.* **2015**, *15*, 82. [CrossRef] [PubMed]
21. Sangwan, V.; Foulds, I.; Singh, J.; Dhindsa, R.S. Cold-activation of *Brassica napus* BN115 promoter is mediated by structural changes in membranes and cytoskeleton and requires Ca^{2+} influx. *Plant J.* **2001**, *27*, 1–12. [CrossRef]
22. Ensminger, I.; Busch, F.; Huner, N.P.A. Photostasis and cold acclimation: Sensing low temperature through photosynthesis. *Physiol. Plant.* **2006**, *126*, 28–44. [CrossRef]
23. Lin, K.H.; Hwang, W.C.; Lo, H.F. Chilling stress and chilling tolerance of sweet potato as sensed by chlorophyll fluorescence. *Photosynthetica* **2007**, *45*, 628–632. [CrossRef]
24. Brestic, M.; Zivcak, M.; Kunderlikova, K.; Sytar, O.; Shao, H.; Kalaji, H.M.; Allakhverdiev, S.I. Low PSI content limits the photoprotection of PSI and PSII in early growth stages of chlorophyll b-deficient wheat mutant lines. *Photosynth. Res.* **2015**, *125*, 151–166. [CrossRef] [PubMed]
25. Liu, Y.-F.; Han, X.-R.; Zhan, X.-M.; Yang, J.-F.; Wang, Y.-Z.; Song, Q.-B.; Chen, X. Regulation of calcium on peanut photosynthesis under low night temperature stress. *J. Integr. Agric.* **2013**, *12*, 2172–2178. [CrossRef]
26. Kołodziejczyk, I.; Dzitko, K.; Szewczyk, R.; Posmyk, M.M. Exogenous melatonin improves corn (*Zea mays* L.) embryo proteome in seeds subjected to chilling stress. *J. Plant Physiol.* **2016**, *193*, 47–56. [CrossRef]
27. Xu, S.; Hu, J.; Li, Y.; Ma, W.; Zheng, Y.; Zhu, S. Chilling tolerance in Nicotiana tabacum induced by seed priming with putrescine. *Plant Growth Regul.* **2010**, *63*, 279–290. [CrossRef]
28. Shi, Q.; Bao, Z.; Zhu, Z.; Ying, Q.; Qian, Q. Effects of different treatments of salicylic acid on heat tolerance, chlorophyll fluorescence, and antioxidant enzyme activity in seedlings of *Cucumis sativa* L. *Plant Growth Regul.* **2006**, *48*, 127–135. [CrossRef]
29. Shi, H.T.; Ye, T.T.; Zhong, B.; Liu, X.; Chan, Z.L. Comparative proteomic and metabolomic analyses reveal mechanisms of improved cold stress tolerance in bermudagrass (*Cynodon dactylon* (L.) Pers.) by exogenous calcium. *J. Integr. Plant Biol.* **2014**, *56*, 1064–1079. [CrossRef]
30. Bodelón, O.G.; Blanch, M.; Sanchez-Ballesta, M.T.; Escribano, M.I.; Merodio, C. The effects of high CO_2 levels on anthocyanin composition, antioxidant activity, and soluble sugar content of strawberries stored at low non-freezing temperatures. *Food Chem.* **2010**, *122*, 673–678. [CrossRef]

31. Bolouri-Moghaddam, M.R.; Le Roy, K.; Xiang, L.; Rolland, F.; Ende, W.V.D. Sugar signaling and antioxidant network connections in plant cells. *FEBS J.* **2010**, *277*, 2022–2037. [CrossRef]
32. Dörffling, K.; Schulenburg, S.; Lesselich, G.; Dörffling, H. Abscisic acid and proline levels in cold hardened winter wheat leaves to variety-specific differences in freezing resistance. *J. Agron. Crop. Sci.* **1990**, *165*, 230–239. [CrossRef]
33. Ludwig, A.A.; Saitoh, H.; Felix, G.; Freymark, G.; Miersch, O.; Wasternack, C.; Boller, T.; Jones, J.D.G.; Romeis, T. Ethylene-mediated cross-talk between calcium-dependent protein kinase and MAPK signaling controls stress responses in plants. *Proc. Natl. Acad. Sci. USA* **2005**, *102*, 10736–10741. [CrossRef]
34. Wang, J.; Song, J.; Wu, X.; Deng, Q.; Zhu, Z.; Ren, M.; Ye, M.; Zeng, R. Seed priming with calcium chloride enhances wheat resistance against wheat aphid *Schizaphis graminum* Rondani. *Pest Manag. Sci.* **2021**, *77*, 4709–4718. [CrossRef]
35. Min, K.; Liu, B.; Lee, S.-R.; Arora, R. Supplemental calcium improves freezing tolerance of spinach (*Spinacia oleracea* L.) by mitigating membrane and photosynthetic damage, and bolstering anti-oxidant and cell-wall status. *Sci. Hortic.* **2021**, *288*, 110212. [CrossRef]
36. Luan, S.; Kudla, J.; Rodriguez-Concepcion, M.; Yalovsky, S.; Gruissem, W. Calmodulins, and calcineurin B-like proteins: Calcium sensors for specific signal response coupling in plants. *Plant Cell.* **2002**, *14*, 14389–14400. [CrossRef]
37. Batistic, O.; Kudla, J. Integration and channeling of calcium signaling through the CBL calcium sensor/CIPK protein kinase network. *Planta* **2004**, *219*, 915–924. [CrossRef] [PubMed]
38. Xiang, Y.; Huang, Y.M.; Xiong, L.Z. Characterization of stress-responsive CIPK genes in rice for stress tolerance improvement. *Plant Physiol.* **2007**, *144*, 1416–1428. [CrossRef]
39. Ruelland, E.; Vaultier, M.-N.; Zachowski, A.; Hurry, V. Cold signaling and cold acclimation in plants. *Adv. Bot. Res.* **2009**, *49*, 35–150.
40. Castellani, J.W.; Tipton, M.J. Cold stress effects on exposure tolerance and exercise performance. *Compr. Physiol.* **2015**, *6*, 443–469.
41. Hwarari, D.; Guan, Y.; Ahmad, B.; Movahedi, A.; Min, T.; Hao, Z.; Lu, Y.; Chen, J.; Yang, L. ICE-CBF-COR Signaling Cascade and Its Regulation in Plants Responding to Cold Stress. *Int. J. Mol. Sci.* **2022**, *23*, 1549. [CrossRef] [PubMed]
42. Ito, Y.; Katsura, K.; Maruyama, K.; Taji, T.; Kobayashi, M.; Seki, M.; Shinozaki, K.; Yamaguchi-Shinozaki, K. Functional analysis of rice DREB1/CBF-type transcription factors involved in cold-responsive gene expression in transgenic rice. *Plant Cell Physiol.* **2006**, *47*, 141–153. [CrossRef]
43. Niu, C.-F.; Wei, W.; Zhou, Q.-Y.; Tian, A.-G.; Hao, Y.-J.; Zhang, W.-K.; Ma, B.; Lin, Q.; Zhang, Z.-B.; Zhang, J.-S.; et al. Wheat WRKY genes TaWRKY2 and TaWRKY19 regulate abiotic stress tolerance in transgenic Arabidopsis plants. *Plant Cell Environ.* **2012**, *35*, 1156–1170. [CrossRef] [PubMed]
44. Shi, Y.; Ding, Y.; Yang, S. Molecular Regulation of CBF Signaling in Cold Acclimation. *Trends Plant Sci.* **2018**, *23*, 623–637. [CrossRef] [PubMed]
45. Ganeshan, S.; Vitamvas, P.; Fowler, D.B.; Chibbar, R.N. Quantitative expression analysis of selected COR genes reveals their differential expression in leaf and crown tissues of wheat (*Triticum aestivum* L.) during an extended low-temperature acclimation regimen. *J. Exp. Bot.* **2008**, *59*, 2393–2402. [CrossRef]
46. Breton, G.; Danyluk, J.; Charron, J.-B.F.; Sarhan, F. Expression profiling and bioinformatic analyses of a novel stress-regulated multispanning transmembrane protein family from cereals and Arabidopsis. *Plant Physiol.* **2003**, *132*, 64–74. [CrossRef]
47. Henriksson, K.N.; Trewavas, A.J. The effect of short-term low-temperature treatments on gene expression in *Arabidopsis* correlates with changes in intracellular Ca^{2+} levels. *Plant Cell Environ.* **2003**, *26*, 485–496. [CrossRef]
48. Cai, Q.; Wang, S.; Cui, Z.; Sun, J.; Ishii, Y. Changes in freezing tolerance and its relationship with the contents of carbohydrates and proline in overwintering centipedegrass (*Eremochloa ophiuroides* (Munro) Hack.). *Plant Prod. Sci.* **2004**, *7*, 421–426. [CrossRef]
49. Cui, Y.; Tian, Z.; Zhang, X.; Muhammad, A.; Han, H.; Jiang, D.; Cao, W.; Dai, T. Effect of water deficit during vegetative growth periods on post-anthesis photosynthetic capacity and grain yield in winter wheat (*Triticum aestivum* L.). *Acta Physiol. Plant.* **2015**, *37*, 196. [CrossRef]
50. Maxwell, K.; Johnson, G.N. Chlorophyll fluorescence—A practical guide. *J. Exp. Bot.* **2000**, *51*, 659–668. [CrossRef]
51. Moya, J.L.; Ros, R.; Picazo, I. Influence of cadmium and nickel on growth, net photosynthesis and carbohydrate distribution in rice plants. *Photosynth. Res.* **1993**, *36*, 75–80. [CrossRef] [PubMed]
52. He, L.; Li, B.; Lu, X.; Yuan, L.; Yang, Y.; Yuan, Y.; Du, J.; Guo, S. The effect of exogenous calcium on mitochondria, respiratory metabolism enzymes and ion transport in cucumber roots under hypoxia. *Sci. Rep.* **2015**, *5*, 11391. [CrossRef] [PubMed]
53. Wu, Y.; Wang, X.; Peng, X.; Ge, J.; Cai, J.; Huang, M.; Zhou, Q.; Zhong, Y.; Jiang, D. Cold priming improves chilling resistance in wheat seedlings: Changing of photosystem II imprints during recovery from priming. *Environ. Exp. Bot.* **2023**, *207*, 105220. [CrossRef]

plants

MDPI

Article

Comparison of Different Machine Learning Algorithms for the Prediction of the Wheat Grain Filling Stage Using RGB Images

Yunlin Song †, Zhuangzhuang Sun †, Ruinan Zhang, Haijiang Min, Qing Li *, Jian Cai, Xiao Wang, Qin Zhou and Dong Jiang *

National Technique Innovation Center for Regional Wheat Production, Key Laboratory of Crop Ecophysiology, Ministry of Agriculture, Nanjing Agricultural University, Nanjing 210095, China; 2018201011@njau.edu.cn (Y.S.); 2020201018@stu.njau.edu.cn (Z.S.); 2020801253@stu.njau.edu.cn (R.Z.); 2023101025@stu.njau.edu.cn (H.M.); caijian@njau.edu.cn (J.C.); xiaowang@njau.edu.cn (X.W.); qinzhou@njau.edu.cn (Q.Z.)

* Correspondence: qingli@njau.edu.cn (Q.L.); jiangd@njau.edu.cn (D.J.)

† These authors contributed equally to this work.

Abstract: Grain filling is essential for wheat yield formation, but is very susceptible to environmental stresses, such as high temperatures, especially in the context of global climate change. Grain RGB images include rich color, shape, and texture information, which can explicitly reveal the dynamics of grain filling. However, it is still challenging to further quantitatively predict the days after anthesis (DAA) from grain RGB images to monitor grain development. Results: The WheatGrain dataset revealed dynamic changes in color, shape, and texture traits during grain development. To predict the DAA from RGB images of wheat grains, we tested the performance of traditional machine learning, deep learning, and few-shot learning on this dataset. The results showed that Random Forest (RF) had the best accuracy of the traditional machine learning algorithms, but it was far less accurate than all deep learning algorithms. The precision and recall of the deep learning classification model using Vision Transformer (ViT) were the highest, 99.03% and 99.00%, respectively. In addition, few-shot learning could realize fine-grained image recognition for wheat grains, and it had a higher accuracy and recall rate in the case of 5-shot, which were 96.86% and 96.67%, respectively. Materials and Methods: In this work, we proposed a complete wheat grain dataset, WheatGrain, which covers thousands of wheat grain images from 6 DAA to 39 DAA, which can characterize the complete dynamics of grain development. At the same time, we built different algorithms to predict the DAA, including traditional machine learning, deep learning, and few-shot learning, in this dataset, and evaluated the performance of all models. Conclusions: To obtain wheat grain filling dynamics promptly, this study proposed an RGB dataset for the whole growth period of grain development. In addition, detailed comparisons were conducted between traditional machine learning, deep learning, and few-shot learning, which provided the possibility of recognizing the DAA of the grain timely. These results revealed that the ViT could improve the performance of deep learning in predicting the DAA, while few-shot learning could reduce the need for a number of datasets. This work provides a new approach to monitoring wheat grain filling dynamics, and it is beneficial for disaster prevention and improvement of wheat production.

Keywords: wheat; grain filling; RGB image; machine learning; deep learning; few-shot learning

1. Introduction

Wheat is the principal supply of energy for mankind, which highlights the fact that wheat yield plays a crucial role in world food security [1]. The grain filling stage is a crucial stage for wheat grain formation and maturation after anthesis. Therefore, grain filling lays the foundation for wheat yield formation, and the characteristics of grain filling determine the final grain weight, a major yield component of wheat [2].

However, grain filling is highly susceptible to adverse factors such as high temperatures, and these are more severe in the context of global climate change. As a result, for every 1 °C increase in the maximum temperature during the wheat growing season, global wheat production decreases by 5.6% [3]. Some studies have also shown that grain filling characteristics include two important factors: the grain filling rate and grain filling duration [4], and their interaction determines the final grain weight of wheat [5]. In addition, other studies have shown that post-anthesis high-temperature stress could accelerate the grain filling rate and decrease the grain filling duration. Still, the increase in the grain filling rate is not enough to compensate for the loss caused by the shortening of the grain filling duration [6]. Therefore, the timely improvement of the grain filling process has guiding significance for the safety of wheat production in terms of disaster prevention and reduction.

The development of the grain shape has an important influence on the wheat grain filling process. The grain length, grain width, and grain thickness are crucial factors in determining the grain shape of wheat. Traditionally, these parameters were measured manually using vernier calipers [7]. However, this method was time-consuming, labor-intensive, prone to human error, and provided results that may not be representative due to the small sample size. RGB images of the grain can provide sufficient color, shape, and texture information [8], which has built a bridge between different inspections such as seed quality and purity. ImageJ 1.3.7 software can be used to recognize RGB images and extract more comprehensive grain morphology parameters. However, the grains should be arranged neatly for accurate identification, and the number of samples that can be processed simultaneously may be limited [9]. Some studies used computer vision technology to extract the texture and color traits of wheat, barley, oats, and rye, which classified each crop with a classification accuracy of 100% [10]. Similarly, four deep convolutional neural network models were proposed for the classification of lightly sprouted wheat kernels and sound wheat kernels, with over 95% accuracy on test sets [11]. By extracting 55 traits of wheat grains, Neethirajan et al. identified healthy and damaged wheat grains with an accuracy of 95% [12]. Although these studies have achieved progress in purity detection, viability detection, and the pest and disease detection of mature seeds, there is still a challenge in predicting the days after anthesis (DAA) of wheat grain due to the lack of an RGB image dataset for the complete grain filling stage.

A machine learning algorithm is an effective way to predict the DAA. Traditional machine learning algorithms, such as Decision Trees (DTs), Support Vector Machines (SVMs), and Random Forest (RF), have achieved satisfactory accuracy for wheat yield prediction [13], leaf area classification [14], and variety classification [15], although the accuracy of traditional machine learning algorithms depends on effective feature selection. However, manual feature selection is complicated, requiring a lot of optimization time and labor. In recent years, deep learning has become the mainstream in the field of computer vision, which can avoid feature engineering based on prior knowledge and achieve end-to-end classification or regression tasks. For example, a two-to-two deep learning model was designed to predict both wheat yield and quality from time-series data without any manual feature selection [16]. Furthermore, a CNN-LSTM framework has also provided evidence of the benefits of deep learning classification over traditional image analysis features for the classification of various plant genotypes [17]. Many studies have proved that deep learning algorithms can achieve satisfactory accuracy based on a large number of training samples, but some agriculture images are very time-consuming and laborious to obtain. To solve the data dependency problem in deep learning, few-shot learning was proposed, which has the ability to learn from a small number of training samples [18]. For example, Liang used few-shot learning to achieve the classification of cotton leaf spots, with the DenseNet algorithm having the best classification accuracy [19]. Similarly, a semi-supervised few-shot learning approach was proposed to recognize plant leaf diseases [20]. Although few-shot learning has many applications in plant disease classification, it is still unclear whether it can predict the wheat grain DAA or not, which requires fine-grained image recognition.

Wheat is one of the three major crops in the world [21], and the accurate prediction of the DAA from grain RGB images is of great significance for wheat development. Therefore, the highlights of this paper are: (1) A novel dataset that was established, with thousands of RGB images, which cover the period from 6 days to 39 days after anthesis; (2) we deciphered the dynamic changes in grain color, shape, and texture, during the grain filling stage, which revealed that RGB images can explicitly show the the dynamics of grain filling; and (3) we analyzed the performance of traditional machine learning, deep learning, and few-shot learning in predicting the DAA.

2. Results

2.1. Grain Filling Dynamics of Color, Shape, and Texture Traits

The color, shape, and texture traits changed during the grain filling process. For color traits, the values of R, G, and B were 243–252, 243–252, and 232–250, respectively, while the values of H, S, and V were 1–4, 2–18, and 243–253, respectively (Figure 1). In addition, R, G, B, and V showed a trend of first decreasing and then increasing with the DAA. At the same time, H and S showed a trend of first increasing and then decreasing.

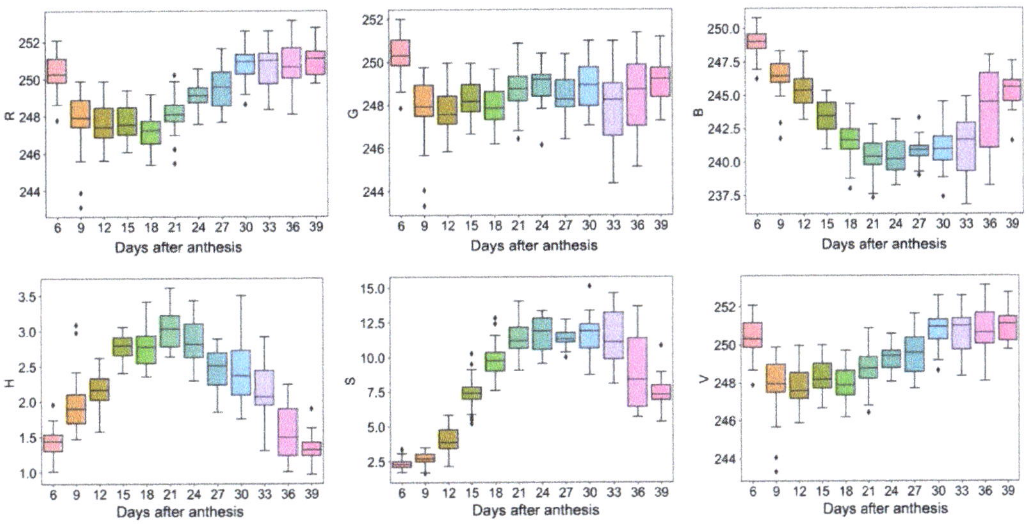

Figure 1. Color traits. Box plots of R, G, B, H, S, and V dynamic after anthesis, where different colored boxes represent different days after anthesis.

For shape traits, the area, perimeter, radius, and equivalent diameter ranged from 9–32 mm^2, 13–30 mm, 4–10 mm, and 3–6 mm, and the eccentric, compact, rectangle degree, and roundness ranged from 0.54–0.94, 16–40, 0.54–0.81, and 0.31–0.75, respectively (Figure 2). Furthermore, most of these traits showed a trend of first increasing and then decreasing, such as the area, perimeter, radius, and equivalent diameter. For example, the grain area increased sharply over days 6–24, slowly increased or was unchanged over days 24–33, and then declined after 33 days.

For texture traits, the homogeneity, dissimilarity, and correlation of the grain ranged from 0.25–0.73, 1–4, and 0.972–0.997, while the entropy, ASM, and energy ranged from 3.3–6.4, 0.001–0.062, and 0.03–0.025, respectively (Figure 3). Moreover, the homogeneity, correlation, ASM, and energy tended to decline with the increase in the DAA, while dissimilarity and entropy tended to rise with the increase in the DAA.

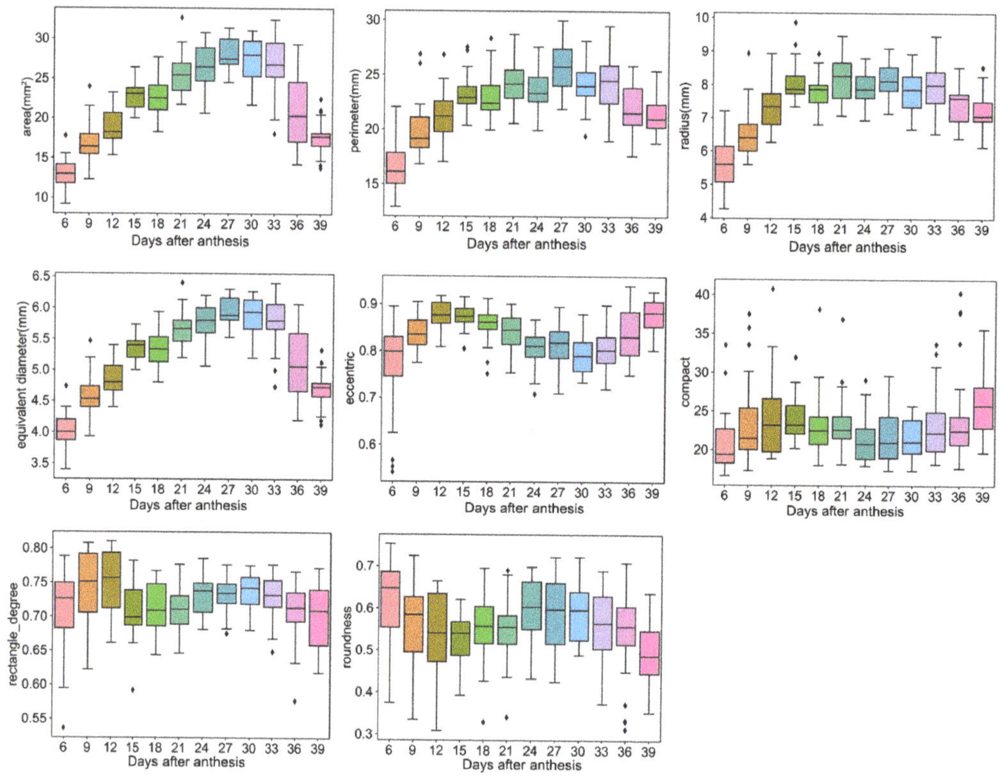

Figure 2. Shape traits. Box plots of area, perimeter, radius, equivalent diameter, eccentric, compact, rectangle degree, and roundness dynamic after anthesis, where different colored boxes represent different days after anthesis.

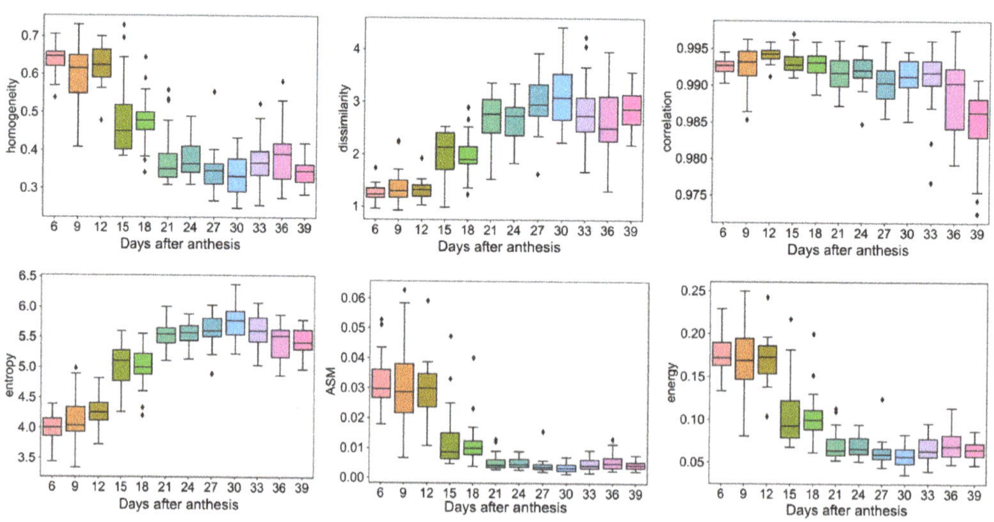

Figure 3. Texture traits. Box plots of homogeneity, dissimilarity, correlation, entropy, ASM, and energy dynamic after anthesis, where different colored boxes represent different days after anthesis.

2.2. *Machine Learning Performance for Predicting DAA*

To verify whether machine learning can predict the DAA, three machine learning algorithms, including Decision Trees (DTs), Support Vector Machines (SVMs), and Random Forest (RF), were applied to predict the DAA. The results showed that the performance of the three machine learning algorithms was ranked as RF > SVM > DT (Figure 4). The average precision and recall of the DT, SVM, and RF were 76.11% and 74.83%, 80.98% and 80.78%, and 88.71% and 87.93%, respectively. In addition, the precision and recall of the three algorithms were higher at 6–15 days and 39 days, while the precision and recall were lower at 21–33 days.

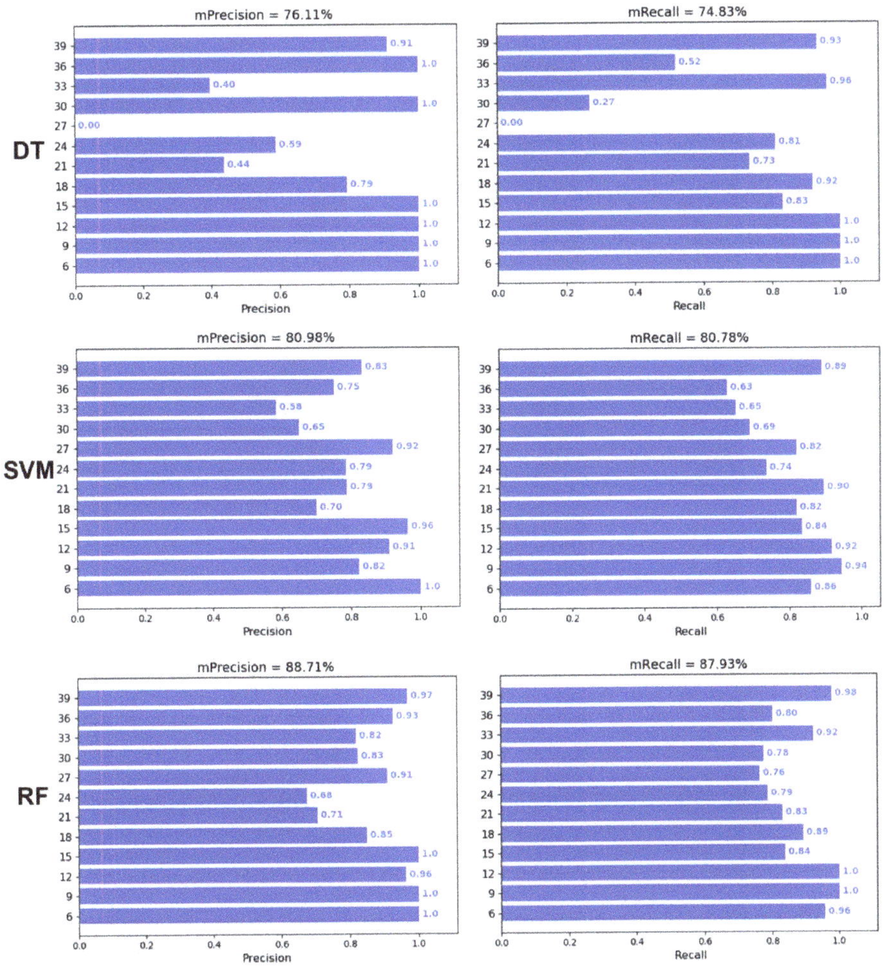

Figure 4. Performance of traditional machine learning.

2.3. *Deep Learning Performance for Predicting DAA*

To verify whether deep learning can predict DAA, an image classification model was constructed with different backbone networks, including HRnet18, Densenet121, Inceptionv3, VGG16, Resnet18, Resnet50, Resnet101, and Vision Transformer (ViT) (Figures 5 and 6). The results showed that the image classification model with a ViT backbone network had the best performance, and its precision and recall were 99.03% and 99.00%, respectively (Table 1). On the other hand, the image classification model with a VGG16 backbone network had the worst performance, and its prediction precision and recall were 94.24% and 94.00%, respectively. However, the precision and recall of VGG16 also far outperformed that of the optimal machine learning algorithm, which had a precision and recall of 83.94% and 83.88%. In addition, all deep learning models also demonstrated a good prediction performance during the middle of the grain filling stage (21–33 days), compared with the machine learning performances. These results illustrated the advantages of deep learning algorithms not only in end-to-end image classification but also in providing higher accuracy.

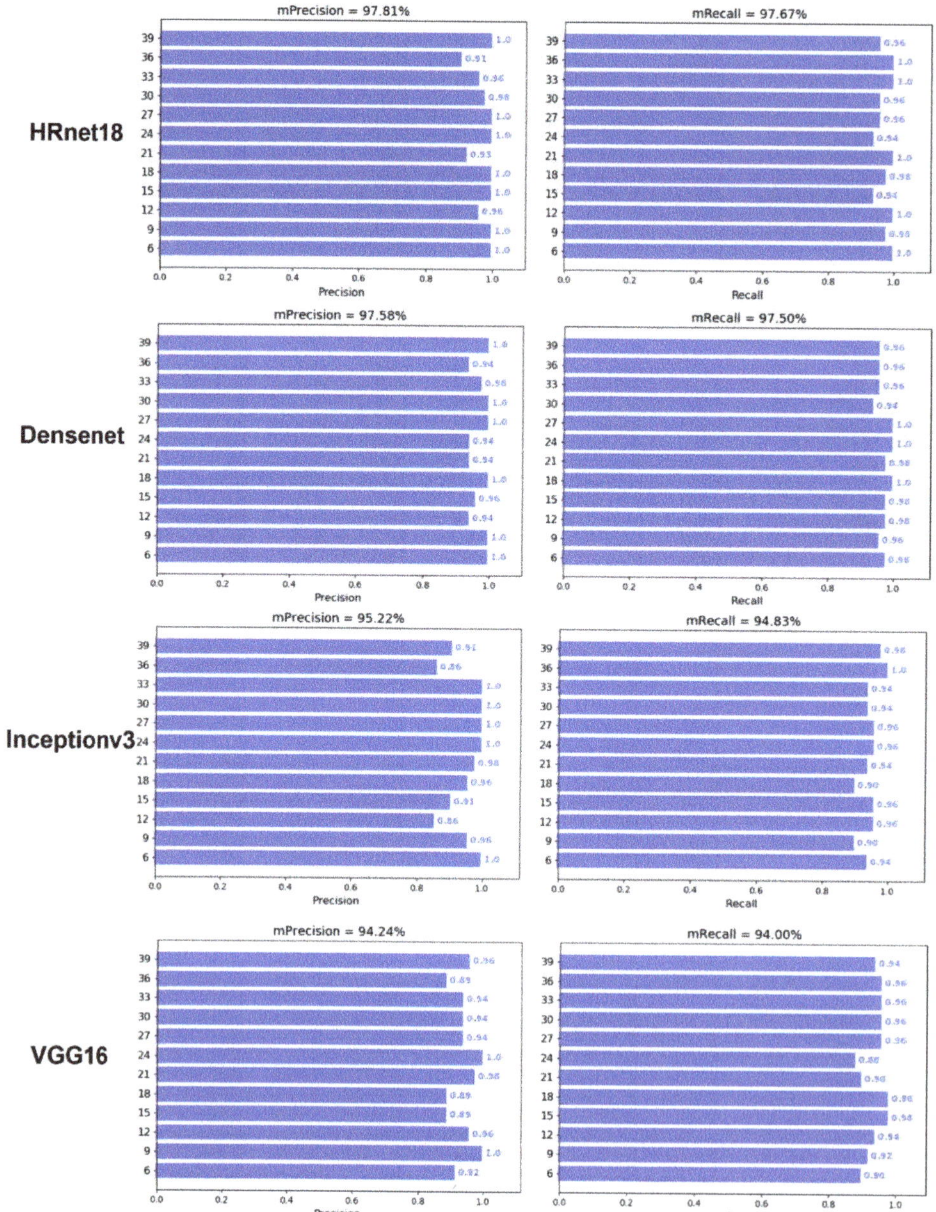

Figure 5. Performance of deep learning with HRnet18, Densenet121, Inception v3, and VGG16.

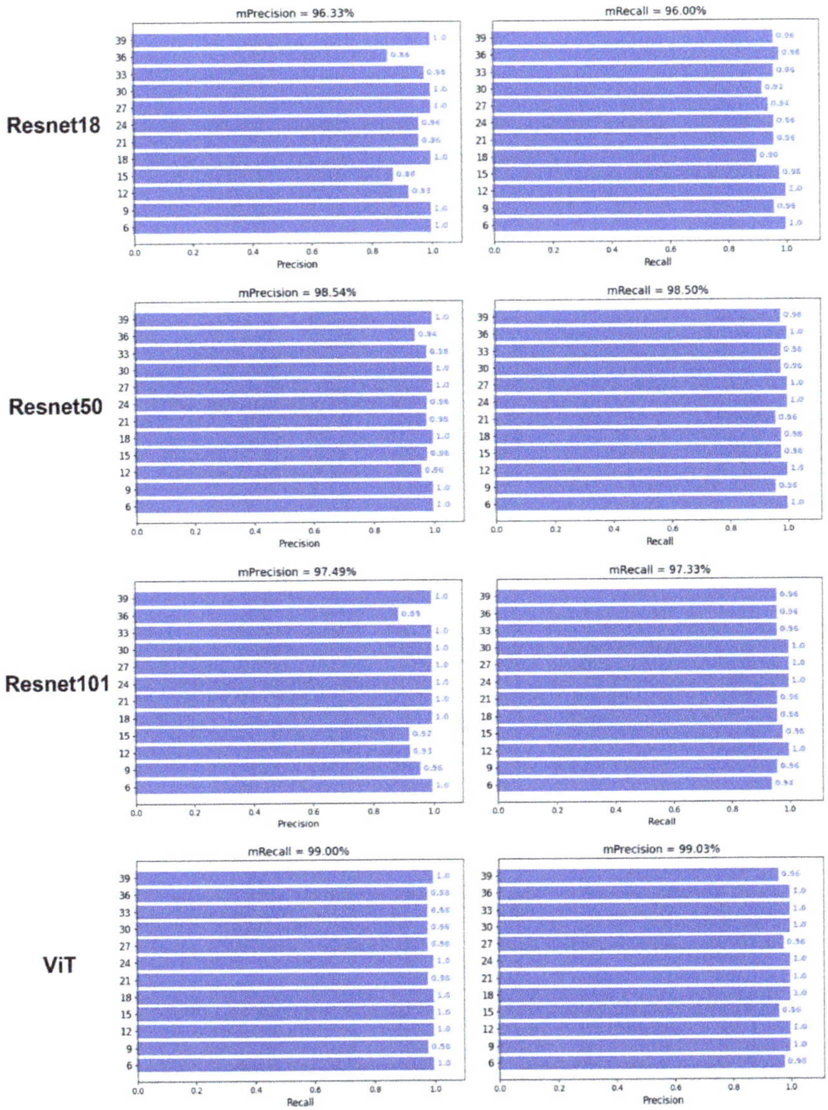

Figure 6. Performance of deep learning with Resnet18, Resnet50, Resnet101, and ViT.

Table 1. Performance of deep learning algorithms.

Network Architecture	mPrecision	mRecall	F1-Score	Model Size (MB)
HRnet18	0.9781	0.9767	0.9774	74.37
Densenet121	0.9758	0.9750	0.9754	27.15
Inception v3	0.9522	0.9483	0.9502	83.51
VGG 16	0.9424	0.9400	0.9412	512.36
Resnet18	0.9633	0.9600	0.96165	42.73
Resnet50	0.9850	0.9854	0.9852	90.07
Resnet101	0.9749	0.9733	0.9741	162.82
ViT	**0.9903**	**0.9900**	**0.9901**	146.01

Note: Bold numbers represent optimal prediction performance.

2.4. Few-Shot Learning Performance for Predicting DAA

In order to reduce our dependence on a number of datasets, we further explored the performance of few-shot learning at predicting DAA. The results showed that the mean precision and mean recall of 1-shot were 92.23% and 91.67%, respectively. In addition, the precision and recall increased as the number of query sets increased. After 5-shot, the gain in the model's performance started to become minuscule. In the end, the precision and recall of 5-shot were 96.86% and 96.67%, which was the optimal shot number (Figure 7).

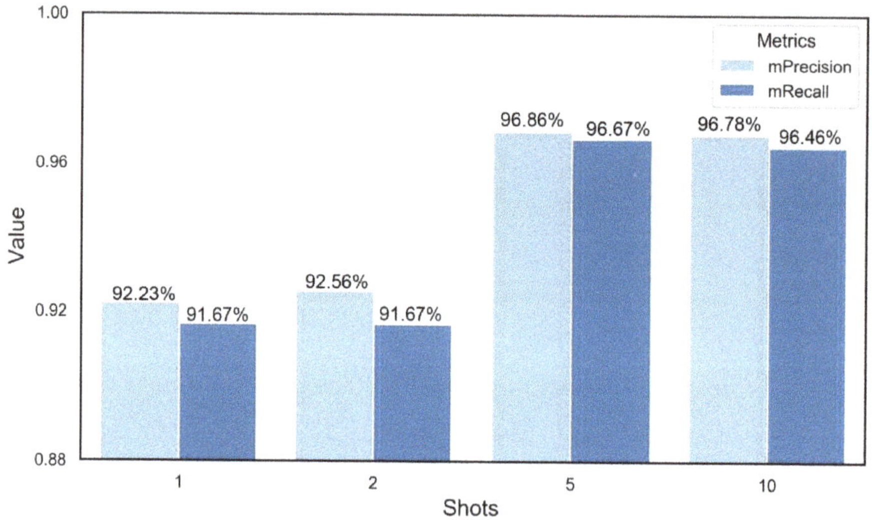

Figure 7. Performance of few-shot learning. The K was set to 1, 2, 5, and 10, which represents the fact that 1, 2, 5, and 10 images were randomly selected from the query dataset.

3. Materials and Methods

3.1. Plant Material and Growth Conditions

The widely cultivated wheat varieties (*Triticum aestivum* L.) Xumai 32, Yangmai 15, and Yangmai 16 were planted at the Yanhu Farm in Tongshan District, Xuzhou City, Jiangsu Province (117°07′ E, 34°28′ N). The plot size was 3.5 m × 5 m, and the seedling density was 2.4×10^6 ha^{-1} with a row space of 0.25 m. Two nitrogen (N) rates (225 and 270 kg N ha^{-1}) were tested; the ratio of basal fertilizer (applied at the seedling stage) to topdressing fertilizer (applied at the jointing stage) was 5:5, respectively. All treatments were given 90 kg P_2O_5 ha^{-1} and 150 kg k_2O ha^{-1} in the form of superphosphate and potassium chloride fertilizers before sowing. Other field management was conducted following local practices.

3.2. RGB Image Acquisition and Processing

In order to obtain RGB images of the 6–39 days of the grain filling period, wheat spike samples were taken from the field every three days, which began at six days after anthesis. After each sampling, 20 wheat grains were manually stripped from the middle of a spike. Twenty grains were neatly placed on a square black background plate for the photographing of each with a mobile phone (iPhone SE, Apple, Cupertino, CA, USA). A fixed device was also used to ensure that each shot was of the same focal length and lighting condition. The resolution of the obtained image was 3024 × 4032, and the image format was JPG. Then, a crop tool was used to crop to the image to the square black background plate with a 2000 × 2000 resolution. Finally, an object detection algorithm was used to detect the location of each grain with a 250 × 250 bounding box (Figure 8).

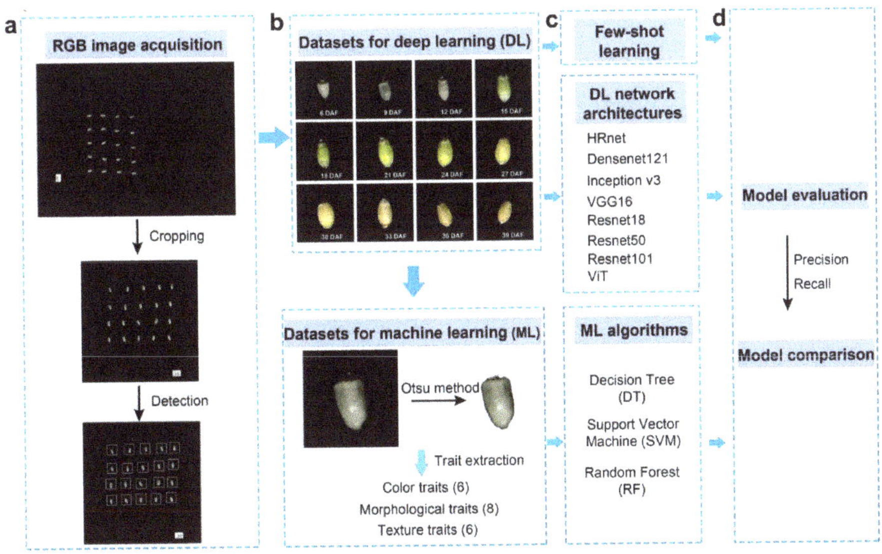

Figure 8. Workflow. (**a**) RGB image acquisition and processing; (**b**) dataset construction; (**c**) model construction; (**d**) model evaluation.

3.3. *Dataset Construction*

In order to compare the performance of different algorithms, the datasets need to be built separately, as required due to their characteristics. Deep learning and few-shot learning belong to an end-to-end model that can directly input raw RGB images, and convolutional neural networks (CNN) are used to automatically extract features; for traditional machine learning, it is necessary to manually carry out feature engineering, which determines the upper limit of the model.

3.3.1. Dataset for Traditional Machine Learning

Machine learning algorithms require manual feature engineering. Therefore, the OTSU method [22] was used for each grain image to remove the black background and obtain a mask of the region of interest. A total of 20 color, shape, and texture traits were extracted for each grain. Specifically, six color traits are the means of R, G, B, H, S, and V; the eight shape traits include the area, perimeter, radius, equivalent diameter, eccentric, compact, rectangle degree, and roundness; the six texture traits include dissimilarity, homogeneity, energy, correlation, ASM, and entropy. These traits were extracted via Python and OpenCV (a Python library), and the codes are available online at https://github.com/shem123456/wheat-grain-traits (accessed on 21 September 2023).

3.3.2. Dataset for Deep Learning

Deep learning is an end-to-end model structure, which can directly use raw RGB images as its input. The coordinate information of each grain is obtained based on the bounding boxes. A crop tool was used to crop the original images and save each grain as a 250 × 250 resolution image. These images were categorized and grouped into different folders according to their DAA (Tables S1 and S2, Figure S1). Finally, these images were randomly split into the training dataset, validation dataset, and testing dataset at proportions of 70%, 10%, and 20%, respectively.

3.3.3. Dataset for Few-Shot Learning

Few-shot learning can learn the image mappings and distance representations from the source domain, so it can classify new categories with few images. Images from 80% of the deep learning dataset were used to fine-tune the pre-training weight of the Omniglot dataset [23]. The remaining 20% of the images were further divided into a query dataset (20%) and a testing dataset (80%). In addition, the typical structure of few-shot learning (N-way and K-shot) means that there are N classes with K samples per class in the target domain. There are 12 classes in this paper, which represent the 6–39 days of the grain filling period. The K is set to 1, 2, 5, and 10, which means that 1, 2, 5, and 10 images are randomly selected from the query dataset.

3.4. Model Construction

For traditional machine learning, DT, SVM, and RF were selected to predict the DAA. A decision tree is a tree structure (it can be a binary tree or a non-binary tree). Each non-leaf node represents a test for a feature attribute; each branch represents the output of this feature attribute over a range; and each leaf node stores a category. Based on the structural risk minimization criterion, SVM constructs the optimal classification hyperplane to maximize the classification interval to improve the generalization ability of the learning machine. The RF is an integrated Bagging-based learning method that can handle classification and regression [24]. All machine learning algorithms were built using Python and Scikit-Learn (a Python library).

For deep learning, convolutional neural networks are used to map one input to another output, which can automatically learn mapping relationships between massive volumes of data without manual feature engineering. The performance of the deep learning depends on the structure of the convolutional neural network, except for the Transformer model, which uses a multi-head attention mechanism. Therefore, HRnet18, Densenet121, Inceptionv3, VGG16, Resnet18, Resnet50, Resnet101, and ViT, were selected in this study, and these models were then evaluated to select the optimal network structure. All models are built using PyTorch (a deep learning framework) and timm (PyTorch Image Models: a collection of image models, layers, utilities, optimizers, schedulers, and data loaders), and trained on a high-performance computer with an Intel Xeon(R) Platinum 8350C CPU, 32 GB memory, and an NVIDIA 3090 GPU. The batch size was 32. The Adam method with a learning rate of 0.001 was set to optimize the model parameters.

For few-shot learning, a Siamese network with contrastive loss was used. There are two subnets in the Siamese network, and each subnet uses VGG16 as its backbone architecture. Meanwhile, both the subnets share the same model weight. When training, a pair of images is fed into the Siamese network, and two images will be mapped to feature vectors with the same dimensions. Then, the contrast loss is applied to calculate the similarity of the two feature vectors. If they belong to the same class, the network will minimize the loss function to shorten the distance between similar classes; otherwise, the network maximizes the loss function to distinguish between the different classes. The details of the Siamese architecture can be found in [25]. The definition of the contrastive loss function is shown in Formula (1). Finally, the Siamese network with contrastive loss was built using PyTorch, and the configuration of its training was consistent with that of the deep learning model described above. The codes are available online at https://github.com/shem123456/grain-filling-classification (accessed on 21 September 2023).

$$Loss = yd^2 + (1-y) \cdot max(0, \text{margin-}d)^2 \tag{1}$$

where d represents the Euclidean distance between the two samples and y is the label of whether the two samples match; $y = 1$ means that the two samples are the same class, and $y = 0$ means that the two samples are of different classes. The margin is the threshold.

3.5. Model Evaluation

We simplified the DAA prediction as a multiclass classification problem, so precision and recall, as common metrics for image classification, were applied to evaluate the models' performance. Precision is the proportion of correct positive predictions to all positive predictions, while recall is the proportion of correct positive predictions to all positive examples. The relevant calculations are provided as follows:

$$precision = \frac{TP}{TP + FP} \quad (2)$$

$$recall = \frac{TP}{TP + FN} \quad (3)$$

where *TP* represents true positives; *FP* represents false positives; and *FN* represents false negatives.

4. Discussion

The traits from RGB images can reflect grain filling dynamics. Previous studies have shown that wheat grain filling follows a slow–fast–slow pattern due to the reactions for protein and starch synthesis [2,26]. Specifically, the development of the wheat grain can be divided into three stages: 0-15 days, 15–27 days, and 27 days to maturation. During the first stage, the grain fresh weight continuously increases because of its division and expansion. Then, the grain dry weight will increase further because of starch and protein accumulation at 15–27 days. Finally, the grain filling rate begins to decline after 27 days, and the grain fresh weight drops rapidly after 33 days due to desiccation [27]. As shown in Figure 2, the grain filling pattern is similar to the trend in shape traits, including the area, perimeter, and equivalent diameter of the grain, suggesting that the shape traits from RGB images can characterize the dynamics of grain filling. Meanwhile, the color and texture traits did not show a similar trend, but these traits improved information abundance for predicting the DAA based on the performance results of traditional machine learning.

The prediction performance of RF is optimal in terms of machine learning, as seen in Figure 4, but the accuracy of traditional machine learning is still limited by feature engineering. Compared with other machine learning methods, RF is an ensemble learning method and is robust to noise and outliers [28]. However, the prediction performance of deep learning is much greater than that of traditional machine learning. The main source of this difference is that the precision and recall of machine learning algorithms were lower at days 21–33 (Figure 4). This might be because the wheat grains at 21–33 days are too similar for the machine learning algorithms to classify. On the one hand, the upper limit of the model for machine learning depends on feature engineering, and although a total of 20 color, shape, and texture traits were extracted in this study, these manually selected traits still have limitations in representing seed properties. On the other hand, the input for deep learning is the complete RGB image, as it is an end-to-end image classification model, and the network structure of the CNN involved can select the optimal feature based on the gradient [29]. Notably, ViT achieved optimal prediction accuracy (Figures 5 and 6), although it uses a network structure that is different from a CNN. ViT is the network construction of an attention mechanism [30], which can be a model that focuses on more useful information while suppressing redundancy, and this structure and its variants have generated state-of-the-art results in many visual tasks [31].

Few-shot learning can also achieve a better performance than traditional machine learning in the prediction of DAA. To explore whether few-shot learning can achieve fine-grained recognition for wheat grains, we evaluated the performance of a Siamese network with contrastive loss in predicting the DAA. As shown in Figure 7, the results showed that the few-shot learning under 1-shot could approximate the performance of a standard deep learning algorithm. Meanwhile, its mean precision and recall increased with the number of shots, which were consistent with other results for disease recognition [18]. These results demonstrate that few-shot learning is an effective strategy for reducing our dependence on the dataset scale, which benefits from metric-based learning to compare and analyze unseen samples [32].

Although a complete wheat grain filling dataset was presented, with a comparison of different algorithms, in this study, there are still some perspectives that need to be considered in the future: (1) Although the deep learning algorithm has great advantages in predictive performance, it is far from practical application in the field. In our next study, we will try some lightweight improvements to make it more suitable for mobile deployment. (2) In this study, only RGB images were used, while multispectral and hyperspectral imaging are also worth exploring to further obtain physiological traits.

5. Conclusions

Predicting the wheat grain DAA provides timely information on wheat grain filling dynamics, which is essential for ensuring the security of the wheat yield. In this work, we presented a complete wheat grain dataset that characterized the dynamic development of wheat grains. This dataset revealed the dynamic changes in traits such as color, shape, and texture during wheat grain development. To predict the DAA from RGB images of wheat grains, we tested the performance of traditional machine learning, deep learning, and few-shot learning on this dataset. The results showed that the deep learning classification model with ViT as the backbone neural network had the highest accuracy and recall, 99.03% and 99.00%, respectively, while few-shot learning had a reduced dependence on the dataset's scale, and had a higher precision and recall in the case of 5-shot learning, 95.08% and 94.83%, respectively. In future work, we expect to propose a grain filling monitoring model compatible with few samples and which has high accuracy in order to achieve its production purpose: disaster prevention and mitigation.

Supplementary Materials: The following supporting information can be downloaded at https://www.mdpi.com/article/10.3390/plants12234043/s1, Figure S1: the WheatGrain dataset; Table S1: details of datasets; Table S2: model size.

Author Contributions: Q.L. and D.J. designed the experiments. Y.S. and Z.S. performed the experiments. Z.S. and R.Z. wrote all the code scripts. Y.S., Z.S. and Q.L. interpreted the data and wrote the manuscript. Q.L., R.Z., H.M., J.C., Q.Z., X.W. and D.J. revised the manuscript. All authors have read and agreed to the published version of the manuscript.

Funding: This work was supported by the Earmarked Fund for CARS (CARS-03), the Fundamental Research Funds for Central Universities (XUEKEN2023013), the National Natural Science Foundation of China (32030076), the National Key Research and Development Program of China (2022YFD2301404), the Jiangsu Agriculture Science and Technology Innovation Fund (CX(22)1006), the Jiangsu Innovation Support Program for International Science and Technology Cooperation Project (BZ2023049), and the Collaborative Innovation Center for Modern Crop Production, co-sponsored by Province and Ministry (CIC-MCP).

Data Availability Statement: The original contributions presented in the study are included in the article/Supplementary Materials. Further inquiries can be directed to the corresponding authors.

Acknowledgments: We thank the editors and the anonymous reviewers for their valuable comments and suggestions.

Conflicts of Interest: The authors declare no conflict of interest.

References

1. Kumar, S.B.; Prabhasankar, P. Low glycemic index ingredients and modified starches in wheat based food processing: A review. *Trends Food Sci. Technol.* **2014**, *35*, 32–41. [CrossRef]
2. Xie, Q.; Mayes, S.; Sparkes, D.L. Carpel size, grain filling, and morphology determine individual grain weight in wheat. *J. Exp. Bot.* **2015**, *66*, 6715–6730. [CrossRef] [PubMed]
3. Djanaguiraman, M.; Narayanan, S.; Erdayani, E.; Prasad, P.V.V. Effects of high temperature stress during anthesis and grain filling periods on photosynthesis, lipids and grain yield in wheat. *BMC Plant Biol.* **2020**, *20*, 268. [CrossRef] [PubMed]
4. Sofield, I.; Evans, L.; Cook, M.; Wardlaw, I. Factors influencing the rate and duration of grain filling in wheat. *Funct. Plant Biol.* **1977**, *4*, 785–797. [CrossRef]
5. Ullah, A.; Nadeem, F.; Nawaz, A.; Siddique, K.H.M.; Farooq, M. Heat stress effects on the reproductive physiology and yield of wheat. *J. Agron. Crop Sci.* **2022**, *208*, 1–17. [CrossRef]
6. Bhullar, S.; Jenner, C. Effects of temperature on the conversion of sucrose to starch in the developing wheat endosperm. *Funct. Plant Biol.* **1986**, *13*, 605–615. [CrossRef]

7. Yang, X.-S.; Liang, Z.-Y.; Li, H.; Shen, W.-N.; Li, M.-X.; Xi, Y.-J.; Wang, Z.-L.; Liu, S.-D. Analysis on Genetic Model of Grain Characteristics in Common Wheat by Mixed Inheritance Model of Major Genes Plus Polygenes. *J. Triticeae Crops* **2013**, *33*, 1119–1127.
8. Patrício, D.I.; Rieder, R. Computer vision and artificial intelligence in precision agriculture for grain crops: A systematic review. *Comput. Electron. Agric.* **2018**, *153*, 69–81. [CrossRef]
9. Breseghello, F.; Sorrells, M.E. QTL analysis of kernel size and shape in two hexaploid wheat mapping populations. *Field Crops Res.* **2007**, *101*, 172–179. [CrossRef]
10. Majumdar, S.; Jayas, D. Classification of bulk samples of cereal grains using machine vision. *J. Agric. Eng. Res.* **1999**, *73*, 35–47. [CrossRef]
11. Zhu, J.; Li, H.; Rao, Z.; Ji, H. Identification of slightly sprouted wheat kernels using hyperspectral imaging technology and different deep convolutional neural networks. *Food Control* **2023**, *143*, 109291. [CrossRef]
12. Neethirajan, S.; Jayas, D.S.; White, N. Detection of sprouted wheat kernels using soft X-ray image analysis. *J. Food Eng.* **2007**, *81*, 509–513. [CrossRef]
13. Li, Q.; Jin, S.; Zang, J.; Wang, X.; Sun, Z.; Li, Z.; Xu, S.; Ma, Q.; Su, Y.; Guo, Q.; et al. Deciphering the contributions of spectral and structural data to wheat yield estimation from proximal sensing. *Crop J.* **2022**, *10*, 1334–1345. [CrossRef]
14. Liu, S.; Jin, X.; Nie, C.; Wang, S.; Yu, X.; Cheng, M.; Shao, M.; Wang, Z.; Tuohuti, N.; Bai, Y.; et al. Estimating leaf area index using unmanned aerial vehicle data: Shallow vs. deep machine learning algorithms. *Plant Physiol* **2021**, *187*, 1551–1576. [CrossRef] [PubMed]
15. Tu, K.; Wen, S.; Cheng, Y.; Xu, Y.; Pan, T.; Hou, H.; Gu, R.; Wang, J.; Wang, F.; Sun, Q. A model for genuineness detection in genetically and phenotypically similar maize variety seeds based on hyperspectral imaging and machine learning. *Plant Methods* **2022**, *18*, 1–17. [CrossRef] [PubMed]
16. Sun, Z.; Li, Q.; Jin, S.; Song, Y.; Xu, S.; Wang, X.; Cai, J.; Zhou, Q.; Ge, Y.; Zhang, R.; et al. Simultaneous prediction of wheat yield and grain protein content using multitask deep learning from time-series proximal sensing. *Plant Phenomics* **2022**, *2022*, 9757948. [CrossRef] [PubMed]
17. Taghavi Namin, S.; Esmaeilzadeh, M.; Najafi, M.; Brown, T.B.; Borevitz, J.O. Deep phenotyping: Deep learning for temporal phenotype/genotype classification. *Plant Methods* **2018**, *14*, 66. [CrossRef]
18. Zhong, F.; Chen, Z.; Zhang, Y.; Xia, F. Zero-and few-shot learning for diseases recognition of Citrus aurantium L. using conditional adversarial autoencoders. *Comput. Electron. Agric.* **2020**, *179*, 105828. [CrossRef]
19. Liang, X. Few-shot cotton leaf spots disease classification based on metric learning. *Plant Methods* **2021**, *17*, 114. [CrossRef]
20. Li, Y.; Chao, X. Semi-supervised few-shot learning approach for plant diseases recognition. *Plant Methods* **2021**, *17*, 68. [CrossRef]
21. Zhao, C.; Liu, B.; Piao, S.; Wang, X.; Lobell, D.B.; Huang, Y.; Huang, M.; Yao, Y.; Bassu, S.; Ciais, P.; et al. Temperature increase reduces global yields of major crops in four independent estimates. *Proc. Natl. Acad. Sci. USA* **2017**, *114*, 9326–9331. [CrossRef] [PubMed]
22. Liu, D.; Yu, J. Otsu method and K-means. In Proceedings of the 2009 Ninth International Conference on Hybrid Intelligent Systems, Shenyang, China, 12–14 August 2009.
23. Lake, B.M.; Salakhutdinov, R.; Tenenbaum, J.B. Human-level concept learning through probabilistic program induction. *Science* **2015**, *350*, 1332–1338. [CrossRef] [PubMed]
24. Tsai, C.-F.; Hsu, Y.-F.; Yen, D.C. A comparative study of classifier ensembles for bankruptcy prediction. *Appl. Soft Comput.* **2014**, *24*, 977–984. [CrossRef]
25. Argüeso, D.; Picon, A.; Irusta, U.; Medela, A.; San-Emeterio, M.G.; Bereciartua, A.; Alvarez-Gila, A. Few-Shot Learning approach for plant disease classification using images taken in the field. *Comput. Electron. Agric.* **2020**, *175*, 105542. [CrossRef]
26. Shewry, P.R.; Underwood, C.; Wan, Y.; Lovegrove, A.; Bhandari, D.; Toole, G.; Mills, E.C.; Denyer, K.; Mitchell, R.A. Storage product synthesis and accumulation in developing grains of wheat. *J. Cereal Sci.* **2009**, *50*, 106–112. [CrossRef]
27. Shewry, P.R.; Mitchell, R.A.; Tosi, P.; Wan, Y.; Underwood, C.; Lovegrove, A.; Freeman, J.; Toole, G.A.; Mills, E.C.; Ward, J.L. An integrated study of grain development of wheat (cv. Hereward). *J. Cereal Sci.* **2012**, *56*, 21–30. [CrossRef]
28. Liaw, A.; Wiener, M. Classification and Regression by randomForest. *R News* **2002**, *23*, 18–22.
29. Mortazi, A.; Bagci, U. Automatically designing CNN architectures for medical image segmentation. In Proceedings of the Machine Learning in Medical Imaging: 9th International Workshop, MLMI 2018, Held in Conjunction with MICCAI 2018, Granada, Spain, 16 September 2018.
30. Han, K.; Xiao, A.; Wu, E.; Guo, J.; Xu, C.; Wang, Y. Transformer in transformer. In Proceedings of the Neural Information Processing Systems, Online, 6–14 December 2021; Volume 34, pp. 15908–15919.
31. Moutik, O.; Sekkat, H.; Tigani, S.; Chehri, A.; Saadane, R.; Tchakoucht, T.A.; Paul, A. Convolutional neural networks or vision transformers: Who will win the race for action recognitions in visual data? *Sensors* **2023**, *23*, 734. [CrossRef]
32. Li, Y.; Chao, X. ANN-based continual classification in agriculture. *Agriculture* **2020**, *10*, 178. [CrossRef]

plants

MDPI

Article

The Role of the Ascorbic Acid–Glutathione Cycle in Young Wheat Ears' Response to Spring Freezing Stress

Yuting Zhang, Chunyang Ni, Yongwen Dong, Xue Jiang, Chang Liu, Weiling Wang *, Can Zhao, Guohui Li, Ke Xu and Zhongyang Huo *

Jiangsu Key Laboratory of Crop Genetics and Physiology, Jiangsu Key Laboratory of Crop Cultivation and Physiology, Jiangsu Co-Innovation Center for Modern Production Technology of Grain Crops, Agricultural College, Yangzhou University, Yangzhou 225009, China; mz120221323@stu.yzu.edu.cn (Y.Z.); mz120221369@stu.yuz.edu.cn (C.N.); mz120221361@stu.yuz.edu.cn (Y.D.); mz120231361@stu.yzu.edu.cn (X.J.); mx120200672@yzu.edu.cn (C.L.); 007148@yzu.edu.cn (C.Z.); lgh@yzu.edu.cn (G.L.); xuke@yzu.edu.cn (K.X.)

* Correspondence: wangwl@yzu.edu.cn (W.W.); zyhuo@yzu.edu.cn (Z.H.)

Abstract: Freezing stress in spring often causes the death and abnormal development of young ears of wheat, leading to a significant reduction in grain production. However, the mechanisms of young wheat ears responding to freezing are largely unclear. In this study, the role of the ascorbic acid–glutathione cycle (AsA–GSH cycle) in alleviating freezing-caused oxidative damage in young wheat ears at the anther connective tissue formation phase (ACFP) was investigated. The results showed that the release rate of reactive oxygen species (ROS) and the relative electrolyte conductivity in young ears of Jimai22 (JM22, freezing-tolerant) were significantly lower than those in young ears of Xumai33 (XM33, freezing-sensitive) under freezing. The level of the GSH pool (231.8~392.3 μg/g FW) was strikingly higher than that of the AsA pool (98.86~123.4 μg/g FW) in young wheat ears at the ACFP. Freezing significantly increased the level of the AsA pool and the activities of ascorbate peroxidase (APX) and monodehydroascorbate reductase (MDHAR) in the young ears of both varieties. The level of the GSH pool increased in the young ears of XM33 under freezing but decreased in the young ears of JM22. The young ears of JM22 showed higher activities of glutathione reductase (GR), glutathione-S-transferase (GST) and glutathione peroxidase (GPX) than the young ears of XM33 under freezing. Collectively, these results suggest that the AsA–GSH cycle plays a positive role in alleviating freezing-induced oxidative damage in young wheat ears. Furthermore, the ability of utilizing GSH as a substrate to scavenge ROS is an important factor affecting the freezing tolerance of young wheat ears. In addition, abscisic acid (ABA), salicylic acid (SA), 3-indolebutyric acid (IBA) and cis-zeatin (cZ) may be involved in regulating the AsA–GSH cycle metabolism in young wheat ears under freezing.

Keywords: wheat; young ears; spring freezing; AsA–GSH cycle; hormone

1. Introduction

Wheat (*Triticum aestivum* L.) is the world's major food crop and plays an important role in food supply security [1]. With the intensification of global warming in recent years, temperature fluctuations in spring increase the risk of wheat plants suffering from frost stress [2,3]. Spring frost is an important meteorological factor that adversely affects plant growth and development and causes serious economic losses to wheat production in China, the United States, Australia and Europe [4–6]. In Australia, even under optimal management conditions, spring frost can reduce wheat yields by approximately 10% [7]. The yield of wheat plants suffering from spring frost are generally reduced by 10% to 30%, and in severe cases, this can be up to 50% or more in Huanghuai and the middle and lower reaches of the Yangtze River of China [8,9]. In China, spring frost mainly occurs from March to April, when winter wheat plants are in the stage from jointing to booting [7,10]. The jointing and booting stages occur during the critical period for the rapid growth and differentiation of young wheat ears, which are very sensitive to freezing temperatures [7,11]. The death and maldevelopment of young ears are the main reasons for wheat yield reductions caused by spring freezing [12–14]. Zhong et al. [7] found that the freezing tolerance of young wheat ears declines as development advances through the jointing stage and drops off particularly abruptly at the point when the anther connective tissue formation phase (ACFP) begins.

Reactive oxygen species (ROS), including hydrogen peroxide (H_2O_2), superoxide anion (O_2^-), singlet oxygen ($^1O_2$) and hydroxyl radical (OH^-), are byproducts of aerobic metabolism [15]. The inhibitory effects of freezing stress on electron transport in the membranes of mitochondria and chloroplasts and the activities of antioxidant enzymes can cause an over-accumulation of ROS in plant cells [16]. ROS are highly reactive and toxic; their over-accumulation can lead to the oxidative destruction of biological macromolecules (nucleic acids, proteins, lipids, etc.) [17]. Therefore, freezing-induced ROS over-accumulation is an important physiological cause of organ death [18–20]. To prevent the over-accumulation of ROS under an ever-changing environment, plants have evolved sophisticated mechanisms for scavenging ROS, including enzymatic and non-enzymatic antioxidants [21]. The ascorbate–glutathione (AsA–GSH) cycle serves as an essential antioxidant defense system in plants, which is composed of two major non-enzymatic antioxidants, AsA and GSH, together with a combination of four enzymes, ascorbate peroxidase (APX), monodehydroascorbate reductase (MDHAR), dehydroascorbate reductase (DHAR) and glutathione reductase (GR) [22,23]. Among these enzymes, APX converts H_2O_2 into water with the help of AsA as an electron donor, which is converted into monodehydroascorbate (MDHA). MDHAR can convert MDHA into AsA by consuming reducing power. A part of MDHA is spontaneously converted into dehydroascorbate (DHA), which can be reverted into AsA by DHAR with GSH as a substrate. GR catalyzes the reduction in GSSG for GSH regeneration [22]. In addition, glutathione peroxidase (GPX) and glutathione-S-transferase (GST) can eliminate ROS with GSH as a substrate [24]. It has been documented that the metabolism of the AsA–GSH cycle plays a crucial role in maintaining redox homeostasis in many species under drought, salt and freezing conditions [25,26]. Hormones can act as signaling molecules to regulate responses to biotic and abiotic stresses in plants [10]. Previous studies found that hormones such abscisic acid (ABA) and salicylic acid (SA) have important regulatory effects on the metabolism of the AsA–GSH cycle under stresses [20,27–29]. However, the responses of the AsA–GSH cycle and hormones to freezing stress in young wheat ears at the ACFP and their relationships are largely unclear.

Therefore, in this study, we analyzed the changes in the AsA–GSH cycle metabolism and hormones levels in young wheat ears' response to freezing stress of two varieties (Jimai22, JM22, freezing-tolerant; Xumai33, XM33, freezing-sensitive) with different spring freezing tolerances at the ACFP. The results provide new insights into understanding the mechanisms of spring freezing tolerance in young wheat ears at the ACFP.

2. Results

2.1. Growth Parameters

The freezing treatment significantly inhibited the growth of wheat plants in both varieties, as indicated by the decreases in the plant height and biomass (dry mass) (Figure 1A–C). However, the decreases in the plant height and biomass in XM33 (14.89% and 27.48%) were significantly higher than that in JM22 (6.129% and 17.49%). Due to the death of the young ears from main stems or tillers caused by the freezing treatment, axillary buds sprouted to form regenerated tillers [30]. As shown in Figure 1A,D, there were more regenerated spikes (refers to those regenerated tillers that can produce grains) in XM33 than JM22 under the freezing treatment at maturity (Figure 1A,D), indicating that the freezing treatment caused more death in the young ears of XM33 than JM22. In addition, the decrease in the grain number per spike of the surviving main stems in JM22 (9.017%) was significantly lower than that in XM33 (20.63%), suggesting that there was more floret abortion in the young ears of the surviving main stems in XM33 compared to JM22 under freezing.

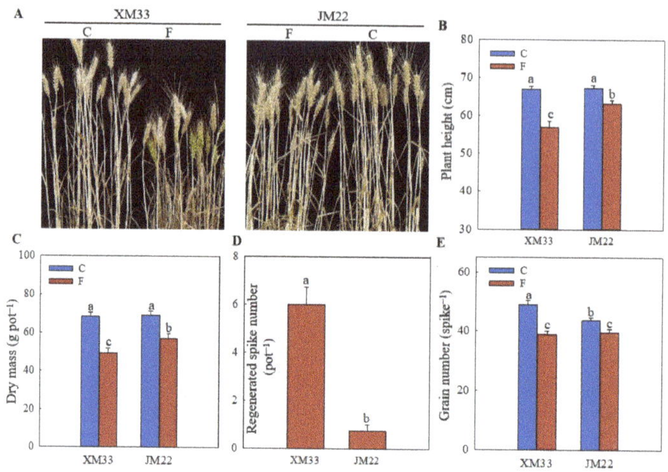

Figure 1. Freezing tolerances of JM22 and XM33 at the ACFP. Note: (**A**) Plant phenotype at maturity. (**B**) Plant height at maturity. (**C**) Dry mass at maturity. (**D**) Regenerated spike number. (**E**) Grain number. C and F indicate the control and freezing treatments, respectively. JM22 and XM33 represent the freezing-tolerant and freezing-sensitive winter wheat varieties Jimai22 and Xumai33, respectively. Each value is the mean ± SE of three biological replicates. The different lowercase letters indicate statistically significant differences at the $p < 0.05$ level.

2.2. ROS Level and Electrolyte Leakage

As shown in Figure 2A, the release rate of ROS was increased in the young ears of both varieties under freezing stress compared to the normal temperature control group, with increases of 5.149% in XM33 and 2.745% in JM22. An over-accumulation of ROS can destroy the cell membrane structure and lead to electrolyte leakage. Thus, the relative electrolyte conductivity (REC) is commonly used to evaluate the freezing tolerance in plants [31]. The freezing treatment significantly increased the REC in the young ears of both varieties, indicating that the freezing treatment caused declining function in the cell membranes. However, the increase in the REC in the young ears of JM22 (45.07%) was significantly lower than that in the young ears of XM33 (100.6%) under freezing stress (Figure 2B).

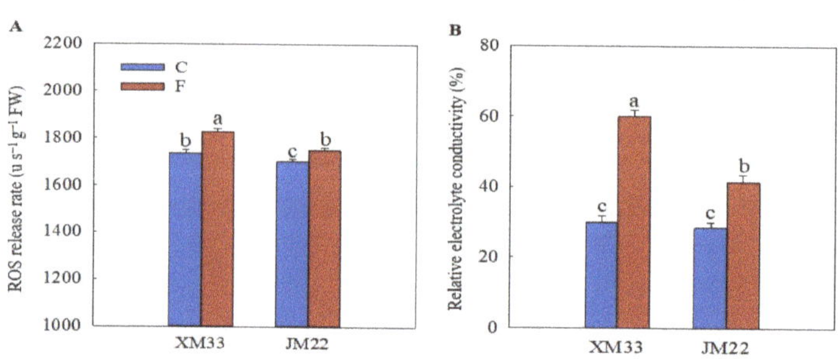

Figure 2. Response of the ROS release rate and relative electrolyte conductivity to freezing stress in the young wheat ears at the ACFP. Note: (**A**) ROS release rate. (**B**) Relative electrolyte conductivity (REC). C and F indicate the control and freezing treatments, respectively. JM22 and XM33 represent the freezing-tolerant and freezing-sensitive winter wheat varieties Jimai22 and Xumai33, respectively. Each value is the mean ± SE of three biological replicates. The different lowercase letters indicate statistically significant differences at the $p < 0.05$ level.

2.3. Ascorbate and Glutathione Content

The young ears of JM22 showed significantly higher GSH (74.49%), oxidized glutthone (GSSG, 43.40%), GSH+GSSG (69.24%) and AsA (45.46%) contents, and higher GSH/GSSG and ascorbic acid/dehydroascorbic acid (AsA/DHA) ratios than those in the young ears of XM33 under normal conditions (Figures 3 and 4). There were no significant differences in the DHA and AsA+DHA contents between the young ears of the two varieties (Figure 4). The freezing treatment obviously increased the contents of GSH, GSSG and GSH+GSSG, and remarkably decreased the GSH/GSSG ratio in the young ears of XM33 (Figure 3). However, the change patterns of the GSH, GSSG and GSH+GSSG contents and GSH/GSSG ratio in the young ears of JM22 under freezing stress were the opposite to those in the young ears of XM33 (Figure 3). Moreover, the freezing treatment caused an increase in the contents of AsA, DHA and AsA+DHA, as well as the AsA/DHA ratio in the young ears of both varieties (Figure 4). The young ears of JM22 exhibited significantly higher GSH (39.53%) and GSH+GSSG (22.07%) contents, as well as a higher GSH/GSSG ratio (112.9%) and a lower GSSG content (42.73%) than those in the young ears of XM33 under freezing stress (Figure 3). However, there were no significant differences in the AsA, DHA and AsA+DHA contents and the AsA/DHA ratio between the young ears of the two varieties under freezing (Figures 3 and 4).

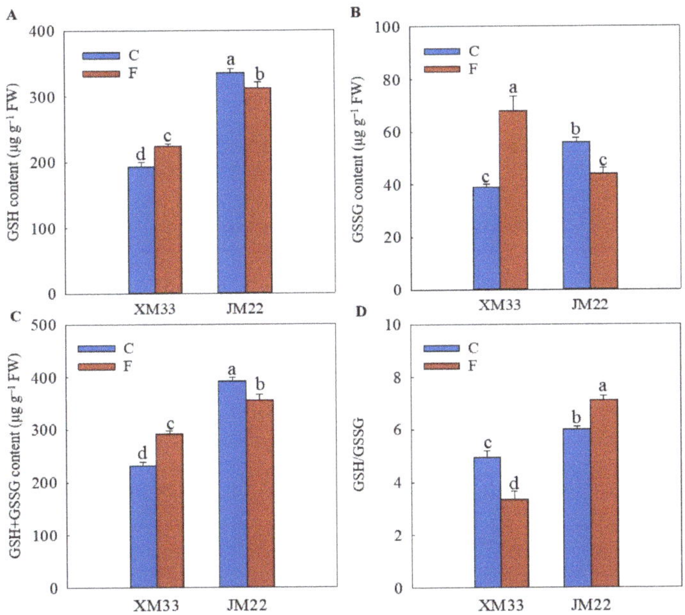

Figure 3. Response of the glutathione pool to freezing stress in the young wheat ears at the ACFP. Note: (**A**,**B**) Contents of reduced glutathione (GSH) and oxidized glutathione (GSSG). (**C**,**D**) Total contents of GSH and GSSG and the GSH/GSSG ratio. C and F indicate the control and freezing treatments, respectively. JM22 and XM33 represent the freezing-tolerant and freezing-sensitive winter wheat varieties Jimai22 and Xumai33, respectively. Each value is the mean ± SE of three biological replicates. The different lowercase letters indicate statistically significant differences at the $p < 0.05$ level.

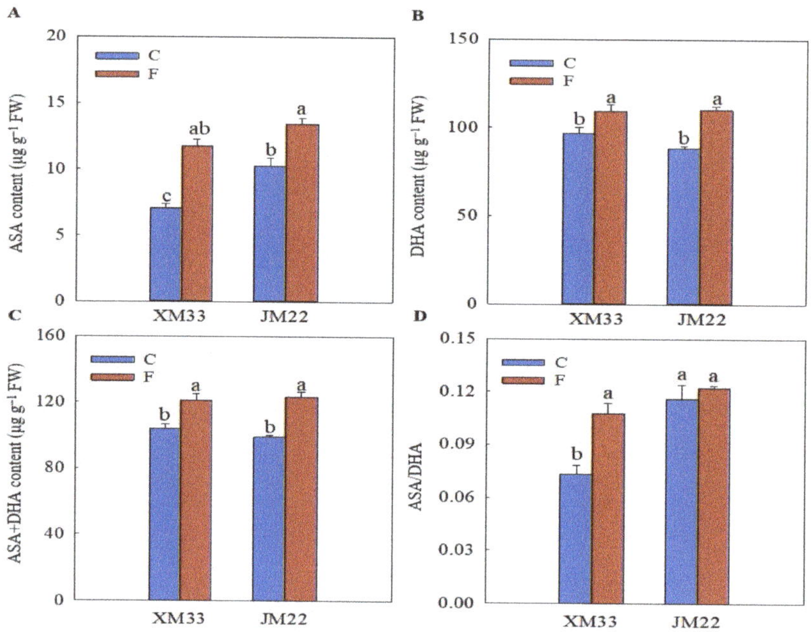

Figure 4. Response of the ascorbate pool to freezing stress in the young wheat ears at the ACFP. Note: (**A**,**B**) Contents of ascorbic acid (AsA) and dehydroascorbic acid (DHA). (**C**,**D**) Total contents of AsA and DHA and the AsA/DHA ratio. C and F indicate the control and freezing treatments, respectively. JM22 and XM33 represent the freezing-tolerant and freezing-sensitive winter wheat varieties Jimai22 and Xumai33, respectively. Each value is the mean ± SE of three biological replicates. The different lowercase letters indicate statistically significant differences at the $p < 0.05$ level.

2.4. AsA–GSH Metabolism Associated Enzymes Activities

As shown in Table 1, the young ears of JM22 exhibited significantly higher activities of APX and GR and a significantly lower activity of GPX compared to the young ears of XM33 under normal conditions. There were no significant differences in the activities of MDHAR, DHAR and GST between the young ears of the two varieties under normal conditions. The freezing treatment enhanced the activities of MDHAR and APX and decreased the activity of GR in the young ears of both varieties. The activities of DHAR significantly increased (73.86%) in the young ears of XM33, whereas it obviously decreased (28.29%) in the young ears of JM22 under freezing stress. The change trend in the activity of GST was the opposite to that of DHAR in the young ears of the two varieties under freezing stress. The freezing treatment had no obvious effect on the GST activity in the young ears of JM22 but caused a significant decline (44.95%) in the GST activity in the young ears of XM33. The young ears of JM22 showed higher activities of GR, GST and GPX and lower activities of MDHAR, DHAR and APX compared to the young ears of XM33 under freezing stress.

Table 1. Response of the activities of the enzymes involved in the AsA–GSH cycle metabolism to freezing stress in the young wheat ears at the ACFP.

Enzymes	XM33		JM22	
	C	F	C	F
MDHAR (U g^{-1} FW)	2.845 ± 0.013 c	3.621 ± 0.031 a	2.822 ± 0.109 c	3.162 ± 0.145 b
DHAR (U g^{-1} FW)	0.153 ± 0.005 b	0.266 ± 0.008 a	0.152 ± 0.005 b	0.109 ± 0.005 c
APX (U g^{-1} FW)	3.233 ± 0.109 c	4.102 ± 0.023 a	3.701 ± 0.085 b	3.892 ± 0.149 ab
GR (U g^{-1} FW)	0.302 ± 0.007 b	0.189 ± 0.010 c	0.382 ± 0.010 a	0.304 ± 0.009 b
GST (U g^{-1} FW)	0.436 ± 0.014 a	0.240 ± 0.008 b	0.437 ± 0.014 a	0.417 ± 0.011 a
GPX (U g^{-1} FW)	1.211 ± 0.0323 a	1.042 ± 0.044 b	0.957 ± 0.014 b	1.204 ± 0.024 a

Note: The measurements were taken at the end of the temperature treatments. MDHAR, monodehydroascorbate reductase; DHAR, dehydroascorbate reductase; APX, ascorbate peroxidase; GR, glutathione reductase; GST, glutathione-S-transferase; GPX, glutathione peroxidase. C refers to the normal temperature control, F refers to the freezing treatment. JM22 and XM33 represent the freezing-tolerant and freezing-sensitive winter wheat varieties Jimai22 and Xumai33, respectively. The data are the means ± SE of three biological replicates. The different letters at the same row indicate significant differences at the $p < 0.05$ level.

2.5. Hormones Levels

As shown in Figure 5, the freezing treatment induced an increase in the contents of ABA (58.84% in XM33 and 50.37% in JM22), SA (82.71% in XM33 and 223.8% in JM22) and jasmonic acid (JA, 111.3% in XM33 and 28.33% in JM22) in the young ears of both varieties compared to their normal temperature control groups. The young ears of JM22 showed higher JA (87.40% and 13.79%, respectively) and SA (42.03% and 151.7%, respectively) contents and a lower ABA (21.75% and 25.92%, respectively) content than those in the young ears of XM33 under both normal and freezing conditions (Figure 5).

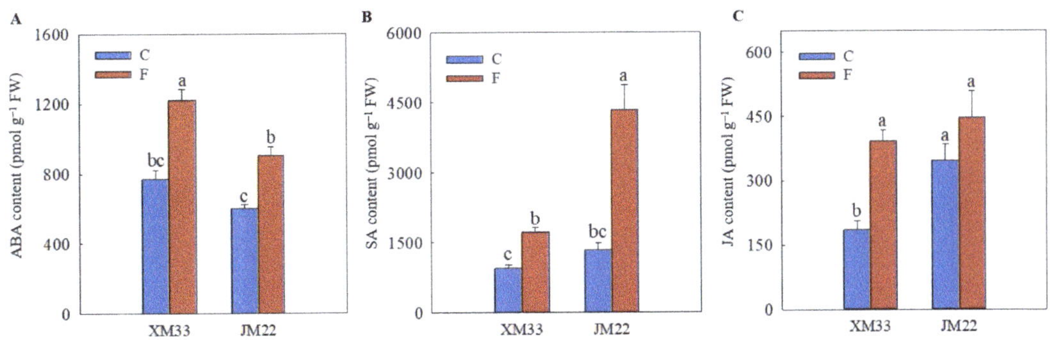

Figure 5. Response of the ABA, SA and JA levels to freezing stress in the young wheat ears at the ACFP. Note: (**A–C**) Contents of abscisic acid (ABA), salicylic acid (SA), jasmonic acid (JA). C and F indicate the control and freezing treatments, respectively. JM22 and XM33 represent the freezing-tolerant and freezing-sensitive winter wheat varieties Jimai22 and Xumai33, respectively. Each value is the mean ± SE of three biological replicates. The different lowercase letters indicate statistically significant differences at the $p < 0.05$ level.

N6-isopentenyladenosine (IPA), trans-zeatin (tZ) and cis-zeatin (cZ) are major cytokinin derivatives in plants. The young ears of JM22 showed significantly higher contents of tZ (42.50%), cZ (146.69%) and IPA (31.04%) compared to the young ears of XM33 under normal conditions (Figure 6). The freezing treatment induced a significant decline in the contents of tZ (21.73% in XM33 and 59.90% in JM22) and IPA (25.06% in XM33 and 29.51% in JM22) in the young ears of both varieties. However, the content of cZ increased by 13.22% in the young ears of JM22 under freezing stress, while it decreased by 22.99% in the young ears of XM33. There were no significant differences in the contents of IPA and tZ between the young ears of the two varieties under freezing stress. However, the content of cZ in the young ears of JM22 was significantly higher than that in the young ears of XM33 under freezing stress (Figure 6).

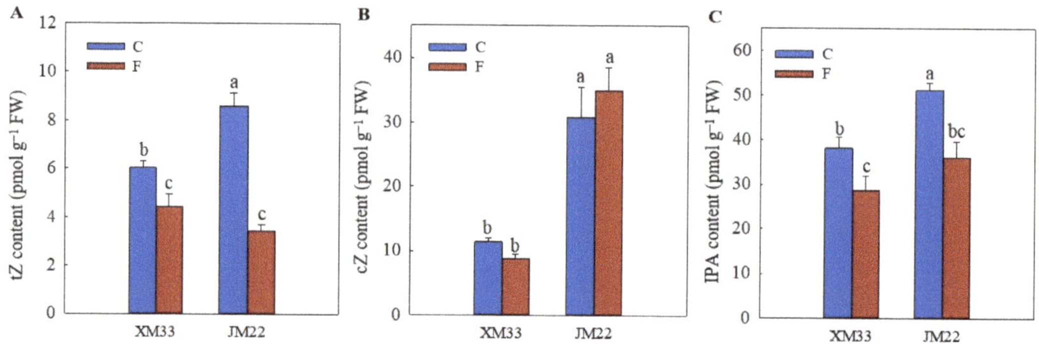

Figure 6. Response of the levels of cytokinin to freezing stress in the young wheat ears at the ACFP. Note: (**A–C**) Contents of trans-zeatin (tZ), cis-zeatin (cZ) and N6-isopentenyladenosine (IPA). C and F indicate the control and freezing treatments, respectively. JM22 and XM33 represent the freezing-tolerant and freezing-sensitive winter wheat varieties Jimai22 and Xumai33, respectively. Each value is the mean ± SE of three biological replicates. The different lowercase letters indicate statistically significant differences at the $p < 0.05$ level.

There was no significant difference in the indole-3-acetic acid (IAA) content between the young ears of the two varieties under normal conditions (Figure 7). However, the young ears of JM22 showed a significantly higher 3-indolebutyric acid (IBA) content (49.60%) than that in the young ears of XM33 under normal conditions. The freezing treatment had little effect on the IAA content in the young ears of XM33, whereas it caused a significant increase in the IAA content (26.42%) in the young ears of JM22. The content of IBA was significantly decreased (48.12%) in the young ears of XM33 under freezing stress but had no significant change in the young ears of JM22. The young ears of JM22 showed higher IAA (10.76%) and IBA (209.9%) contents compared to the young ears of XM33 under freezing stress (Figure 7).

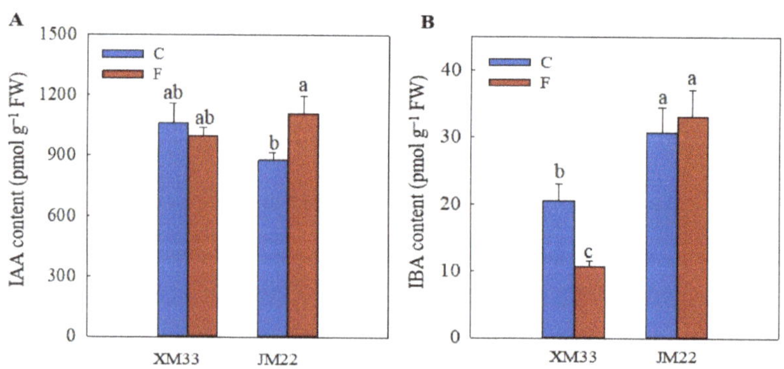

Figure 7. Response of the auxin levels to freezing stress in the young wheat ears at the ACFP. Note: (**A**,**B**) Contents of indoleacetic acid (IAA), indolebutyric acid (IBA). C and F indicate the control and freezing treatments, respectively. JM22 and XM33 represent the freezing-tolerant and freezing-sensitive winter wheat varieties Jimai22 and Xumai33, respectively. Each value is the mean ± SE of three biological replicates. The different lowercase letters indicate statistically significant differences at the $p < 0.05$ level.

2.6. Correlation Analysis

As shown in Figure 8, there was a significant positive correlation between the ROS release rate and the REC in the young wheat ears under freezing, suggesting that freezing stress-induced ROS over-accumulation caused oxidative injury in the young wheat ears. The ROS release rate and REC were significantly and negatively related to the contents of AsA and GSH and the activities of GR, GST and GPX, while they were positively correlated with the activities of APX, DHAR and MDHAR in the young wheat ears under freezing. The ABA content was positively correlated with the activities of APX, MDHAR and DHA, while it was negatively related to the activities of GR, GST and GPX. In contrast, the contents of SA, IBA and cZ were positively correlated with the activities of GR, GST and GPX, while they were negatively correlated with the activities of APX, MDHAR and DHAR in the young wheat ears under freezing (Figure 8).

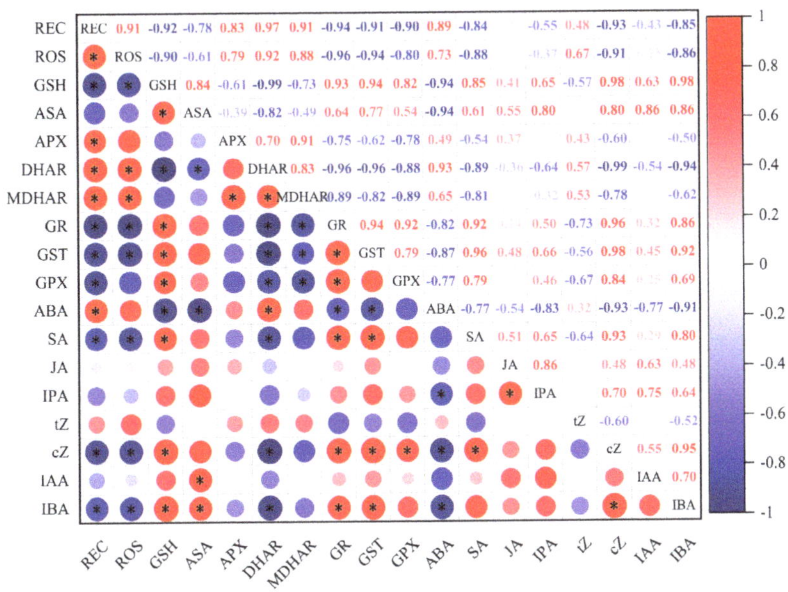

Figure 8. Correlation heat map for the freezing tolerance, AsA–GSH cycle metabolism and hormones levels in the young wheat ears under freezing stress at the ACFP. Note: REC, relative electrolyte conductivity; ROS, ROS release rate; GSH, the content of reduced glutathione; AsA, the content of ascorbic acid; APX, the activity of ascorbate peroxidase; DHAR, the activity of dehydroascorbate reductase; MDHAR, the activity of monodehydroascorbate reductase; GR, the activity of glutathione reductase; GST, the activity of glutathione-S-transferase; GPX, the activity of glutathione peroxidase; ABA, the content of abscisic acid; SA, the content of salicylic acid; JA, the content of jasmonic acid; IPA, the content of N6-isopentenyladenosine; tZ, the content of trans-zeatin; cZ, the content of cis-zeatin; IAA, the content of indoleacetic acid; IBA, the content of indolebutyric acid. *: Significant at the $p < 0.05$ level.

3. Discussion

The adverse effects of spring freezing on the growth and grain yield of wheat plants have been widely investigated [16,30,32,33]. Freezing stress seriously decreases the photosynthetic rates and causes a significant reduction in biomass accumulation, which can lead to declining yields [27]. Moreover, the death and abnormal development of young ears under freezing stress, resulting in reductions in the effective spike numbers and grain numbers per spike, are also important factors of spring freezing-caused yield loss [2,34,35]. The results in this study were consistent with the findings of previous studies (Figure 1). The JM22 plants showed smaller decreases in the plant height, biomass and grain numbers per spike and lower numbers of regenerated spikes compared to the XM33 plants under freezing stress (Figure 1), indicating that the JM22 plants (including the leaves and ears) were more tolerant to freezing stress than the XM33 plants. Moreover, the increase in the relative electrolyte conductivity (REC) in the young ears of JM22 was lower compared to that in the young ears of XM33 under freezing stress (Figure 2B), further demonstrating that the ears of JM22 had a better freezing tolerance than the ears of XM33 at the ACFP. Jiang et al. [3] and Wang et al. [17] reported that the content of malondialdehyde (MDA) was obviously increased in the young wheat ears at the ACFP. In the present study, the freezing treatment significantly increased the ROS release rates in the young ears of both varieties (Figure 2A). These results indicated that freezing stress caused oxidative injury to the young wheat ears, which was similar to what was observed in the wheat leaves [27,35]. The young ears of JM22 showed lower ROS release rates compared to the young ears of XM33 under freezing stress (Figure 2A), indicating that the young ears of JM22 had a stronger antioxidant capacity than the young ears of XM33. The significant positive correlation between the ROS release rate and REC (Figure 8) suggests that enhancing the antioxidant capacity is an important way to improve the freezing tolerance in young wheat ears.

Liang et al. [36] indicated that the AsA–GSH cycle was involved in maintaining redox homeostasis in wheat florets under cold stress at the booting stage. In the present study, the activity of APX in the young ears of XM33 was significantly increased under freezing stress (Table 1), suggesting that APX played a positive role in alleviating freezing-caused oxidative damage in young wheat ears at the ACFP. In addition, the activities of MDHAR and DHAR as well as the AsA/DHA ratio were significantly enhanced in the young ears of XM33 under freezing stress compared to the normal temperature control (Figure 4 and Table 1), indicating that the regeneration of AsA in the young ears of XM33 was accelerated in response to freezing stress. This result demonstrated that the regeneration of AsA plays an important role in young wheat ears resisting freezing stress. However, the generation of GSH in the young ears of XM33 was inhibited by freezing stress, as demonstrated by the decline in the GR activity and GSH/GSSG ratio (Figure 3 and Table 1). Moreover, the activities of GPX and GST in the young ears of XM33 were significantly decreased under the freezing treatment (Table 1). These results suggest that the young ears of XM33 mainly relied on AsA as a substrate to scavenge ROS rather than GSH under freezing stress at the ACFP. The increase in the GSH content in the young ears of XM33 under freezing stress might have been mainly used for the regeneration of AsA catalyzed by the enzyme DHAR [22].

The contents of AsA and DHA and the activities of MDHAR and APX were enhanced in the young ears of JM22 under freezing (Figure 4 and Table 1), implying that the young ears of JM22 also utilized AsA as a substrate to scavenge ROS under freezing. However, the young ears of JM22 displayed lower activities of MDHAR, DHAR and APX, and higher activities of GR, GST and GPX compared to the young ears of XM33 under freezing (Table 1), suggesting that the young ears of JM22 mainly depended more on GSH rather than AsA to alleviate oxidative damage caused by freezing stress compared to the young ears of XM33 at the ACFP. It is worth noting that the level of the GSH pool (231.8~392.3 μg/g FW) in the young wheat ears was strikingly higher than that of the AsA pool (98.86~123.4 μg/g FW) at the ACFP (Figures 3 and 4), which was generally consistent with the results of a study on wheat florets at the booting stage [35]. These results indicated that GSH may play a more important role in maintaining redox homeostasis than AsA in young wheat ears, especially under stress conditions. In addition, the young ears of JM22 showed a significantly higher GSH content and GSH/GSSG ratio than the young ears of XM33 under freezing stress (Figure 3). It was speculated that the difference in the ability to scavenge ROS using GSH may have been an important physiological reason for the difference in the freezing tolerance between the young ears of JM22 and XM33 at the ACFP.

Although the regulatory roles of hormones in wheat leaves and rhizomes in freezing tolerance have been widely studied [37,38], their roles in the response of young wheat ears to freezing are unclear. In this study, the levels of ABA, JA and SA were significantly increased in the young wheat ears of both varieties under freezing (Figure 5), suggesting that ABA, JA and SA play important roles in regulating the freezing response of young wheat ears at the ACFP. These results were consistent with previous studies that reported that the exogenous application of ABA, SA and JA could effectively enhance the freezing tolerance of wheat plants [37,39,40]. In addition, the response patterns of cZ, IAA and IBA were different in the young ears of JM22 (increased) and XM33 (decreased) under freezing (Figures 6 and 7). Zhang et al. [34] indicated that the freezing-induced abortion of florets was due to a decline in the level of IAA. The exogenous application of 6-benzylamino adenine (6-BA, a type of cytokinin) significantly increased the grain number per spike in wheat under chilling stress at the booting stage [41]. Our results indicated that cZ, IAA and IBA were positively related to the freezing tolerance of the young wheat ears at the ACFP, in accordance with the previous studies from other authors.

Kosová et al. [39] found that exogenous ABA significantly increased the AsA and GSH contents by regulating the expression levels and activities of AsA–GSH cycle metabolism-related enzymes in wheat leaves under freezing. In this study, the content of ABA was positively correlated with the activities of APX, DHAR and MDHAR, and was negatively correlated with the activities of GR, GST and GPX (Figure 8), suggesting that ABA plays opposing roles in regulating the regeneration of AsA and GSH in young wheat ears under freezing (Figure 9). These results were similar to those of Repkina et al. [42], who reported that exogenous ABA inhibited the activity of GR in pepper leaves under cold stress but increased the activity of DHAR. A previous study showed that SA, 6-BA and IAA treatments could upregulate the activities of the AsA–GSH cycle under stress conditions in plants [23,27,43]. In the present study, the contents of SA, IBA and cZ were positively correlated with the activities of GR, GST and GPX, and were negatively related to the activities of APX, DHAR and MDHAR, indicating that SA, IBA and cZ promoted the regeneration of GSH, but not AsA, in the young wheat ears under freezing stress. It is worth noting that the regulating roles of SA, IBA and cZ in the regeneration of GSH and AsA seem to be the opposite compared to ABA (Figure 8), suggesting that the metabolism of the AsA–GSH cycle is co-regulated by multiple hormones in the response of young wheat ears to freezing stress (Figure 9).

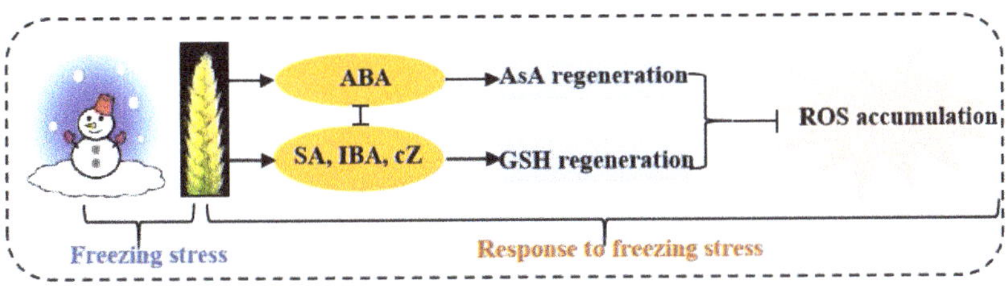

Figure 9. Summary showing the mechanisms of young wheat ears responding to freezing at the ACFP. Note: ABA, abscisic acid; SA, salicylic acid; IBA, indolebutyric acid; cZ, cis-zeatin; AsA, ascorbic acid; GSH, reduced glutathione; ROS, reactive oxygen species. The arrows indicate a positive regulation, whereas the lines ending with a bar show a negative regulation.

4. Materials and Methods

4.1. Experimental Design and Plant Material

This experiment was conducted during the wheat growing season of 2021–2022 at the Experimental Station of Yangzhou University, Yangzhou City, Jiangsu Province, China. Two winter wheat varieties, Xumai33 (XM33, freezing-sensitive) and Jimai22 (JM22, freezing-tolerant), were selected as the experimental materials. The wheat plants were planted in plastic pots (25 cm in diameter × 22 cm in height) filled with 9.5 kg of clay soil. The planting density was 10 plants per pot. The fertilizer and water management were carried out as described in our previous study [17]. The plants were grown outdoors until the young ears of the main stems developed to the ACFP. The development stage of the ears of the main stems in the two varieties was dynamically monitored using a microscope.

When the ears of the main stem of each variety developed to the ACFP, the wheat plants were randomly divided into two groups and moved into growth chambers (Eshengtaihe Co., Ltd., Beijing, China) for the temperature treatments. One group was transferred to a growth chamber with a temperature of 15 °C/5 °C (day/night, set according to the ambient temperature) as a control treatment (C). The other group was transferred to another growth chamber for freezing treatment (F). The freezing treatment was performed by cooling the temperature at a rate of 2 $°C \cdot h^{-1}$ and then holding it at −4 °C for 24 h. At the end of the freezing treatment, the ears of the main stems in each treatment were collected for further physiological and metabolic analyses. After 24 h of freezing treatment, the chamber temperature was increased to 4 °C (12 h) for thawing, and then the plants in the two growth chambers were moved out and grown under natural conditions until harvest. During the temperature treatments, the photosynthetically active radiation (PAR) in the growth chambers was set to 300 $\mu mol\ m^{-2}\ s^{-1}$, the photoperiod was 10 h and the relative air humidity was maintained at approx. 70%.

4.2. Measurements

4.2.1. Plant Height and Dry Mass

At maturity, fifteen plants were sampled and divided into three groups as three biological replicates for the measurements of the plant height. Plant height refers to the distance from the base of the stem to the top of the spike (excluding the length of the awns). Three pots of plants from each treatment were separately harvested as three biological replicates for the measurements of the aboveground biomass. The plants were killed at 105 °C for 30 min and then dried (80 °C) to a constant weight for the determination of the dry mass accumulation.

4.2.2. Regenerated Tiller and Grain Number

The regenerated tillers of the three pots from each treatment were marked with small tags to distinguish them from the original tillers. The regenerated tillers that could produce grains were counted as regenerated spikes. The number of regenerated spikes and original spikes were counted upon harvest. Fifteen spikes from the main stems were collected from each treatment at maturity and were equally divided into three groups as three biological replicates for the measurements of the grain numbers per spike. Threshing and counting the number of grains per spike were carried out manually.

4.2.3. ROS Release Rate and REC

The determination of the ROS release rates was accomplished using assay kits (Item No. ROS-1-Y) from Suzhou Keming Biotechnology Co., Ltd. (Suzhou, Jiangsu, China), using the method of specific probe fluorescence (2′,7′-dichlorodi-hydrofluorescin diacetate). The rate of change in the fluorescence intensity (u) was used to indicate the ROS release rate.

The REC was measured according to the method of Singh et al. [44]. Briefly, 0.1 g of fresh young ears was immersed in deionized water in test tubes for 2 h at room temperature, and the initial conductivity (E1) was measured using a conductivity bridge (DDS-307A, LEX Instruments Co., Ltd., Shanghai, China). Then, the tubes containing the young ears were boiled for 15 min to release all the electrolytes. They were cooled to room temperature, and the final conductivity (E2) was measured. The REC was calculated according to the formula REC = (E1/E2) × 100. Three replications were performed for each treatment.

4.2.4. AsA–GSH Cycle

The determination of AsA, GSH, GR and GPX were accomplished using assay kits (item No. A009-1-1, item No. A006-2-1, item No. A062-1-1 and item No. A005-1, respectively) from Nanjing Jiancheng Biotechnology Co., Ltd. (Nanjing, Jiangsu, China). The measurement of DHAR, MDHAR and GSSG were accomplished using assay kits (item No. DHAR-2-W, item No. MDHAR-2-W and item No. GSSG-2-W, respectively) from Suzhou Keming Biotechnology Co., Ltd. (Suzhou, Jiangsu, China). The contents of DHA and the activity of GST were determined using kits (item No. BC1240 and item No. AKPR013U) produced by Beijing Solarbio Science and Technology Co., Ltd. (Beijing, China) and Beijing Box Bio-engineering Technology Co., Ltd. (Beijing, China), respectively. The determination of the above indicators strictly followed the manufacturer's instructions.

The activity of APX (APX; EC1.11.1.11) was measured as described by Xia et al. [45]. In brief, 0.3 g of fresh young wheat ears were placed into a mortar, 4 mL of HEPES buffer (pH = 7.0) was added to make a homogenate, and then freeze centrifuging was applied for 20 min (10,000 r/min, 4 °C), taking 0.1 mL of the supernatant for use. To the supernatant, 2.7 mL of a reaction buffer (containing 0.1 mmol/L EDTA and 0.3 mmol/L AsA) was added sequentially, and finally 0.15 mL of H_2O_2 was added to initiate the enzymatic reaction, which was mixed immediately. The APX activity was calculated by recording the change in the absorbance value at 290 nm.

4.2.5. Endogenous Hormone Contents

The endogenous hormone content was determined with reference to the method of Simura et al. [46]. As follows, young wheat ears were lyophilized and ground into powder. An appropriate amount of the sample was then weighed, and 1 mL of the extract (50% acetonitrile in water, precooled at −40 °C, containing isotopically labeled internal standard mixture) was added to the sample and vortexed. Then, it was mixed to make a homogenate. The homogenate was allowed to stand at −40 °C for 1 h, and then cryocentrifuged for 15 min (12,000 r/min, 4 °C). The supernatant was taken as 850 μL, which was blown dry with nitrogen and then resolubilized with 85 μL of 10% aqueous acetonitrile to obtain the endogenous hormone solution to be measured. Chromatographic separation of the various hormones was performed on a Waters ACQUITY UPLC CSH C18 (150 × 2.1 mm, 1.7 μm, Waters, Milford, MA, USA) liquid chromatography column using an ultra-high-performance liquid chromatograph (EXIONLC system, SCIEX, Framingham, MA, USA), followed by qualitative and quantitative analyses using liquid chromatography–mass spectrometry (LC–MS) in the multiple reaction monitoring (MRM) mode. In the liquid chromatography, phase A was an aqueous solution containing 0.01% formic acid, phase B was acetonitrile containing 0.01% formic acid, the temperature of the column oven was 50 °C, the sample plate was set at 4 °C and the injection volume was 5 μL.

4.3. Statistical Analysis

The data calculations and graphing were conducted using Microsoft Excel 2021 and Origin 2021, respectively, while the analysis of variance (ANOVA) involved SPSS 20.0. The significant differences were determined using Duncan's multiple range test at the $p < 0.05$ level.

5. Conclusions

The results in this study indicated that the metabolism of the AsA–GSH cycle was involved in alleviating freezing-caused oxidative damage in young wheat ears at the ACFP. Compared to AsA, the regeneration and utilization of GSH were more important for maintaining redox homeostasis in young wheat ears under freezing stress. The hormones of SA, IBA and cZ may play a positive role in regulating the regeneration of GSH, and ABA may be implicated in controlling the regeneration of AsA in young wheat ears under freezing stress (Figure 9). However, the specific regulatory role of the various hormones in the response of young wheat ears to freezing stress and the synergistic mechanism between them need to be studied further.

Author Contributions: W.W. and Z.H. conceived and designed the study. W.W., C.L., Y.D. and X.J. participated in the experiments. W.W., C.N., C.L. and Y.Z. performed the data analysis. W.W. and Y.Z. wrote the manuscript. K.X., G.L. and C.Z. guided the experiments. All authors have read and agreed to the published version of the manuscript.

Funding: This research was funded by the National Natural Science Foundation of China (32301937), China Postdoctoral Science Foundation (2021M692721), Jiangsu Postdoctoral Research Funding Program (2021K342C), Jiangsu Province Higher Education Basic Science (Natural Science) Research Project (23KJB210016), Jiangsu Agricultural Science and Technology Innovation Fund (CX(22)1001), Jiangsu Provincial Key Research and Development Program (BE2020319), Lv Yang Jin Feng Research Plan and a project funded by the Priority Academic Program Development of Jiangsu Higher Education Institutions (PAPD).

Institutional Review Board Statement: Not applicable.

Informed Consent Statement: Not applicable.

Data Availability Statement: The data presented in this study are available upon request from the authors.

Conflicts of Interest: The authors declare no conflict of interest.

References

1. Wang, K.; Gong, Q.; Ye, X. Recent developments and applications of genetic transformation and genome editing technologies in wheat. *Theor. Appl. Genet.* **2020**, *133*, 1603–1622. [CrossRef]
2. Zhong, X.; Wang, D.; Zhao, P.; Yan, X.; Su, C. Occurrence of frost temperature in Huanghuai wheat production zone after winter wheat elongation. *Chin. J. Eco-Agri.* **2007**, *15*, 17–20.
3. Jiang, G.; Hassan, M.A.; Muhammad, N.; Arshad, M.; Chen, X.; Xu, Y.H.; Xu, H.; Ni, Q.Q.; Liu, B.B.; Yang, W.K.; et al. Comparative Physiology and Transcriptome Analysis of Young Spikes in Response to Late Spring Coldness in Wheat (*Triticum aestivum* L.). *Front. Plant Sci.* **2022**, *13*, 811–884. [CrossRef]
4. Lin, F.F.; Cheng, L.; Xu, B.; Chen, J.; Chen, A.H.; Hassan, M.A.; Liu, B.B.; Xu, H.; Chen, X.; Sun, J.Q.; et al. Late spring cold reduces grain number at various spike positions by regulating spike growth and assimilate distribution in winter wheat. *Crop J.* **2023**, *11*, 1272–1278. [CrossRef]
5. Xiao, L.J.; Liu, L.L.; Asseng, S.; Xia, Y.M.; Tang, L.; Liu, B.; Cao, W.X.; Zhu, Y. Estimating spring frost and its impact on yield across winter wheat in China. *Agric. For. Meteorol.* **2018**, *260–261*, 154–164. [CrossRef]
6. Crimp, S.; Zheng, B.; Khimashia, N.; Gobbett, D.; Chapman, S.; Howden, M.; Nicholls, N. Recent changes in Southern Australian frost occurrence: Implications for wheat production risk. *Crop Pasture Sci.* **2016**, *8*, 801–811. [CrossRef]
7. Zhong, X.; Mei, X.; Li, Y.; Yoshida, H.; Zhao, P.; Wang, X.; Han, L.; Hu, X.; Huang, S.; Huang, J. Changes in frost resistance of wheat young ears with development during jointing stage. *J. Agron. Crop Sci.* **2008**, *194*, 343–349. [CrossRef]
8. Saeidi, M.; Eliasi, P.; Abdoly, M.; Sasani, S. Freezing tolerance of wheat cultivars at the early growing season after winter. *Afr. J. Biotechnol.* **2012**, *11*, 4045–4052. [CrossRef]
9. Shimono, H. Earlier rice phenology as a result of climate change can increase the risk of cold damage during reproductive growth in northern Japan. *Agric. Ecosyst. Environ.* **2011**, *144*, 201–207. [CrossRef]
10. Li, L.; Han, C.L.; Yang, J.W.; Tian, Z.Q.; Jiang, R.Y.; Yang, F.; Jiao, K.M.; Qi, M.L.; Liu, L.L.; Zhang, B.Z.; et al. Comprehensive Transcriptome Analysis of Responses during Cold Stress in Wheat (*Triticum aestivum* L.). *Genes* **2023**, *14*, 844. [CrossRef] [PubMed]
11. Fuller, M.P.; Fuller, A.M.; Kaniouras, S.; Christophers, J.; Fredericks, T. The freezing characteristics of wheat at ear emergence. *Eur. J. Agron.* **2007**, *26*, 435–441. [CrossRef]
12. Gabriela, P.; Eugen, P.; Cătălin, D.S.; Cătălin, M. Basal sterility of wheat ears—Index of sensitivity to delayed spring frozen and its relationship with yield and productivity elements, in the climatic conditions of 2020 year at ards caracal. *Ann. Univ. Craiova Agri. Montan. Cad. Ser.* **2020**, *50*, 180–184.
13. Zhang, L.L.; Chen, X.; Ke, Y.Y.; Liu, B.B.; Muhammad, A.H.; Zhang, Y.; Xu, H.; Liu, J.C. An Inclusive Comparison of Identification Methods and Indices of Winter Wheat Tolerance against Late Spring Coldness. *Chin. J. Agrom.* **2021**, *42*, 146–157.
14. Yue, Y.J.; Zhou, Y.; Wang, J.A.; Ye, X.Y. Assessing Wheat Frost Risk with the Support of GIS: An Approach Coupling a Growing Season Meteorological Index and a Hybrid Fuzzy Neural Network Model. *Sustainability* **2016**, *8*, 1308. [CrossRef]
15. Choudhury, S.; Panda, P.; Sahoo, L.; Panda, S.K. Reactive oxygen species signaling in plants under abiotic stress. *Plant Signal. Behav.* **2013**, *8*, e23681. [CrossRef]
16. Li, X.N.; Cai, J.; Liu, F.L.; Dai, T.B.; Cao, W.X.J.D. Cold priming drives the sub-cellular antioxidant systems to protect photosynthetic electron transport against subsequent low temperature stress in winter wheat. *Plant Physiol. Biochem.* **2014**, *82*, 34–43. [CrossRef] [PubMed]

17. Wang, W.L.; Zhang, Y.T.; Liu, C.; Dong, Y.W.; Jiang, X.; Zhao, C.; Li, G.H.; Xu, K.; Huo, Z.Y. Label-Free Quantitative Proteomics Reveal the Mechanisms of Young Wheat (*Triticum aestivum* L.) Ears' Response to Spring Freezing. *Int. J. Mol. Sci.* **2023**, *24*, 15892. [CrossRef]
18. Apostoloval, P.; Yordanoval, R.; Popova, L. Response of antioxidative defence system to low temperature stress in two wheat cultivars. *Gen. Appl. Plant Physiol.* **2008**, *34*, 281–294.
19. Ruelland, E.; Vaultier, M.N.; Zachowski, A.; Hurry, V. Cold signalling and cold acclimation in plants. *Adv. Bot. Res.* **2009**, *49*, 35–150.
20. Wang, W.L.; Wang, X.; Lv, Z.S.; Khanzada, A.; Huang, M.; Cai, J.; Zhou, Q.; Huo, Z.Y.; Jiang, D. Effects of cold and salicylic acid priming on free proline and sucrose accumulation in winter wheat under freezing stress. *J. Plant Growth Regul.* **2022**, *41*, 2171–2184. [CrossRef]
21. Tiwari, Y.K.; Yadav, S.K. Effect of High-Temperature Stress on Ascorbate–Glutathione Cycle in Maize. *Agric. Res.* **2020**, *9*, 179–187. [CrossRef]
22. Hasanuzzaman, M.; Bhuyan, M.H.M.B.; Anee, T.I.; Parvin, K.; Nahar, K.; Mahmud, J.A.; Fujita, M. Regulation of ascorbate-glutathione pathway in mitigating oxidative damage in plants under abiotic stress. *Antioxidants* **2019**, *8*, 384. [CrossRef]
23. Mittler, R.; Vanderauwera, S.; Gollery, M.; Van Breusegem, F. Reactive oxygen gene network of plants. *Trends Plant Sci.* **2004**, *9*, 490–498. [CrossRef] [PubMed]
24. Wang, J.; Wang, D.Y.; Zhu, M.; Li, F.H. Exogenous 6-benzyladenine improves waterlogging tolerance in maize seedlings by mitigating oxidative stress and upregulating the ascorbate-glutathione cycle. *Front. Plant Sci.* **2021**, *12*, 680376. [CrossRef] [PubMed]
25. Iwaniuk, P.; Kaczyński, P.; Pietkun, M.; Łozowicka, B. Evaluation of titanium and silicon role in mitigation of fungicides toxicity in wheat expressed at the level of biochemical and antioxidant profile. *Chemosphere* **2022**, *308*, 136284. [CrossRef]
26. Hasanuzzaman, M.; Parvin, K.; Bardhan, K.; Nahar, K.; Anee, T.I.; Masud, A.A.C.; Fotopoulos, V. Biostimulants for the regulation of reactive oxygen species metabolism in plants under abiotic stress. *Cells* **2021**, *10*, 2537. [CrossRef] [PubMed]
27. Wang, W.L.; Wang, X.; Zhang, J.; Huang, M.; Cai, J.; Zhou, Q.; Dai, T.B.; Jiang, D. Salicylic acid and cold priming induce late-spring freezing tolerance by maintaining cellular redox homeostasis and protecting photosynthetic apparatus in wheat. *Plant Growth Regul.* **2021**, *90*, 109–121. [CrossRef]
28. Yu, J.; Cang, J.; Lu, Q.W.; Fan, B.; Xu, Q.H.; Li, W.N.; Wang, X.T. ABA enhanced cold tolerance of wheat 'dn1'via increasing ROS scavenging system. *Plant Signal. Behav.* **2020**, *15*, 1780403. [CrossRef]
29. Zhang, L.R.; Zhang, N.; Wang, S.S.; Tian, H.Y.; Liu, L.; Pei, D.; Yu, X.D.; Zhao, L.; Chen, F. A TaSnRK1 α Modulates TaPAP6L-Mediated Wheat Cold Tolerance through Regulating Endogenous Jasmonic Acid. *Adv. Sci.* **2023**, *10*, 2303478. [CrossRef]
30. Liu, L.L.; Xia, Y.M.; Liu, B.; Chang, C.Y.; Xiao, L.J.; Shen, J.; Tang, L.; Cao, W.X. Individual and combined effects of jointing and booting low-temperature stress on wheat yield. *Eur. J. Agron.* **2020**, *113*, 125989. [CrossRef]
31. Roxas, V.P.; Lodhi, S.A.; Garrett, D.K.; Mahan, J.R.; Allen, R.D. Stress tolerance in transgenic tobacco seedlings that overexpress glutathione S-transferase/glutathione peroxidase. *Plant Cell Physiol.* **2000**, *41*, 1229–1234. [CrossRef]
32. Han, Q.X.; Kang, G.Z.; Guo, T.C. Proteomic analysis of spring freeze-stress responsive proteins in leaves of bread wheat (*Triticum aestivum* L.). *Plant Physiol Biochem.* **2013**, *63*, 236–244.
33. Cheong, B.E.; Onyemaobi, O.; Wing Ho Ho, W.W.; Biddulph, T.B.; Rupasinghe, T.W.; Roessner, U.; Dolferus, R. Phenotyping the Chilling and Freezing Responses of Young Microspore Stage Wheat Spikes Using Targeted Metabolome and Lipidome Profiling. *Cells* **2020**, *9*, 1309. [CrossRef]
34. Zhang, W.J.; Wang, J.Q.; Huang, Z.L.; Mi, L.; Xu, K.; Wu, J.J.; Fan, Y.H.; Ma, S.Y.; Jiang, D.G. Effects of low temperature at booting stage on sucrose metabolism and endogenous hormone contents in winter wheat spikelet. *Front. Plant Sci.* **2019**, *10*, 498. [CrossRef] [PubMed]
35. Li, X.N.; Pu, H.C.; Liu, F.L.; Zhou, Q.; Cai, J.; Dai, T.B.; Cao, W.X.; Jiang, D. Winter wheat photosynthesis and grain yield responses to spring freeze. *Agron. J.* **2015**, *107*, 1002–1010. [CrossRef]
36. Liang, Z.; Luo, J.; Wei, B.; Liao, Y.; Liu, Y. Trehalose can alleviate decreases in grain number per spike caused by low-temperature stress at the booting stage by promoting floret fertility in wheat. *J. Agron. Crop Sci.* **2021**, *207*, 717–732. [CrossRef]
37. Wang, W.L.; Wang, X.; Huang, M.; Cai, J.; Zhou, Q.; Dai, T.B.; Cao, W.X.; Jiang, D. Hydrogen peroxide and abscisic acid mediate salicylic acid-induced freezing tolerance in wheat. *Front. Plant Sci.* **2018**, *9*, 1137. [CrossRef]
38. Si, T.; Wang, X.; Wu, L.; Zhao, C.Z.; Zhang, L.N.; Huang, M.; Cai, J.; Zhou, Q.; Dai, T.B.; Zhu, J.K.; et al. Nitric oxide and hydrogen peroxide mediate wounding-induced freezing tolerance through modifications in photosystem and antioxidant system in wheat. *Front. Plant Sci.* **2017**, *8*, 1284. [CrossRef]
39. Kosová, K.; Prášil, I.T.; Vítámvás, P.; Dobrev, P.; Motyka, V.; Floková, K.; Vanková, R. Complex phytohormone responses during the cold acclimation of two wheat cultivars differing in cold tolerance, winter Samanta and spring Sandra. *J. Plant Physiol.* **2012**, *169*, 567–576. [CrossRef]
40. Zhang, N.; Li, L.j.; Zhang, L.R.; Li, J.; Fang, Y.H.; Zhao, L.; Ren, Y.; Chen, F. Abscisic acid enhances tolerance to spring freeze stress and regulates the expression of ascorbate-glutathione biosynthesis-related genes and stress-responsive genes in common wheat. *Mol. Breed.* **2020**, *40*, 108. [CrossRef]
41. Wang, W.L.; Wang, X.; Huang, M.; Cai, J.; Zhou, Q.; Dai, T.B.; Jiang, D. Alleviation of field low-temperature stress in winter wheat by exogenous application of salicylic acid. *J. Plant Growth Regul.* **2021**, *40*, 811–823. [CrossRef]

42. Repkina, N.; Ignatenko, A.; Holoptseva, E.; MiszalskI, Z.; Kaszycki, P.; Talanova, V. Exogenous methyl jasmonate improves cold tolerance with parallel induction of two cold-regulated (COR) genes expression in *Triticum aestivum* L. *Plants* **2021**, *10*, 1421. [CrossRef] [PubMed]
43. Guo, W.L.; Chen, R.G.; Gong, Z.H.; Yin, Y.X.; Ahmed, S.S.; He, Y.M. Exogenous abscisic acid increases antioxidant enzymes and related gene expression in pepper (*Capsicum annuum*) leaves subjected to chilling stress. *Genet. Mol. Res.* **2012**, *11*, 4063–4080. [CrossRef]
44. Singh, H.P.; Batish, D.R.; Kohli, R.K.; Arora, K. Arsenic-induced root growth inhibition in mung bean (*Phaseolus aureus* Roxb.) is due to oxidative stress resulting from enhanced lipid peroxidation. *Plant Growth Regul.* **2007**, *53*, 65–73. [CrossRef]
45. Xia, J.X.; Wang, Y.J.M.; Zhou, Y.H.; Tao, Y.; Mao, W.H.; Shi, K.; Asami, T.; Chen, Z.X.; Yu, J.Q. Reactive oxygen species are involved in brassinosteroid-induced stress tolerance in cucumber. *Plant Physiol.* **2009**, *150*, 801–814. [CrossRef] [PubMed]
46. Simura, J.; Antoniadi, I.; Široká, J.; Tarkowská, D.; Strnad, M.; Ljung, K.; Novák, O. Plant Hormonomics: Multiple Phytohormone Profiling by Targeted Metabolomics. *Plant Physiol.* **2018**, *177*, 476–489. [CrossRef] [PubMed]

plants

Article

Enhanced Stomatal Conductance Supports Photosynthesis in Wheat to Improved NH_4^+ Tolerance

Jinling Hu †, Qiaomei Zheng †, Chaofeng Dong, Zhihui Liang, Zhongwei Tian and Tingbo Dai *

Key Laboratory of Crop Physiology Ecology and Production Management of Ministry of Agriculture, Nanjing Agricultural University, Nanjing 210095, China; 2018201007@njau.edu.cn (J.H.); 2020201015@stu.njau.edu.cn (Q.Z.); 2021101015@stu.njau.edu.cn (C.D.); 2015101019@njau.edu.cn (Z.L.); zhwtian@njau.edu.cn (Z.T.)

* Correspondence: tingbod@njau.edu.cn; Tel.: +86-025-84395033

† These authors contributed equally to this work.

Abstract: The impact of ammonium (NH_4^+) stress on plant growth varies across species and cultivars, necessitating an in-depth exploration of the underlying response mechanisms. This study delves into elucidating the photosynthetic responses and differences in tolerance to NH_4^+ stress by investigating the effects on two wheat (*Triticum aestivum* L.) cultivars, Xumai25 (NH_4^+-less sensitive) and Yangmai20 (NH_4^+-sensitive). The cultivars were grown under hydroponic conditions with either sole ammonium nitrogen (NH_4^+, AN) or nitrate nitrogen (NO_3^-, NN) as the nitrogen source. NH_4^+ stress exerted a profound inhibitory effect on seedling growth and photosynthesis in wheat. However, these effects were less pronounced in Xumai25 than in Yangmai20. Dynamic photosynthetic analysis revealed that the suppression in photosynthesis was primarily attributed to stomatal limitation associated with a decrease in leaf water status and osmotic potential. Compared to Yangmai20, Xumai25 exhibited a significantly higher leaf K^+ concentration and *TaAKT1* upregulation, leading to a stronger stomatal opening and, consequently, a better photosynthetic performance under NH_4^+ stress. In conclusion, our study suggested stomatal limitation as the primary factor restricting photosynthesis under NH_4^+ stress. Furthermore, we demonstrated that improved regulation of osmotic substances contributed to higher stomatal conductance and enhanced photosynthetic performance in Xumai25.

Keywords: ammonium stress; photosynthesis; wheat; stomatal conductance; osmatic potential

1. Introduction

It is well known that ammonium (NH_4^+), though a principal nitrogen source, can induce plant toxicity, particularly when used as the sole or predominant nitrogen source [1–3]. Several studies have investigated the manifestations of NH_4^+ stress in plants, including diminished biomass, elevated reactive oxygen species (ROS) production, disruptions in pH equilibrium, and ion regulation [2,4]. The sensitivity of plants to NH_4^+ stress is inherently variable, contingent upon factors such as species, genotypes, and soil conditions [1,4,5]. Though NH_4^+ assimilation heavily relies on the sugar generated during photosynthesis in the leaves, extant research has predominantly concentrated on the mechanisms underlying NH_4^+ uptake and assimilation in plant roots [6,7]. A discernible knowledge gap exists in how NH_4^+ stress affects photosynthesis, especially in susceptible plant species.

During photosynthesis, the diffusion of CO_2 from the atmosphere to the chloroplast sites for carboxylation encounters multiple resistances within leaves. The regulatory mechanism underlying stomatal opening is well elucidated [8]. It comprises an initial influx of K^+, which is later replaced by sucrose, culminating in a reduction of sucrose levels [9]. Recent investigations in rice plants have demonstrated that K^+ deficiency can reduce stomatal conductance, restricting gas exchange [10]. Furthermore, NH_4^+ absorption has been shown to impede the uptake of other cations, such as K^+ or Ca^{2+}, and elevate

ABA content [4,11]. Thus, NH_4^+ stress might instigate a reduction in stomatal conductance, adversely impacting photosynthesis.

While the regulation of stomatal conductance (g_s) and transpiration by nitrogen nutrition is a well-established tenet in plant physiology, the impact of excess NH_4^+ on transpiration and water use efficiency (WUE) remains unclear. Intriguingly, distinct nitrogen forms can impart divergent effects on plant water uptake, this phenomenon is intricately linked to the nitrogen preference of various plant species [12]. Some studies posit that excessive NH_4^+ can inhibit plant water uptake [13] or induce symptoms akin to water deficit, termed "ammoniacal syndrome" [14]. Conversely, other studies argue that compared to NO_3^- nutrition, NH_4^+ nutrition can further strengthen water stress resistance in rice via marked upregulation of AQP genes [12,15]. Thus, the influence of NH_4^+ stress on the water status of plants and its intricate relationship to stomatal opening and subsequent photosynthetic capacity remain unclear.

Chlorophyll fluorescence, closely linked to chlorophyll content, emerges as a crucial indicator of the photosynthetic response to stress [16]. NH_4^+ stress adversely impacts the chlorophyll content of plants, however, the degree of impact depends on the severity of stress and the plant species [17,18]. Previous studies have postulated that NH_4^+ stress can disrupt the stability of leaf membrane lipids, hindering electron transfer and consequently impeding photosynthesis [19]. In contrast, a previous study in *Arabidopsis* showed that NH_4^+ stress elevates mitochondrial ROS levels without significant impacts on photosynthesis [20]. Such divergent findings warrant the need to comprehensively elucidate the regulatory mechanisms governing chlorophyll content and chlorophyll fluorescence under NH_4^+ stress.

Wheat, a key part of global food production, is well known for its high susceptibility to NH_4^+ toxicity [21,22]. In wheat seedlings, NH_4^+ stress is manifested as stunted seedling growth and the onset of oxidative stress [21]. The pivotal role of nitrogen in the metabolism of photosynthetic pigments and the efficient functioning of the photosynthetic apparatus underscores its profound impact on overall plant growth and development, including in wheat. Despite the indispensable role of nitrogen in photosynthesis, the specific photosynthetic response of wheat seedlings to NH_4^+ stress remains unknown. The primary objective of this study was to delve into the mechanism underlying the tolerance and photosynthetic response of wheat seedlings to NH_4^+ stress. Here, we conducted a comparative analysis of two wheat cultivars: NH_4^+-less sensitive Xumai25 and NH_4^+-sensitive Yangmai20. We focused on growth responses, gas exchange, leaf water status, and adjustments in osmotic balance in both these cultivars in response to NH_4^+ stress.

2. Results

2.1. Plant Growth and Nitrogen Concentration

The impact of NH_4^+ stress (AN) on plant growth was evident from the significantly lower dry weight of AN-treated plants than NN-treated plants at 10 DAT, with a more pronounced decrease in Yangmai20 (25%) than Xumai25 (7%). Similarly, the AN-treated plants exhibited reduced leaf area, with a more pronounced reduction in Yangmai20 than Xumai25 (Table 1). In contrast, the AN-treated plants showed higher specific leaf weight (SLW) and nitrogen concentration than the NN-treated plants, with a more pronounced increase in Yangmai20.

2.2. Photosynthesis and Related Attributes

The AN-treated plants exhibited a progressive decline in the net photosynthetic rate (A) (Figure 1A) from 1 to 10 DAT, with a more pronounced decline observed in NH_4^+-sensitive Yangmai20. At 5 DAT, the value of A was significantly reduced in Yangmai20 under NH_4^+ stress compared to NN-treated plants, but this phenomenon was not observed in Xumai25. Furthermore, AN significantly reduced stomatal conductance (g_s), sub-stomatal CO_2 concentration (C_i), and transpiration rate (Tr) in Yangmai20 (Figure 1B,D,E), compared to the NN (17%, 11%, and 13%, respectively). In contrast, AN substantially increased stomatal limitation (l) in both cultivars, with consistently higher stomatal limitation, evidenced by lower g_s, in Yangmai20 than in Xumai25 (Figure 1B,F). Additionally, AN did not significantly affect leaf mesophyll conductance (g_m) (Figure 1C), carboxylation efficiency (CE), electron transfer rate (J_e), maximum carboxylation rate (V_{max}), or maximum electron transport rate (J_{max}) in either cultivar (Figure 2A–D).

Table 1. Effects of ammonium stress on dry weight, leaf area, leaf nitrogen concentration, and specific leaf weight of wheat seedling at 10 days after treatment.

Cultivar	Treatment	Plant Dry Weight (mg plant^{-1})	Plant Leaf Area (cm^2 plant^{-1})	Leaf Nitrogen Concentration (mg g^{-1})	Specific Leaf Weight of 4th Leaf (mg cm^{-2} DW)
Yangmai20	NN	1266 ± 39.7 a	100 ± 4.6 a	69 ± 1.9 b	0.30 ± 0.008 b
	AN	980 ± 31.2 c	82 ± 5.4 c	77 ± 2.5 a	0.38 ± 0.017 a
Xumai25	NN	1128 ± 50.3 b	97 ± 4.7 ab	65 ± 3.1 b	0.29 ± 0.013 b
	AN	1033 ± 50.2 c	86 ± 3.4 bc	75 ± 2.2 a	0.37 ± 0.006 a

Note: NN, nitrate treatment; AN, ammonium treatment. Data are means ± standard deviation (SD) of three biological replicates, and different letters indicate significant differences (p < 0.05) according to ANOVA.

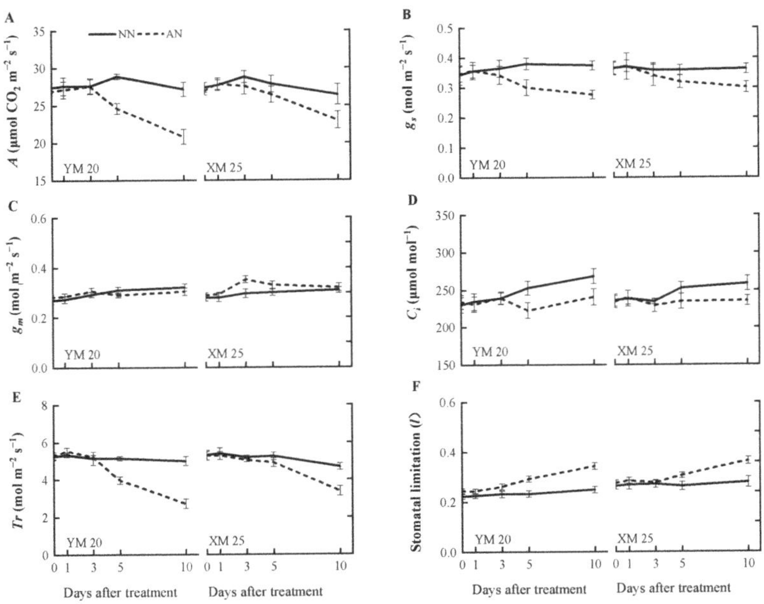

Figure 1. Effects of ammonium stress on (**A**) net photosynthetic rate (A), (**B**) stomatal conductance (g_s), (**C**) leaf mesophyll conductance (g_m), (**D**) sub-stomatal CO_2 concentration (C_i), (**E**) transpiration rate (Tr), and (**F**) stomatal limitation (l) of wheat seedlings at 0, 1, 3, 5, and 10 days after treatment. NN, nitrate treatment; AN, ammonium treatment. YM20, Yangmai20 (NH_4^+-sensitive); XM25, Xumai25 (NH_4^+-less sensitive).

Furthermore, at 10 DAT, AN led to a significant decline in CE, V_{max}, and J_e of Yangmai20 but did not markedly impact the corresponding parameters in Xumai25 (Figure 2A–C).

2.3. *Chlorophyll Content and Chlorophyll Fluorescence Parameters*

Changes in chlorophyll content and chlorophyll fluorescence were concurrently measured alongside photosynthetic parameters. The AN-treated Yangmai20 exhibited slightly elevated total chlorophyll content at 3 and 5 DAT, while this phenomenon was not observed in Xumai25. However, there was no significant difference in the total chlorophyll content of the AN- and NN-treated plants at 10 DAT (Figure 3A). Notably, no significant differences were observed in the actual photochemical efficiency ($Y(II)$) of AN- and NN-treated Xumai25 (Figure 3B–D). In contrast, the $Y(II)$ value of AN-treated Yangmai20 was significantly lower than NN-treated Yangmai20 at 10 DAT (Figure 3B). Furthermore, no significant differences were observed in the maximum quantum yield (Fv/Fm) of the AN- and NN-treated plants. The non-photochemical quenching (NPQ) of AN-treated plants was significantly higher than NN-treated plants at 10 DAT.

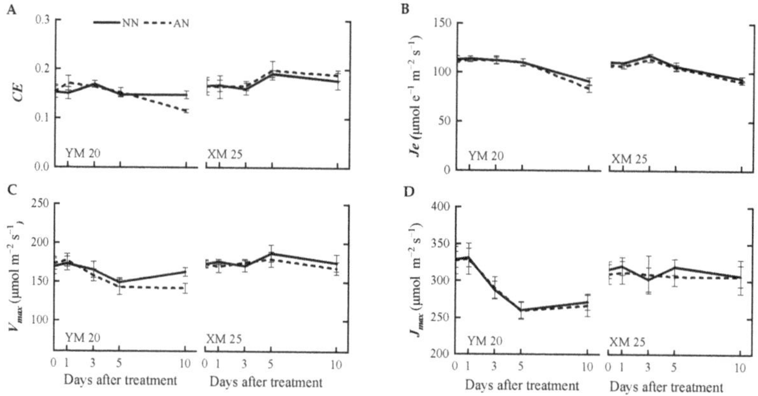

Figure 2. Effects of ammonium stress on the (**A**) carboxylation efficiency (CE), (**B**) electron transfer rate (J_e), (**C**) maximum carboxylation rate (V_{max}), and (**D**) the maximum electron transport rate (J_{max},) at 0, 1, 3, 5, and 10 days after treatment NN, nitrate treatment; AN, ammonium treatment. YM20, Yangmai20 (NH_4^+-sensitive); XM25, Xumai25 (NH_4^+-less sensitive).

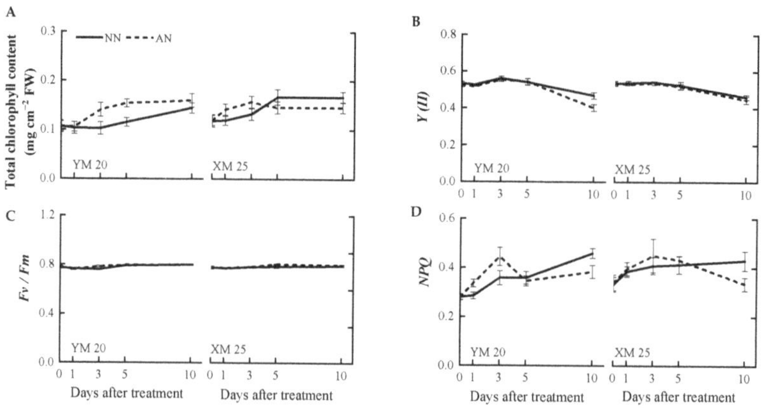

Figure 3. Effects of ammonium stress on (**A**) total chlorophyll content, (**B**) actual photochemical efficiency ($Y(II)$), (**C**) maximum quantum yield (Fv/Fm), and (**D**) non-photochemical quenching (NPQ) of wheat seedlings at 0, 1, 3, 5, and 10 days after treatment. NN, nitrate treatment; AN, ammonium treatment. YM20, Yangmai20 (NH_4^+-sensitive); XM25, Xumai25 (NH_4^+-less sensitive).

2.4. *Leaf Moisture Status*

AN treatment adversely impacted the leaf moisture status compared to the NN treatment. At 5 DAT, AN-treated Yangmai20 exhibited significantly reduced relative water content (RWC), pre-dawn water potential (ψp), osmotic potential (ψs), and midday water potential (ψm) compared to NN-treated Yangmai20, while no significant differences were observed between AN- and NN-treated Xumai25. At 10 DAT, the AN-treated plants exhibited a decrease in RWC, ψs, ψp, and ψm, with a more pronounced decrease in Yangmai20 (Figure 4A–D).

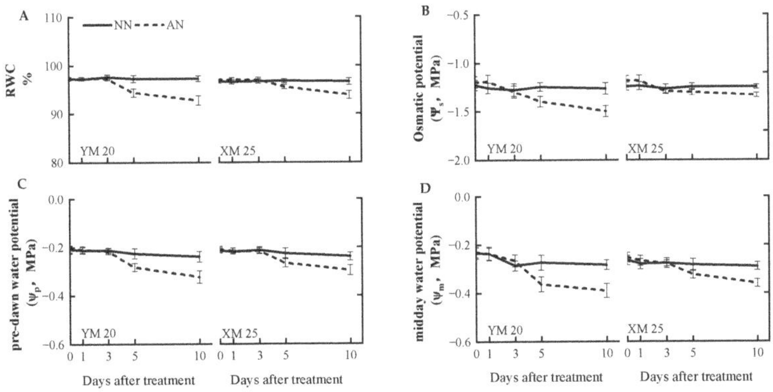

Figure 4. Effects of ammonium stress on the (**A**) relative water content (RWC), (**B**) leaf osmotic potential (ψs), (**C**) pre-dawn water potential (ψp), and (**D**) midday water potential (ψm) of wheat seedlings at 0, 1, 3, 5, and 10 days after treatment (DAT). NN, nitrate treatment; AN, ammonium treatment. YM20, Yangmai20 (NH_4^+-sensitive); XM25, Xumai25 (NH_4^+-less sensitive).

2.5. Concentration of K^+, Sucrose and ABA

AN treatment significantly impacted the parameters related to osmotic homeostasis and stomatal opening. All AN-treated plants exhibited consistently decreasing K^+ and sucrose levels (Figure 5A,B). Although, Xumai25 exhibited substantially less pronounced K^+ reduction than Yangmai20 (Figure 5A). In addition, AN-treated Yangmai20 exhibited higher ABA concentration in the leaves than NN-treated plants at 5 DAT, while no significant differences were observed between AN- and NN-treated Xumai25. At 10 DAT, ABA concentration was significantly increased in both cultivars under AN treatment (Figure 5C).

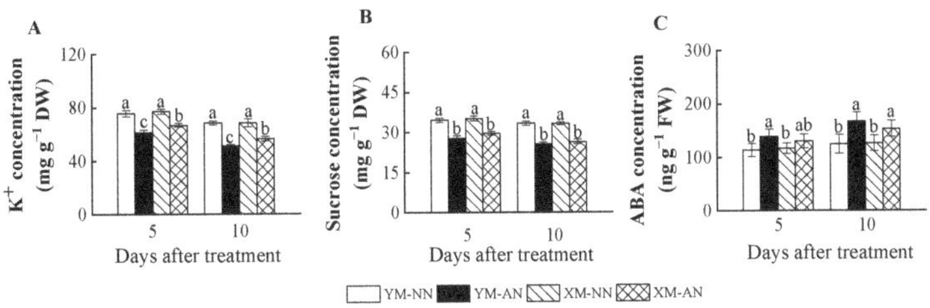

Figure 5. Effects of ammonium stress on (**A**) K^+, (**B**) sucrose, and (**C**) ABA concentration of wheat seedlings at 5 and 10 days after treatment. Different letters indicate significant differences ($p < 0.05$) according to ANOVA. NN, nitrate treatment; AN, ammonium treatment. YM, Yangmai20 (NH_4^+-sensitive); XM, Xumai25 (NH_4^+-less sensitive).

2.6. Relative Gene Expression

In leaves, the AN-treated plants exhibited higher *TaHA1* and *TaAKT1* expression than NN-treated plants in both cultivars (Figure 6A,B). Furthermore, AN-treated Yangmai20 exhibited significantly lower *TaKOR1* expression than NN-treated Yangmai20, while no significant difference was observed between AN- and NN-treated Xumai25 (Figure 6C). However, the *TaKAT1* expressions of AN- and NN-treated Yangmai20 were comparable to AN- and NN-treated Xumai25, respectively (Figure 6D).

In roots, AN-treated Yangmai20 exhibited significantly lower *TaTIP2.3* and *TaPIP1.1* expression than NN-treated Yangmai20, while this phenomenon is not observed in Xumai25 (Figure 6E,F). Moreover, the AN-treated plants exhibited higher *TaAKT1* expression than NN-treated plants, with significantly higher expression in AN-treated Xumai25 than NN-treated Xumai25 (Figure 6G).

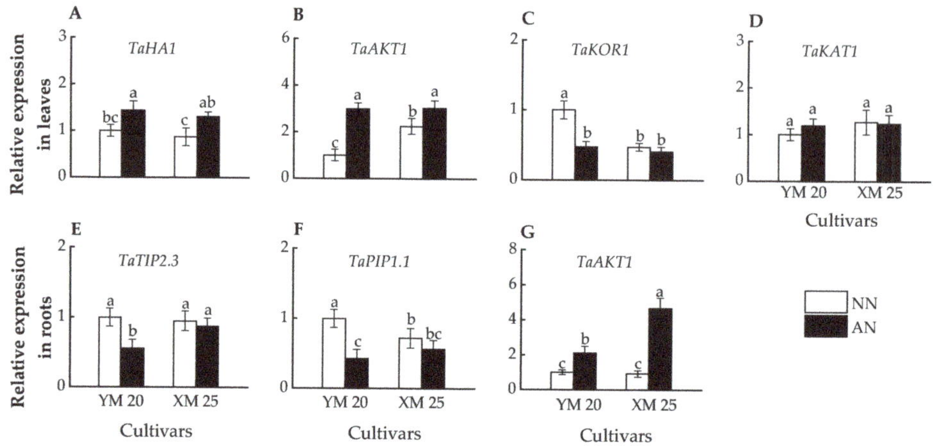

Figure 6. Effects of ammonium stress on the expressions of genes encoding K^+ channel, water channel, and proton transport protein in leaves and roots of wheat seedlings at 5 days after treatment, including (**A**) *TaHA1*, (**B**) *TaAKT1*, (**C**) *TaKOR1*, (**D**) *TaKAT1*, (**E**) *TaTIP2.3*, (**F**) *TaPIP1.1*, (**G**) *TaAKT1*. Data are expressed as means of three biological replicates, and different letters indicate significant differences ($p < 0.05$) according to ANOVA. NN, nitrate treatment; AN, ammonium treatment. YM, Yangmai20 (NH_4^+-sensitive); XM, Xumai25 (NH_4^+-less sensitive).

2.7. The Relationships between Photosynthetic Parameters and Physiological Traits with Nitrogen Forms

Principal component analysis (PCA) demonstrated that PC1 and PC2 explained 74.3% of the total trait variance for the two wheat cultivars under the two treatments (AN and NN) (Figure 7). The eigenvalue and cumulative contribution of PC1 and PC2 of photosynthetic parameter and physiological traits and their corresponding loading are shown in Supplementary Table S1. PC1 explained 56.8% of total variation and had a positive association with ψm, ψs, ψp, K^+, g_s, and *A*, and a negative association with *l*. Thus, PC1 tended to represent traits like leaf moisture status, osmotic substance, and photosynthetic parameters. Alternatively, PC2 explained 17.5% of the total variation and was positively associated with *NPQ* and *Y(II)*, representing the chlorophyll fluorescence. These results indicated that photosynthetic traits might be positively related to leaf moisture status and osmotic substance, and were negatively correlated to *l*.

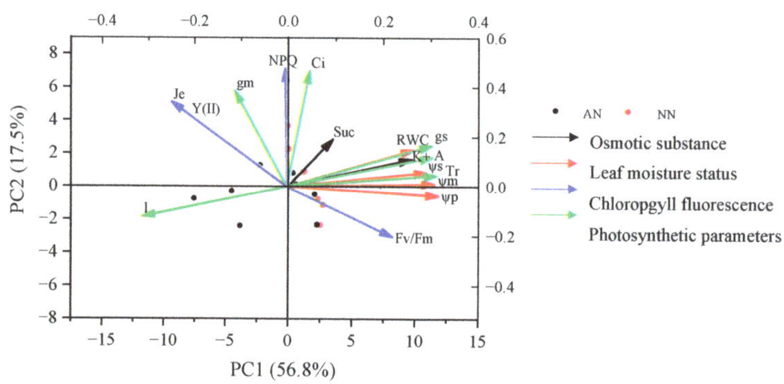

Figure 7. *Cont.*

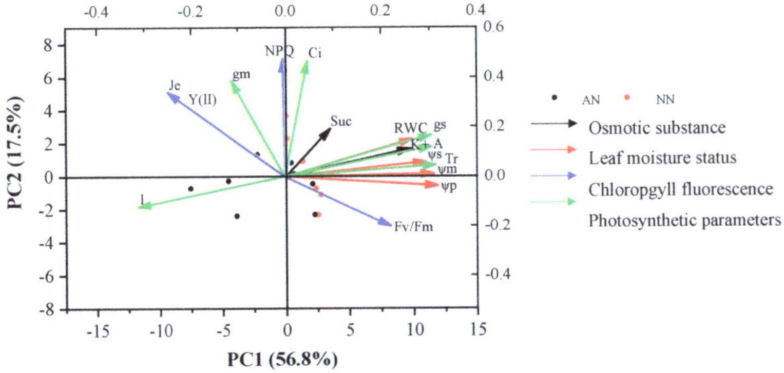

Figure 7. Principal component analysis (PCA) of photosynthetic parameters and other physiological traits of two wheat cultivars (Yangmai20 and Xumai25) under AN and NN treatments. NN, nitrate treatment; AN, ammonium treatment.

3. Discussion

Diverse plant species and cultivars manifest distinctive responses to NH_4^+ stress, as demonstrated by studies on peas [23], tomatoes [24], and *Arabidopsis* [25]. Several studies have extensively reported the NH_4^+ sensitivity of wheat [21,26]. In this study, the deleterious impact of NH_4^+ stress on the biomass and photosynthesis of wheat seedlings (Table 1 and Figure 1A) is similar to the effects reported in other plant species [2,27]. Notably, upon NH_4^+ stress, the NH_4^+-less sensitive cultivar Xumai25 demonstrated a mitigated reduction in net photosynthetic rate (A) compared to the NH_4^+-sensitive cultivar Yangmai20, resulting in superior dry matter production (Table 1, Figure 1A).

3.1. Variation in Photosynthetic Response and Stomatal Limitation

Photosynthesis, intricately linked to NH_4^+ tolerance [28], involves limitation and biochemical limitations influencing photosynthetic capacity. Our findings indicated that Yangmai20 exhibited a notable reduction in A, g_s, C_i, and an increase in l (Figure 1A–D). However, parameters related to the photosynthetic dark reaction, such as CE, J_e, V_{max}, and J_{max}, did not differ significantly between AN- and NN-treated plants (Figure 2A–D). Previous studies have demonstrated l to be a critical factor in reducing A when both C_i and g_s decrease [29,30]. Thus, in the present study, l might predominantly contribute to the reduced photosynthesis in wheat under NH_4^+ stress. Consistent with our findings, NH_4^+ stress has previously been found to induce stomatal closure in tomatoes [28], and reduction in rice [31] and wheat [32]. Notably, Xumai25 exhibited a less pronounced decrease in g_s at 5 DAT, indicating its superior ability to maintain the open state of stomata, contributing to enhanced photosynthetic performance under NH_4^+ stress. Furthermore, at 10 DAT, the dark reaction parameters (CE, V_{max}, and J_e) significantly decreased in Yangmai20 under prolonged NH_4^+ stress (Figure 2A,C,D), illustrating a delayed response, which might be attributed to initial CO_2 assimilation induced by reduced g_s [33].

3.2. Leaf Mesophyll Conductance (g_m) under NH_4^+ Stress

The impact of NH_4^+ on g_m varies based on the plant species and severity of the stress. For instance, Liu et al. (2021) [34] found a significantly lower g_m in female Populous cathayana plants after being supplied with NH_4^+ under salt stress than after being supplied with NO_3^-. In contrast, Li et al. (2012) [35] reported a significantly higher g_m in NH_4^+- supplied rice than NO_3^--supplied rice under drought stress. In the present study, however, we did not observe a significant difference in g_m between AN- and NN-treated plants (Figure 1C). The observed variations in g_m can be attributed to two factors. First, increased NH_4^+ assimilation products augmented the specific leaf weight (SLW) in both cultivars (Table 1), influencing CO_2 partial pressure inside chloroplasts, which can increase g_m [36]. Second, heightened leaf nitrogen content (Table 1) after NH_4^+ stress possibly led to increased cell wall thickness [37].

3.3. Chlorophyll Concentration and Fluorescence under NH_4^+ Stress

The impact of NH_4^+ stress on chlorophyll content varies based on stress severity, plant species, and cultivars [18]. For instance, several studies have reported a decrease in the chlorophyll content of *Arabidopsis* in the presence of high NH_4^+ levels [17,38]. In contrast, some studies have reported increased and decreased chlorophyll content after mild and severe NH_4^+ stress, respectively [18]. In the present study, the chlorophyll content in both cultivars was not markedly different between NN-treated and AN-treated plants (Figure 3A), suggesting that 5 mM NH_4^+ was considered a moderate stress level for wheat seedlings. Additionally, chlorophyll content is closely associated with the efficiency of electron transport and photosynthesis. For instance, Chen et al. (2023) [27] demonstrated that NH_4^+ exposure disrupts the electron transport chain in citrus plants. Interestingly, in this study, we did not observe any significant difference in J_e as A decreased in Yangmai20 at 5 DAT and in Xumai25 at 10 DAT. Chlorophyll fluorescence measurements indicated no significant damage to the PSII reaction centers during the treatment period, as evident by the unchanged Fv/Fm in both cultivars (Figure 3C). The reduction in $Y(II)$ in AN-treated Yangmai20 (Figure 3B) might be responsible for the decrease in J_e (Figure 2B). In addition, the increase in NPQ in AN-treated plants was possibly related to photorespiration [39].

3.4. Decline in Stomatal Conductance and Osmotic Potential under NH_4^+ Stress

Several factors, such as stomatal morphology, distribution, and movement, influence g_s, a pivotal determinant of photosynthesis [40]. The focus of the current investigation was the nuanced impact of NH_4^+ stress on g_s, with specific attention to the intricate regulation of stomatal movement. We meticulously examined the newest fully expanded leaves before treatment initiation to circumvent potential confounding effects from stomatal morphology and distribution. Stomatal movement, primarily orchestrated by the changes in the turgor pressure and volume of the guard cells, is governed by major osmotic entities, such as K^+ and sucrose [8]. Our findings revealed a marked reduction in the leaf osmotic potential (Ψs) of Yangmai20 since 5 DAT (Figure 4B), aligning with the decline in g_s. This finding indicates a direct correlation between diminishing Ψs under NH_4^+ stress and the consequential reduction in stomatal aperture.

Recent research emphasizes the pivotal role of sucrose and K^+ as primary solutes influencing the guard cell osmotic potential [8,9]. K^+ deficiency has been shown to reduce leaf water potential and stomatal area in rice, resulting in a decline in photosynthesis [10]. In the current study, a notable reduction in leaf K^+ concentration was observed for both wheat cultivars under NH_4^+ stress, with a more pronounced reduction in Yangmai20 (Figure 5A). Intriguingly, our study unveiled a higher *TaAKT1* upregulation under NH_4^+ stress in the roots of Xumai25 than Yangmai20 (Figure 6G), potentially contributing to the higher K^+ concentration in Xumai25. Moreover, we observed a significant decrease in the sucrose level of both cultivars under NH_4^+ stress (Figure 1A), which potentially decreases osmotic potential. The decreased sucrose concentration is potentially linked to inhibited photosynthesis (Figure 1A) and increased carbon skeleton requirements due to ammonium assimilation in the roots [2]. Taken together, the reduction in K^+ and sucrose levels under NH_4^+ stress led to a decrease in leaf osmotic potential, which in turn resulted in a decrease in g_s. Xumai25, benefiting from its higher leaf K^+ concentration, sustained an elevated Ψs under NH_4^+ stress, thus maintaining an enhanced g_s.

Additionally, meticulous K^+ flux regulation in guard cells, facilitated by inward (KAT) and outward rectifying K^+ channels (KOR), also plays a crucial role [41]. Our study revealed a significant *TaKOR1* downregulation in Yangmai20 in response to NH_4^+ stress (Figure 6C), with no significant response in Xumai25. In contrast, no significant alteration in *TaKAT1* expression was observed in any of the cultivars (Figure 6D). This finding indicates that the high K^+ concentration and influx in NH_4^+-less sensitive cultivars contributed to the consistently high osmotic potential under NH_4^+ stress. Additionally, both cultivars exhibited *TaHA1* upregulation (Figure 6A), indicating the crucial role of K^+ flow [42] in NH_4^+ stress responses. Further, ABA has been demonstrated to activate anion channels in the plasma membrane of guard cells, leading to anion efflux and subsequent release of K^+ from guard cells, resulting in stomatal closure under drought stress [42]. The elevated ABA level observed in our study (Figure 5C) further contributes to decreased g_s. However, the nuanced interplay of NH_4^+ stress with ion channels governing osmotic potential in stomatal guard cells warrants further exploration. Our results provide an initial foundation for understanding this intricate relationship.

3.5. Water Uptake under NH_4^+ Stress

In the current study, the impact of NH_4^+ exposure on water uptake in the plants remained non-definitive. Some studies suggested that NH_4^+ nutrition enhances water absorption [15,43] while some studies proposed that exposure to NH_4^+ significantly inhibits water uptake in plants [13]. The present study observed a significant reduction in RWC and ψm in both cultivars, with a more pronounced effect in Yangmai20 (Figure 4A,D). Moreover, NH_4^+ stress prompted a substantial downregulation of *TaTIP2.3* and *TaPIP1.1* in Yangmai20 (Figure 6E,F). These findings indicate a more pronounced impact of NH_4^+ stress on water uptake in Yangmai20 than in Xumai25. Based on the existing literature, we postulate that this phenomenon might be attributed to two factors. First, the downregulation of AQPs under NH_4^+ stress to mitigate NH_3/NH_4^+ absorption and transport, as some studies have reported that aquaporins (AQPs) can transport NH_3/NH_4^+ [12,44]. Second, the cellular acidification induced by NH_4^+ stress [2], as evidenced by *TaHA1* upregulation (Figure 6A), adversely impacts the activity of AQPs [45].

4. Materials and Methods

4.1. Plant Materials and Experimental Design

Two wheat cultivars, Xumai25 (NH_4^+-less sensitive) and Yangmai20 (NH_4^+-sensitive), were subjected to hydroponic experiments. The seeds were surface-sterilized using a 10% (v/v) H_2O_2 solution for 15 min and subsequently rinsed thoroughly with sterile distilled water. Subsequently, the seeds were germinated under dark conditions in Petri dishes until the seed bud length reached approximately 1 cm (typically 36 h). Then, the seedlings were transplanted into water-filled opaque plastic containers (45 cm × 32 cm × 25 cm, volume 36 L). When the seedlings reached the two-leaf stage, they were transferred to a modified 50% Hoagland nutrient solution and grown until they reached the four-leaf stage. After this pre-treatment, the seedlings were divided into two groups, each group comprising seedlings from both cultivars. One group was treated with nitrate nitrogen (NN, 5 mM NO_3^--N) nutrient solution and the other with NH_4^+ nitrogen (AN, 5 mM NH_4^+-N) nutrient solution. The macronutrient compositions of both treatment solutions are detailed in Supplementary Table S2. The micronutrient composition was kept consistent in both treatment solutions, as described previously [21]. The solutions were refreshed every three days to ensure a consistent nitrogen supply in each solution and were continuously aerated to prevent anoxia. The pH of each solution was adjusted daily to 5.8 using either 0.1 mM H_2SO_4 or 0.1 mM NaOH. The entire experiment was conducted in a controlled greenhouse environment with a 16 h/8 h light/dark cycle and temperature maintained at 18 °C during the day and 8.5 °C at night. The light intensity and relative air humidity in the greenhouse were set at 400 μmol m^{-2} s^{-1} and 60%, respectively. We adopted a completely randomized block design, and each experiment was replicated three times. Each replication consisted of 3 containers, and each container housed 60 plants.

The random seedlings (belonging to both cultivars) from both groups were collected at 0, 1, 3, 5, and 10 days after treatment (DAT). The fourth leaf (the newest fully expanded leaf before treatment) was used for the measurement of photosynthetic, fluorescence, and leaf water status parameters. The fourth leaves, other leaves, stems, and roots of the collected seedlings were separated and divided into two segments. One segment was oven-dried at 105 °C for 20 min, followed by drying at 85 °C, to determine the dry weight and nitrogen concentration measurements. The other segment was promptly frozen in liquid nitrogen and stored at −80 °C for subsequent analyses.

4.2. Gas Exchange Measurements

Gas exchange measurements were constructed using a gas exchange system (Li-Cor 6800, Li-Cor Inc. Lincoln, NE, USA) equipped with a standard chamber (2 cm^2) and a Multiphase Flash Fluorometer (6800-01F). The parameters were set as follows: leaf temperature, 25.0 ± 0.5 °C; steady-state photosynthetic photon flux density (PPFD), 1200 µmol photons m^{-2} s^{-1}; vapor pressure deficit (VPD), 1.1 ± 0.05 kPa; and relative humidity, 55–65%. Leaves were introduced into the chamber at a reference CO_2 concentration of 400 µmol mol^{-1} for 10 min for stabilization before measurement. Parameters including net photosynthetic rate (*A*), sub-stomatal CO_2 concentration (C_i), stomatal conductance (g_s), transpiration rate (*Tr*), and electron transport rate (J_e) were recorded after stabilization.

For the A–Ci curves, the reference CO_2 concentration was systematically adjusted to the following levels: 400, 200, 150, 100, 50, 400, 400, 600, 800, 1000, 1200, and 1500 µmol mol^{-1}. Data were recorded, following a stabilization period of 2–3 min at each level. Six leaves from each treatment group were selected for this experiment. Carboxylation efficiency (*CE*) was calculated as the initial slope of the A–Ci curves when C_i was <200 µmol mol^{-1} [46]. Parameters such as maximum carboxylation rate (V_{max}), RuBP regeneration (J_{max}), mesophyll conductance (g_m), and stomatal limitation (*l*) were determined using a modified equation derived from the works of S. P. Long and C. J. Bernacchi [47] and Li et al. [46].

The photosynthetic rate (*A*) was expressed mathematically as:

$$A = v_c - 0.5v_o - \text{Rd} = v_c\,(1 - \Gamma^*/C_i) - \text{Rd} \tag{1}$$

Here, v_c is the Rubisco carboxylation rate, v_o is the Rubisco oxygenation rate, Rd is the mitochondrial respiration rate in the light, and Γ* is the CO_2 compensation point related to the C_i [46]. Here, v_c is determined by the minimum of three potential rates: potential Rubisco carboxylation rate, RuBP regeneration rate, and triose phosphate utilization (TPU) rate, (w_c, w_j, and w_p, respectively):

$$v_c = \min\,(w_c, w_j, w_p) \tag{2}$$

Also:

$$w_c = (V_{max} \times C_i)/(C_i + Kc \times (1 + O/Ko)) \tag{3}$$

$$w_j = (J_{max} \times C_i)/(4.5\,C_i + 10.5\,\Gamma^*) \tag{4}$$

Here, *O* is the O_2 concentration (210 mmol mol^{-1}). Furthermore,

$$Kc = \exp^{(38.05\,-\,79.43/(R\,\times\,(T\,+\,273.15)))} \tag{5}$$

$$Ko = \exp^{(20.30\,-\,36.38/(R\,\times\,(T\,+\,273.15)))} \tag{6}$$

Here, *Kc* and *Ko* are the Michaelis constants of carboxylation and oxygenation, respectively. T is the leaf temperature (°C).

The A–Ci curve comprises three phases. In the first phase, Rubisco is limiting and Je increases with the increase in C_i. Thus Equation (3) can be fitted to Equations (1) and (2) as follows:

$$A = V_{max} \times C_i \times (1 - \Gamma^*/C_i)/(C_i + Kc \times (1 + O/Ko)) - \text{Rd} \tag{7}$$

Setting f as a variable changing with C_i:

$$\text{f} = C_i \times (1 - \Gamma^*/C_i)/(C_i + Kc \times (1 + O/Ko)) \tag{8}$$

Equation (8) can then be rewritten as follows:

$$A = V_{max} \times f - \text{Rd} \tag{9}$$

Thus, A can be plotted as a linear function of f, where V_{max} is the slope and Rd is the intercept.

In the second phase, RuBP regeneration is limiting and J_e remains constant with increasing C_i. Thus, Equation (4) can be fitted to Equations (1) and (2) as follows:

$$A = J_{max} \times C_i \times (1 - \Gamma^*/C_i)/(4.5\,C_i + 10.5\,\Gamma^*) - \mathrm{Rd} \quad (10)$$

Setting g as a variable changing with C_i:

$$g = C_i \times (1 - \Gamma^*/C_i)/(4.5\,C_i + 10.5\,\Gamma^*) \quad (11)$$

Equation (11) can then be written as follows:

$$A = J_{max} \times g - \mathrm{Rd} \quad (12)$$

Thus, A can be plotted as a linear function of g, where J_{max} is the slope and Rd is the intercept.

Stomatal limitation (l) was calculated as follows:

$$l = (\mathrm{A}'' - \mathrm{A}')/\mathrm{A}'' \quad (13)$$

Here, A′ is the net photosynthetic rate at the ambient atmospheric CO_2 concentration (400 μmol mol^{-1}) and A″ is the net photosynthetic rate when C_i is equal to ambient atmospheric CO_2 concentration (C_i = 400 μmol mol^{-1}).

The leaf mesophyll conductance (g_m) was calculated using the constant J method [47,48]:

$$g_m = A/(C_i - \Gamma^*\,(J_e + 8(A + \mathrm{Rd}))/(J_e - 4(A + \mathrm{Rd}))) \quad (14)$$

Here, J_e is the electron transport rate.

4.3. Leaf Area and Chlorophyll Fluorescence Measurements

Plant leaf areas were measured using an LI-3100 AREA METER (Li-Cor, Inc., Lincoln, NE, USA).

After the photosynthetic measurements, the fluorescence characteristics of the fourth leaves were determined using a PAM-2500 portable chlorophyll fluorescence apparatus (PAM-2500, Walz, Germany). The leaves were dark-adapted for 30 min before measurements. The minimum fluorescence, Fo, and the maximal fluorescence yield, Fm, were recorded after executing a saturation pulse. Then the actinic light (497 μmol photons m^{-2} s^{-1}) was opened, and the saturation pulses were generated. The initial fluorescence Fo′, the maximum fluorescence Fm′, and the steady fluorescence Ft were recorded during this progress. The maximum quantum yield of the PSII reaction center (*Fv/Fm*), the actual photochemical efficiency (*Y(II)*), and the non-photochemical quenching (*NPQ*) were calculated using a modified equation developed by Christof Klughammer (2008) [49]. The equation is as follows:

$$Fv/Fm = (\mathrm{Fm} - \mathrm{Fo})/\mathrm{Fm} \quad (15)$$

$$Y(II) = (\mathrm{Fm}' - \mathrm{F})/\mathrm{Fm}' \quad (16)$$

$$NPQ = \mathrm{Fm}/\mathrm{Fm}' - 1 \quad (17)$$

4.4. Leaf Water Status

The pre-dawn water potential (ψp) and midday water potential (ψm) of the fourth leaf were determined during the pre-dawn period (between 04:00 and 05:00) and noon (between 12:00 and 13:00), respectively, using a Model 1505D-EXP Pressure Chamber Instrument (1505D-EXP, Decagon, Albany, OR, USA). Briefly, the leaf veins were vertically cut using a sharp blade and the leaf veins were immediately placed into the pressure chamber. The water potential was recorded as the pressure at which blisters emerged on the cross-section of the leaf.

The osmotic potential (Ψs) of the fourth leaf was determined using a vapor pressure osmometer (Wescor 5600, Wescor Inc., Logan, UT, USA) at 25 °C. Briefly, the leaf was flash-frozen in liquid nitrogen at 9:00–10:00 a.m. Then, cell sap was extracted through maceration and filtration using fine nylon mesh and a syringe. The osmotic potential was then calculated from the instrument, as per the manufacturer's instructions.

The leaf relative water content (RWC) of the fourth leaf was measured using a previously described protocol [50]. Briefly, the leaves were weighed immediately after harvest and the weight was recorded as Wf. Then, the leaves were immersed in distilled water for 12 h and weighed to obtain the saturated fresh weight (Wt). Subsequently, the leaves were dried to a constant weight in an oven at 85 °C and weighed to obtain the dry weight (Wd). The RWC was calculated using the formula: (Wf − Wd) / (Wt − Wd) × 100%.

4.5. Determination of N, K+ and Sucrose Concentration

Fresh fourth leaves were freeze-dried and then ground into a powder.

For N and K concentration analyses, approximately 0.1 g of the sample powder was mixed with 5 mL of H_2SO_4. The resulting mixture was heated to 200 °C until a clear solution was obtained. Subsequently, the reaction was terminated by adding H_2O_2. The resulting solutions were then analyzed using inductively coupled plasma optical emission spectrometry (ICP-OES, Optima 8000, Perkin Elmer, Waltham, MA, USA).

The sucrose concentration was determined using the resorcinol method as described by Zeng et al. (2014) [51]. Briefly, 0.1 g of the sample powder was weighed and extracted with a sugar extraction solution. The mixture was centrifuged (15,000× *g*, 15 min), and the supernatant was collected. Next, 0.3 mL of the supernatant was mixed with 0.1 mL of 2 M NaOH, and the solution was incubated at 95 °C for 10 min. Next, 1 mL of 0.1 M resorcinol and 3.5 mL of 10 M HCl were added to the mixture, and the solution was incubated at 80 °C for 60 min. The OD of the solution was measured at 500 nm using UV/VIS spectrophotometer (Pharmacia, Cambridge, UK). The sucrose concentration was calculated using the standard curve.

4.6. Determination of ABA Concentration

ABA concentration was determined using the method described by Greco et al. (2012) [52], with some modifications. Briefly, approximately 0.5 g of the fresh leaf sample was weighed and added to a pre-cooled mortar with 5 mL of 50% chromatographic grade methanol (*v*/*v*) and ground into a slurry in an ice bath. The slurry was then extracted at 4 °C in the dark for 12 h. Afterward, the mixture was centrifuged at 10,000 r/min for 10 min at 4 °C, and the supernatant was collected and stored in a refrigerator. The residue from the first extraction was subjected to two more extractions. For the second extraction, 4 mL pre-cooled 80% chromatographic grade methanol was added to the residue, and the mixture was again extracted for 12 h, followed by centrifugation. The same progress was repeated for the third extraction using 2 mL of pre-cooled 100% chromatographic grade methanol. All supernatants from the three extractions were combined. To adsorb phenols and pigments, PVPP (0.2 g/g FW) was added to the supernatants, which were then shaken at 4 °C for 60 min and centrifuged as mentioned above. The resulting supernatant was slowly passed through a prepared C18 column, and the effluent was collected and freeze-dried in the dark. After freeze-drying, the sample was dissolved in 3 mL of 50% chromatographic grade methanol, filtered through a 0.22 μm membrane, and finally injected into an ultra-performance liquid chromatography (UPLC) system (ACQUITY UPLC H-Class system, Waters, Milford, MA, USA) for analysis. The UPLC analysis was performed using an ACQUITY UPLC HSS T3 column (100 mm × 2.1 mm × 1.8 μm, Waters).

4.7. RT-PCR Analysis

Total RNA was extracted from root and leaf samples using TRIzol reagent (Vazyme Bio, Nanjing, China). The HiScript III Q RT SuperMix (Vazyme Bio, Nanjing, China) was used for cDNA synthesis as per the manufacturer's instructions. The cDNA samples were then diluted 5-fold and subjected to qPCR analysis. qRT-PCR was conducted using the CFX Connect Real-Time PCR Detection System (Bio-Rad, Hercules, CA, USA) with ChamQ SYBR qPCR Master Mix (Vazyme Bio, China).

The primer sequence for *TaHa1* was sourced from Jiang et al. [53]. The primers for aquaporins (*TaTIP2.3*, and *TaPIP1.1*) were sourced from Wu et al. [54]. The primers for potassium transport channel (AKT1) were designed using Primer 5 software. The primers for potassium ion inflow channel (KAT1), and potassium ion outflow channel (KOR1) were sourced from the work of Yang et al. [55]. ACT and ADP genes were used as internal references. All primer sequence is listed in Supplementary Table S3. Relative expression levels were determined using a previously described method [56].

4.8. Statistical Analysis

Physiological data derived from both dry and fresh samples were calculated using three biological replicates, while the data pertaining to photosynthetic, fluorescence, and water status parameters of the leaves were derived from six leaf replicates. Analysis of variance (ANOVA), followed by the Tukey HSD test was used for multiple comparisons. SPSS 26 (IBM, Armonk, NY, US) was used for all statistical analyses. Graphs and tables were generated using Origin 2021 software (OriginLab, Northampton, MA, USA) and Microsoft Excel.

5. Conclusions

In conclusion, our study showed that 5 mM NH_4^+ stress significantly disrupted the crucial photosynthetic processes in wheat plants. Our findings highlight the predominant stomatal limitation in photosynthesis under NH_4^+ stress conditions. In this process, K^+ mediated Ψs decrease played a vital role in the decline of g_s. In addition, the NH_4^+-less sensitive cultivar exhibited a robust ability to maintain osmotic homeostasis, resulting in higher g_s and improved photosynthetic performance and growth under NH_4^+ stress. Further studies are still needed to elucidate how high NH_4^+ concentrations might impact, at the molecular and electrophysiological levels, the stomatal opening in higher plants.

Supplementary Materials: The following supporting information can be downloaded at: https://www.mdpi.com/article/10.3390/plants13010086/s1. Table S1. Eigenvalue and cumulative contribution of PCs of photosynthetic parameter and their corresponding loading. Table S2. The concentration and components of macronutrients in both treatments. Table S3. The primer sequence in this study.

Author Contributions: J.H.: experimentalize, writing—original draft preparation. Q.Z.: experimentalize, data analysis. C.D.: experimentalize. Z.L.: revision. Z.T.: supervision. T.D.: methodology, supervision. All authors have read and agreed to the published version of the manuscript.

Funding: This research was funded by the National Natural Science Foundation of China (Grant No. 32272215) and the National Key R&D Program of Jiangsu (BE2021361-1).

Data Availability Statement: The data presented in this study are available on request from the corresponding author. The data are not publicly available due to we have an in-depth research project on this topic in the future.

Acknowledgments: We express our appreciation to the other researchers and staff members involved in this project for their valuable contributions, expertise, and assistance in project management.

Conflicts of Interest: The authors declare that they have no known competing financial interests or personal relationships that could have appeared to influence the work reported in this paper.

References

1. Britto, D.T.; Kronzucker, H.J. NH_4^+ toxicity in higher plants: A critical review. *J. Plant Physiol.* **2002**, *159*, 567. [CrossRef]
2. Esteban, R.; Ariz, I.; Cruz, C.; Moran, J.F. Review: Mechanisms of ammonium toxicity and the quest for tolerance. *Plant Sci.* **2016**, *248*, 92. [CrossRef] [PubMed]
3. Liu, Y.; von Wirén, N. Ammonium as a signal for physiological and morphological responses in plants. *J. Exp. Bot.* **2017**, *68*, 2581. [CrossRef] [PubMed]
4. Bittsánszky, A.; Pilinszky, K.; Gyulai, G.; Komives, T. Overcoming ammonium toxicity. *Plant Sci.* **2015**, *231*, 184. [CrossRef] [PubMed]
5. Marino, D.; Moran, J.F. Can Ammonium Stress Be Positive for Plant Performance? *Front. Plant Sci.* **2019**, *10*, 1–5. [CrossRef] [PubMed]
6. Bejarano, I.; Marino, D.; Coleto, I. Arabidopsis MYB28 and MYB29 transcription factors are involved in ammonium-mediated alterations of root-system architecture. *Plant Signal Behav.* **2021**, *16*, 1879532. [CrossRef] [PubMed]
7. Robert, G.; Yagyu, M.; Koizumi, T.; Naya, L.; Masclaux Daubresse, C.; Yoshimoto, K. Ammonium stress increases microautophagic activity while impairing macroautophagic flux in Arabidopsis roots. *Plant J.* **2021**, *105*, 1083. [CrossRef]
8. Emily, L.; Harrison, L.A.C.J. The influence of stomatal morphology and distribution on photosynthetic gas exchange. *Plant J.* **2020**, *101*, 768.
9. Cochrane, T.T.; Cochrane, T.A. Differences in the way potassium chloride and sucrose solutions effect osmotic potential of significance to stomata aperture modulation. *Plant Physiol. Biochem.* **2009**, *47*, 205. [CrossRef]
10. Yang, C.; Zhang, J.; Zhang, G.; Lu, J.; Ren, T.; Cong, R.; Lu, Z.; Zhang, Y.; Liao, S.; Li, X. Potassium deficiency limits water deficit tolerance of rice by reducing leaf water potential and stomatal area. *Agric. Water Manag.* **2022**, *271*, 107744. [CrossRef]
11. Li, B.; Li, G.; Kronzucker, H.J.; Baluška, F.; Shi, W. Ammonium stress in Arabidopsis: Signaling, genetic loci, and physiological targets. *Trends Plant Sci.* **2014**, *19*, 107. [CrossRef] [PubMed]
12. Gao, L.; Lu, Z.; Ding, L.; Guo, J.; Wang, M.; Ling, N.; Guo, S.; Shen, Q. Role of Aquaporins in Determining Carbon and Nitrogen Status in Higher Plants. *Int. J. Mol. Sci.* **2018**, *19*, 35. [CrossRef] [PubMed]
13. Naku, M.; Kambizi, L.; Matimati, I. Functional roles of ammonium (NH_4^+) and nitrate (NO_3^-) in regulation of day- and night-time transpiration in Phaseolus vulgaris. *Funct. Plant Biol.* **2019**, *46*, 806. [CrossRef] [PubMed]
14. Cramer, M.D.; Hawkins, H.; Verboom, G.A. The importance of nutritional regulation of plant water flux. *Oecologia* **2009**, *161*, 15. [CrossRef] [PubMed]
15. Ding, L.; Gao, C.; Li, Y.; Li, Y.; Zhu, Y.; Xu, G.; Shen, Q.; Kaldenhoff, R.; Kai, L.; Guo, S. The enhanced drought tolerance of rice plants under ammonium is related to aquaporin (AQP). *Plant Sci.* **2015**, *234*, 14. [CrossRef] [PubMed]
16. Hu, H.; Wang, L.; Wang, Q.; Jiao, L.; Hua, W.; Zhou, Q.; Huang, X. Photosynthesis, chlorophyll fluorescence characteristics, and chlorophyll content of soybean seedlings under combined stress of bisphenol A and cadmium. *Environ. Toxicol. Chem.* **2014**, *33*, 2455. [CrossRef] [PubMed]
17. Rah, M.; Helali, S.; Nebli, H.; Kaddour, R.; Mahmoudi, H.; Lachaâl, M.; Ouerghi, Z. Influence of nitrate—Ammonium ratio on growth and nutrition of *Arabidopsis thaliana*. *Plant Soil* **2010**, *336*, 65.
18. Sanchez-Zabala, J.; Gonzalez-Murua, C.; Marino, D. Mild ammonium stress increases chlorophyll content in *Arabidopsis thaliana*. *Plant Signal Behav.* **2015**, *10*, e991596. [CrossRef]
19. Patterson, K.; Cakmak, T.; Cooper, A.; Lager, I.D.; Rasmusson, A.G.; Escobar, M.A. Distinct signalling pathways and transcriptome response signatures differentiate ammonium- and nitrate-supplied plants. *Plant Cell Environ.* **2010**, *33*, 1486–1501. [CrossRef]
20. Podgórska, A.; Gieczewska, K.; Łukawska-Kuźma, K.; Rasmusson, A.G.; Gardeström, P.; Szal, B. Long-term ammonium nutrition of Arabidopsis increases the extra chloroplastic NAD(P)H/NAD(P)$^+$ ratio and mitochondrial reactive oxygen species level in leaves but does not impair photosynthetic capacity. *Plant Cell Environ.* **2013**, *36*, 2034. [CrossRef]
21. Liu, Y.; Li, Y.; Li, Y.; Tian, Z.; Hu, J.; Adkins, S.; Dai, T. Changes of oxidative metabolism in the roots of wheat (*Triticum aestivum* L.) seedlings in response to elevated ammonium concentrations. *J. Integr. Agric.* **2021**, *20*, 1216. [CrossRef]
22. Ijato, T.; Porras Murillo, R.; Ganz, P.; Ludewig, U.; Neuhäuser, B. Concentration-dependent physiological and transcriptional adaptations of wheat seedlings to ammonium. *Physiol. Plant.* **2021**, *171*, 328. [CrossRef] [PubMed]
23. Cruz, C.; Domínguez-Valdivia, M.D.; Aparicio-Tejo, P.M.; Lamsfus, C.; Bio, A.; Martins-Loução, M.A.; Moran, J.F. Intra-specific variation in pea responses to ammonium nutrition leads to different degrees of tolerance. *Environ. Exp. Bot.* **2011**, *70*, 233. [CrossRef]
24. Vega-Mas, I.; Rossi, M.T.; Gupta, K.J.; González-Murua, C.; Ratcliffe, R.G.; Estavillo, J.M.; González-Moro, M.B. Tomato roots exhibit in vivo glutamate dehydrogenase aminating capacity in response to excess ammonium supply. *J. Plant Physiol.* **2019**, *239*, 83. [CrossRef] [PubMed]
25. Hachiya, T.; Inaba, J.; Wakazaki, M.; Sato, M.; Toyooka, K.; Miyagi, A.; Kawai-Yamada, M.; Sugiura, D.; Nakagawa, T.; Kiba, T.; et al. Excessive ammonium assimilation by plastidic glutamine synthetase causes ammonium toxicity in *Arabidopsis thaliana*. *Nat. Commun.* **2021**, *12*, 4944. [CrossRef] [PubMed]
26. Vega-Mas, I.; Cukier, C.; Coleto, I.; González-Murua, C.; Limami, A.M.; González-Moro, M.B.; Marino, D. Isotopic labelling reveals the efficient adaptation of wheat root TCA cycle flux modes to match carbon demand under ammonium nutrition. *Sci. Rep.* **2019**, *9*, 8925. [CrossRef]

27. Chen, H.; Hu, W.; Wang, Y.; Zhang, P.; Zhou, Y.; Yang, L.; Li, Y.; Chen, L.; Guo, J. Declined photosynthetic nitrogen use efficiency under ammonium nutrition is related to photosynthetic electron transport chain disruption in citrus plants. *Sci. Hortic.* **2023**, *308*, 111594. [CrossRef]
28. Vega-Mas, I.; Marino, D.; Sánchez-Zabala, J.; González-Murua, C.; Estavillo, J.M.; González-Moro, M.B. CO_2 enrichment modulates ammonium nutrition in tomato adjusting carbon and nitrogen metabolism to stomatal conductance. *Plant Sci.* **2015**, *241*, 32. [CrossRef]
29. Chen, G.; Zheng, Z.; Bai, M.; Li, Q. Chronic effects of microcystin-LR at environmental relevant concentrations on photosynthesis of Typha angustifolia Linn. *Ecotoxicology* **2020**, *29*, 514. [CrossRef]
30. Farquhar, G.D.; Caemmerer, S.V.; Berry, J.A. biochemical model of photosynthetic CO_2 assimilation in leaves of C_3 species. *Planta* **1980**, *149*, 78. [CrossRef]
31. Alencar, V.T.C.B.; Lobo, A.K.M.; Carvalho, F.E.L.; Silveira, J.A.G. High ammonium supply impairs photosynthetic efficiency in rice exposed to excess light. *Photosynth. Res.* **2019**, *140*, 321. [CrossRef] [PubMed]
32. Torralbo, F.; González-Moro, M.B.; Baroja-Fernández, E.; Aranjuelo, I.; González-Murua, C. Differential Regulation of Stomatal Conductance as a Strategy to Cope with Ammonium Fertilizer under Ambient versus Elevated CO_2. *Front. Plant Sci.* **2019**, *10*, 597. [CrossRef] [PubMed]
33. Li, H.; Wang, Y.; Xiao, J.; Xu, K. Reduced photosynthetic dark reaction triggered by ABA application increases intercellular CO_2 concentration, generates H_2O_2 and promotes closure of stomata in ginger leaves. *Environ. Exp. Bot.* **2015**, *113*, 11. [CrossRef]
34. Liu, M.; Liu, X.; Du, X.; Korpelainen, H.; Niinemets, Ü.; Li, C. Anatomical variation of mesophyll conductance due to salt stress in *Populus cathayana* females and males growing under different inorganic nitrogen sources. *Tree Physiol.* **2021**, *41*, 1462. [CrossRef] [PubMed]
35. Li, Y.; Ren, B.; Yang, X.; Xu, G.; Shen, Q.; Guo, S. Chloroplast Downsizing under Nitrate Nutrition Restrained Mesophyll Conductance and Photosynthesis in Rice (*Oryza sativa* L.) under Drought Conditions. *Plant Cell Physiol.* **2012**, *53*, 892. [CrossRef] [PubMed]
36. Onoda, Y.; Wright, I.J.; Evans, J.R.; Hikosaka, K.; Kitajima, K.; Niinemets, Ü.; Poorter, H.; Tosens, T.; Westoby, M. Physiological and structural tradeoffs underlying the leaf economics spectrum. *New Phytol.* **2017**, *214*, 1447. [CrossRef] [PubMed]
37. Flexas, J.; Clemente-Moreno, M.J.; Bota, J.; Brodribb, T.J.; Gago, J.; Mizokami, Y.; Nadal, M.; Perera-Castro, A.V.; Roig-Oliver, M.; Sugiura, D.; et al. Cell wall thickness and composition are involved in photosynthetic limitation. *J. Exp. Bot.* **2021**, *72*, 3971. [CrossRef]
38. Jian, S.; Liao, Q.; Song, H.; Liu, Q.; Lepo, J.E.; Guan, C.; Zhang, J.; Ismail, A.M.; Zhang, Z. NRT1.1-Related NH_4^+ Toxicity Is Associated with a Disturbed Balance between NH_4^+ Uptake and Assimilation. *Plant Physiol.* **2018**, *178*, 1473. [CrossRef]
39. Mao, L.Z.; Lu, H.F.; Wang, Q.; Cai, M.M. Comparative photosynthesis characteristics of *Calycanthus chinensis* and *Chimonanthus praecox*. *Photosynthetica* **2007**, *45*, 601. [CrossRef]
40. Brownlee, C. Stomatal Physiology: Cereal Successes. *Curr. Biol.* **2018**, *28*, R551. [CrossRef]
41. Andrés, Z.; Pérez-Hormaeche, J.; Leidi, E.O.; Schlücking, K.; Steinhorst, L.; McLachlan, D.H.; Schumacher, K.; Hetherington, A.M.; Kudla, J.; Cubero, B.; et al. Control of vacuolar dynamics and regulation of stomatal aperture by tonoplast potassium uptake. *Proc. Natl. Acad. Sci. USA* **2014**, *11*, E1806–E1814. [CrossRef] [PubMed]
42. Daszkowska-Golec, A.; Szarejko, I. Open or Close the Gate—Stomata Action under the Control of Phytohormones in Drought Stress Conditions. *Front. Plant Sci.* **2013**, *4*, 138. [CrossRef] [PubMed]
43. Gao, Y.; Li, Y.; Yang, X.; Li, H.; Shen, Q.; Guo, S. Ammonium nutrition increases water absorption in rice seedlings (*Oryza sativa* L.) under water stress. *Plant Soil* **2010**, *331*, 193. [CrossRef]
44. Kirscht, A.; Kaptan, S.S.; Bienert, G.P.; Chaumont, F.; Nissen, P.; de Groot, B.L.; Kjellbom, P.; Gourdon, P.; Johanson, U. Crystal Structure of an Ammonia-Permeable Aquaporin. *PLoS Biol.* **2016**, *14*, e1002411. [CrossRef] [PubMed]
45. Kapilan, R.; Vaziri, M.; Zwiazek, J.J. Regulation of aquaporins in plants under stress. *Biol. Res.* **2018**, *51*, 4. [CrossRef] [PubMed]
46. Li, Y.; Gao, Y.; Xu, X.; Shen, Q.; Guo, S. Light-saturated photosynthetic rate in high-nitrogen rice (*Oryza sativa* L.) leaves is related to chloroplastic CO_2 concentration. *J. Exp. Bot.* **2009**, *60*, 2351. [CrossRef] [PubMed]
47. Long, S.P. Gas exchange measurements, what can they tell us about the underlying limitations to photosynthesis? Procedures and sources of error. *J. Exp. Bot.* **2003**, *54*, 2393. [CrossRef]
48. Harley, P.C.; LFMG. Theoretical Considerations when Estimating the Mesophyll Conductance to CO_2 Flux by Analysis of the Response of Photosynthesis to CO_2. *Plant Physiol.* **1992**, *98*, 1429. [CrossRef]
49. Klughammer, C.; Schreiber, U. Complementary PS II quantum yields calculated from simple fluorescence parameters measured by PAM fluorometry and the Saturation Pulse method. *PAM Appl. Notes* **2008**, *1*, 27–35.
50. Nadal, M.; Brodribb, T.J.; Fernández Marín, B.; García Plazaola, J.I.; Arzac, M.I.; López Pozo, M.; Perera Castro, A.V.; Gulías, J.; Flexas, J.; Farrant, J.M. Differences in biochemical, gas exchange and hydraulic response to water stress in desiccation tolerant and sensitive fronds of the fernAnemia caffrorum. *N. Phytol.* **2021**, *231*, 1415. [CrossRef]
51. Zeng, Y.; Yu, J.; Cang, J.; Liu, L.; Mu, Y.; Wang, J.; Zhang, D. Detection of Sugar Accumulation and Expression Levels of Correlative Key Enzymes in Winter Wheat (*Triticum aestivum*) at Low Temperatures. *Biosci. Biotechnol. Biochem.* **2014**, *75*, 681. [CrossRef] [PubMed]
52. Greco, M.; Chiappetta, A.; Bruno, L.; Bitonti, M.B. In Posidonia oceanica cadmium induces changes in DNA methylation and chromatin patterning. *J. Exp. Bot.* **2012**, *63*, 695. [CrossRef] [PubMed]

53. Jiang, S.; Sun, J.; Tian, Z.; Hu, H.; Michel, E.J.S.; Gao, J.; Jiang, D.; Cao, W.; Dai, T. Root extension and nitrate transporter up-regulation induced by nitrogen deficiency improves nitrogen status and plant growth at the seedling stage of winter wheat (*Triticum aestivum* L.). *Environ. Exp. Bot.* **2017**, *141*, 28. [CrossRef]
54. Wu, S.; Sun, X.; Tan, Q.; Hu, C. Molybdenum improves water uptake via extensive root morphology, aquaporin expressions and increased ionic concentrations in wheat under drought stress. *Environ. Exp. Bot.* **2019**, *157*, 241. [CrossRef]
55. Yang, M.; He, J.; Sun, Z.; Li, Q.; Cai, J.; Zhou, Q.; Wollenweber, B.; Jiang, D.; Wang, X. Drought priming mechanisms in wheat elucidated by in-situ determination of dynamic stomatal behavior. *Front. Plant Sci.* **2023**, *14*, 1138494. [CrossRef]
56. Livak, K.J.; Schmittgen, T.D. Analysis of Relative Gene Expression Data Using Real-Time Quantitative PCR and the $2^{-\Delta\Delta CT}$ Method. *Methods* **2001**, *25*, 402. [CrossRef]

plants

MDPI

Article

A Combination of Three Genomic Regions Conditions High Level of Adult Plant Stripe Rust Resistance in Australian Wheat Cultivar Sentinel

Bosco Chemayek [1,2], William Wagoire [2], Urmil Bansal [1] and Harbans Bariana [1,3,*]

1 Plant Breeding Institute, School of Life and Environmental Sciences, Faculty of Science, The University of Sydney, 107 Cobbitty Road, Cobbitty, NSW 2570, Australia; bchemayek@gmail.com (B.C.); urmil.bansal@sydney.edu.au (U.B.)

2 Buginyanya Zonal Agricultural Research Institute, National Agricultural Research Organisation, Mbale P.O. Box 1356, Uganda; wwwagoire@gmail.com

3 School of Science, Hawkesbury Campus, Western Sydney University, Bourke Street, Richmond, NSW 2753, Australia

* Correspondence: h.bariana@westernsydney.edu.au

Abstract: A seedling susceptible Australian common wheat cultivar Sentinel showed resistance to stripe rust under field conditions. A Sentinel/Nyabing3 (Nyb3)-derived recombinant inbred line (RIL) population was phenotyped. A DArTseq marker-based linkage map of the Sentinel/Nyb3 RIL population was used to determine the chromosomal location of the adult plant stripe rust resistance possessed by Sentinel. Three consistent quantitative trait loci (QTL); *QYr.sun-1BL*, *QYr.sun-2AS* and *QYr.sun-3BS* were detected, and they on an average explained 18%, 15.6% and 10.6% of the variation in stripe rust response, respectively. All three QTL were contributed by Sentinel. *QYr.sun-1B* corresponded to the previously characterized gene *Yr29*. Sentinel expressed resistance at the four-leaf stage at 21 ± 2 °C in the greenhouse. Monogenic segregation among the RIL population was observed when screened at the four-leaf stage at 21 ± 2 °C in the greenhouse, and the underlying resistance locus was temporarily named *YrSen*. *QYr.sun-3BS* peaked on *YrSen*. *QYr.sun-2AS* was mendelized by generating and phenotyping a mongenically sgregating F_6 RIL population, and it was temporarily designated *YrSen2*. RILs carrying *Yr29*, *YrSen* and *YrSen2* in combination exhibited responses like the parent Sentinel. Based on a comparison of the genomic locations and resistance expression with stripe rust resistance genes previously located in their respective chromosomes, *QYr.sun-2AS* (*YrSen2*) and *QYr.sun-3BS* (*YrSen*) were concluded to represent new loci.

Keywords: APR; QTL mapping; QTL interaction; stripe rust; wheat

1. Introduction

Stripe rust of wheat, caused by *Puccinia striiformis* f. sp. *tritici* (*Pst*), is one of the most devastating diseases of wheat and threatens the global food supply [1]. New pathogen genotypes are more aggressive and can infect previously resistant wheat varieties, leading to rapid pathogen migration across the continents [2]. *Pst* infects leaf tissue and significantly reduces grain yield and quality in susceptible cultivars. More than 88% of the global wheat cultivars are now susceptible to stripe rust, and yield losses are severe if infection occurs at the early stages of growth [3]. Stripe rust was first detected in eastern Australia in 1979 [4]. The wheat growing areas of Western Australia remained free from stripe rust for more than two decades, but a new exotic pathotype, 134 E16A+, with virulence for *Yr6*, *Yr7*, *Yr8*, *Yr9* and *YrA* and avirulence for *Yr3* and *Yr4*, was detected in 2002 [5]. This pathotype spread to all wheat growing regions in Australia in 2003 and caused epidemics in eastern Australia [6]. The pathotype 134 E16A+ acquired virulence for stripe rust resistance genes *Yr10*, *Yr17*, *Yr24*, *Yr27*, *YrJ* and *YrT* [6].

The new pathotypes of *Pst* have also caused significant wheat yield losses in most parts of the world in recent years and resulted in global losses estimated to be at least 5.5 million tons per year [3]. In 2009–2010, the outbreak of a pathotype of *Pst* with virulence for *Yr27* caused significant yield losses in Azerbaijan, Ethiopia, Iran, Iraq, Kenya, Morocco, Syria, Turkey and Uzbekistan, threatening the food security and livelihood of resource-poor farmers [7,8]. Consequently, there has been significant use of chemicals to control stripe rust epidemics [9]. Genetic resistance is the most economical and environmentally safe means of stripe rust control; hence, the deployment of combinations of stripe rust resistance genes in new wheat cultivars is the best strategy to mitigate potential yield losses [10,11].

Resistance to stripe rust is categorized into two groups: qualitative/all-stage resistance (ASR) and quantitative/adult plant resistance (APR) [12,13]. The ASR is controlled by genes with major effects, and a single gene can provide a high level of protection against avirulent pathotypes of the target pathogen. In contrast, minor effect genes condition APR, and combinations of two or more genes are needed to achieve a commercially acceptable level of resistance [14,15]. Of the more than 80 stripe rust resistance genes formally named so far, only a few represent the APR category [https://wheat.pw.usda.gov/GG3/wgc; 25 October 2023]. The ease of transfer of ASR genes into commercial cultivars has made them a popular choice among wheat breeders; however, new pathotypes often overcome this type of resistance in due course. To achieve long-lasting control of stripe rust, the deployment of combinations of ASR and APR genes in future wheat cultivars is essential. This highlights the need to identify, characterize and map more APR genes.

Cytogenetic techniques were successfully employed to determine the genomic locations of many ASR genes for stripe rust in wheat [16]; however, their role in determining the chromosomal locations of APR genes has been limited due to the poor expression of APR genes in the heterozygous state. Recent advances in high-throughput (next-generation) sequencing platforms such as genotyping by sequencing (GBS) have led to affordable options for whole-genome sequencing using a large number of markers [17,18]. The rapid technological advances in sequencing facilitated the detection of new QTL for stripe rust resistance in common wheat and durum wheat [19,20].

The Australian common wheat cultivar Sentinel has remained resistant to stripe rust under field conditions since its release in 2005, and the genetic basis of this resistance was unknown. This study focused on molecular mapping of APR to stripe rust in Sentinel.

2. Results

2.1. Stripe Rust Response Assessments

2.1.1. Field Assessment

The resistant parent Sentinel produced adult plant stripe rust response '2', whereas the susceptible parent Nyabing 3 (Nyb3) was scored '9' in the field, when tested with *Pst* pathotype 134 E16A+17+27+ (Figure 1a). The adult plant stripe rust responses of Sentinel/Nyb3 RILs varied from '2' to '9' in all experiments and showed a normal distribution with both parents falling at the tail ends of the curve (Figure 2). Stripe rust response data of the Sentinel/Nyb3 RIL population were subjected to estimation of the number of resistance loci involved using the modified Wright's method. The involvement of two to three loci in controlling stripe rust response variation among Sentinel/Nyb3 RILs was estimated.

2.1.2. Greenhouse Assessment

Sentinel expressed high level of resistance in the field, and to discount the possibility of involvement of any ASR gene(s), Sentinel and Nyb3 were tested at the two-leaf stage in the greenhouse. Both cultivars produced high infection type (IT) 3+. These results demonstrated the absence of seedling resistance in both genotypes. Both parents were then tested at the two-leaf, three-leaf and four-leaf stages, and post-inoculation incubations were performed under two temperature regimes (17 ± 2 °C and 21 ± 2 °C). Both Sentinel and Nyb3 produced IT3+ at the post-inoculation temperature 17 ± 2 °C at all growth stages. In the 21 ± 2 °C post-inoculation experiment, Nyb3 produced susceptible IT3+ at all growth stages. In contrast, Sentinel exhibited susceptible responses (IT3+) at the two-leaf and three-leaf growth stages, and it displayed IT23C at the four-leaf stage (Figure 1b). The entire RIL population was tested at the four-leaf stage with 21 ± 2 °C as the post-inoculation temperature, and it showed monogenic segregation (58 homozygous resistant (HR; IT23C): 59 homozygous susceptible (HS; IT3+), $\chi^2_{1:1} = 0.01$, non-significant at $p = 0.05$ and 1 *d.f.*) (Table 1). The resistance locus was tentatively named *YrSen*.

Figure 1. Stripe rust responses of P1-Sentinel and P2-Nyb3 (**a**) in the field and (**b**) at the four-leaf stage in the greenhouse.

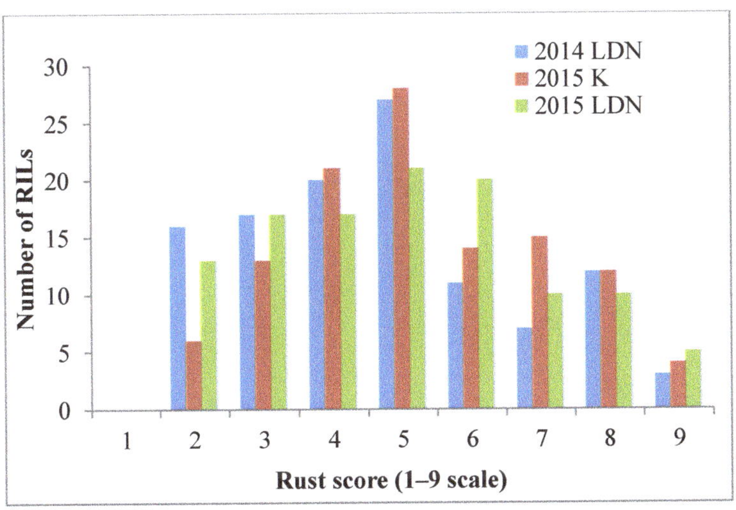

Figure 2. Frequency distribution of FigureSentinel/Nyb3 RILs with respect to adult plant stripe rust response variation (LDN—Lansdowne, K—Karalee).

Table 1. Frequency distribution of Sentinel/Nyb3 F_6 RILs when tested against the *Pst* pathotype 134 E16A+17+27+ at the four-leaf stage at 21 ± 2 °C.

Response Category	Number of RILs		
	Observed	Expected	$\chi^2_{1:1}$
Homozygous Resistant (HR)	58	58.5	0.004
Homozygous Susceptible (HS)	59	58.5	0.004
Total	117	117	0.008 *

* Table value of χ^2 at $p = 0.05$ and 1 *d.f.* = 3.84.

2.2. Construction of Linakge Map and QTL Analysis

2.2.1. Detection of APR Genes *Yr18* and *Yr29*

APR genes *Yr18* and *Yr29* are widespread in wheat germplasm. To check whether Sentinel carries any of these genes, linked markers were used. The *Yr18*-linked STS marker *csLV34* amplified a 150 bp PCR product in the positive control Janz and a 229 bp product in Sentinel and Nyb3. Six *Yr18* gene-specific markers (*ccsfr1-cssfr6*) were also tested on Sentinel and Nyb3 DNA to further confirm the absence of *Yr18* in Sentinel. The dominant marker *cssfr1* resulted in amplification of a 517 bp amplicon in Janz and no amplicon in Sentinel and Nyb3, while *cssfr2* amplified a 523 bp amplicon in Sentinel and Nyb3, and no amplicon was recorded in Janz. The co-dominant marker *cssfr3* produced two bands (150 bp and 517 bp) in Janz and a single band (229 bp) in Sentinel and Nyb3. Another such marker, *cssfr4*, amplified a 150 bp PCR product in Janz and two amplicons (229 bp and 523 bp) in Sentinel and Nyb3. The third co-dominant marker, *cssfr5*, resulted in amplification of a 751 bp product in Janz and a 523 bp amplicon in both Sentinel and Nyb3. CAPS marker *cssfr6* produced three amplicons (63 bp/135 bp/451 bp) in positive control Janz and two (63 bp and 589 bp) in Sentinel and Nyb3, following digestion with the restriction enzyme *Fnu4HI*. These results demonstrated the absence of *Yr18* in both Sentinel and Nyb3. The *Yr29*-linked SNP marker *Lr46_SNP1G22* amplified allele 'A' in Sentinel and the positive control Lalbahadur + *Lr46*, whereas the alternate allele 'G' was detected in Nyb3 and the negative control Lalbahadur. The entire RIL population was tested with *Lr46_SNP1G22*, and monogenic segregation (68 'A': 49 'G'; $\chi^2_{1:1} = 3.09$, non-significant at $p = 0.05$ and 1 *d.f.*) was noted.

2.2.2. Linkage Map Construction

The RIL population was subjected to whole-genome profiling with 16,815 DArTseq markers. A total of 4891 DArTseq markers exhibited segregation distortion, 1502 had more than 10% missing data, and 2079 were monomorphic between parental genotypes and were therefore not included in the linkage map. Of the remaining 8343 markers, chromosome locations were known only for 3605, whereas 4738 markers were not assigned to any chromosomes and were labeled UK (unknown). Redundant markers were filtered out, and only 1619 markers were suitable for linkage map construction. The 4738 unassigned markers were redistributed in the linkage map, and 111 markers were assigned to known chromosomes in the linkage map. Hence, a total of 1730 markers were used for map construction.

A set of 1589 polymorphic DArTseq markers was mapped to 34 discrete linkage groups, and 141 markers remained unlinked. The linkage map covered 4327.1 cM with an average marker density of one marker per 2.7 cM. Marker density was highest for the B genome, and it covered 1783.1 cM. The A genome map included 1552.2 cM, and the coverage of the D genome was low with 912.5 cM. There was an average marker density of one marker per 2.5 cM in the A genome and one marker per 3.7 cM in the D genome. *YrSen* was also included in the linkage map.

2.2.3. QTL Analysis

Composite interval mapping (CIM) was used to scan the Sentinel/Nyb3 RIL population to detect quantitative trait loci (QTL) associated with stripe rust resistance. CIM analyses detected three consistent QTL on chromosomes 1B, 2A and 3B (Figure 3). All QTL were contributed by Sentinel and were named *QYr.sun-1BL*, *QYr.sun-2AS* and *QYr.sun-3BS*. QTL *QYr.sun-1BL*, *QYr.sun-2AS* and *QYr.sun-3BS* on an average explained 18%, 15.3% and 10.6%, respectively, of the phenotypic variation in adult plant stripe rust response among the Sentinel/Nyb3 RIL population (Table 2).

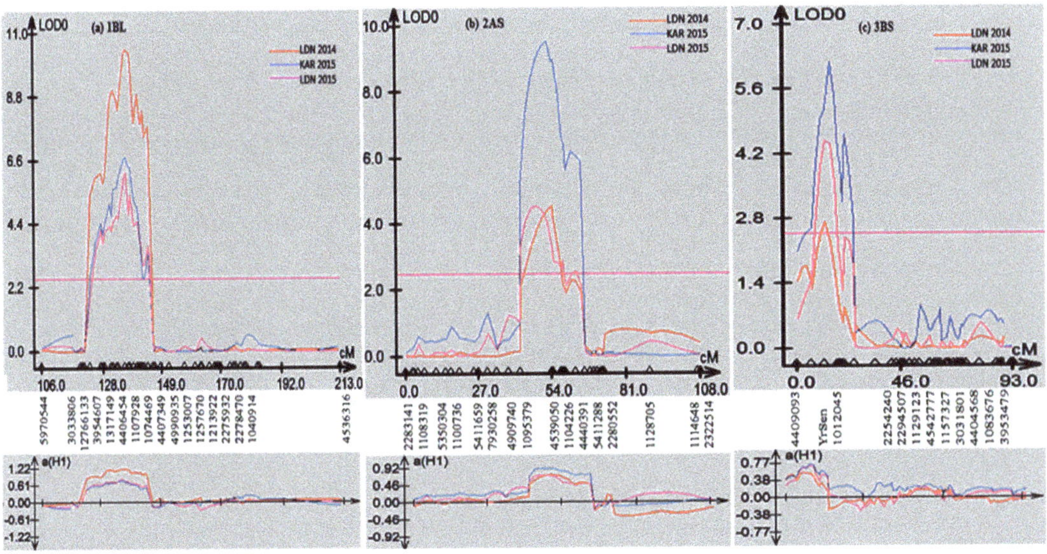

Figure 3. Stripe rust QTL detected on chromosomes (**a**) 1BL, (**b**) 2AS and (**c**) 3BS of Sentinel.

Table 2. Stripe rust QTL detected in the Sentinel/Nyb3 RIL population.

QTL	Season/Site	Peak Marker	Flanking Markers	LOD	R^2
QYr.sun-1BL	2014 LDN	4406454	12766133 & 1074469	10.45	26
	2015 K	4406454	12766133 &1074469	6.69	13
	2015 LDN	4406454	12766133 &1074469	6.18	15
	Mean				18
QYr.sun-2AS	2014 LDN	4539050	1095379 & 4440391	4.54	11.4
	2015 K	4539050	1095379 & 4440391	9.54	22
	2015 LDN	4539050	1095379 & 4440391	4.54	13
	Mean				15.3
QYr.sun-3BS	2014 LDN	*YrSen*	4409093 & 1012045	2.73	6
	2015 K	*YrSen*	4409093 & 1012045	6.17	13.6
	2015 LDN	*YrSen*	4409093 &1012045	4.47	12.1
	Mean				10.6

QYr.sun-1BL explained 13–26% of the variation in stripe rust response, peaked at the DArTseq marker 4406454, and was flanked by markers 12766133 and 1074469 (Table 2). Marker 4539050 was the closest to the *QYr.sun-2AS* peak and contributed 11–22% towards phenotypic variation (Table 2). Markers 1095379 and 4440391 defined the QTL interval. *QYr.sun-3BS* corresponded to *YrSen*, and markers 4409093 and 1012045 flanked it (Table 2). This QTL explained 6 to 14% of the phenotypic variation in stripe rust response in the Sentinel/Nyb3 RIL population.

2.2.4. Contribution of Sentinel Alleles

To assess the individual contribution of each QTL towards adult plant stripe rust severity reduction, rust response data collected on a 1–9 scale were converted to percent rust severity (0–100). Comparisons of the mean stripe rust responses of RILs carrying alternative alleles at the peak marker locus were made to confirm the phenotypic contributions of each QTL (Table 3). Detected QTL were considered real when the mean rust response of RILs carrying the positive allele was less than the mean of those possessing the alternate allele. The contributions of peak markers 4406454 (*QYr.sun-1BL*), 4539050 (*QYr.sun-2AS*) and *YrSen* (*QYr.sun-3BS*) are listed in Table 3. *QYr.sun-1BL*, *QYr.sun-2AS* and *QYr.sun-3BS* reduced stripe rust severities by 23–43%, 27–33% and 25–51%, respectively (Table 3).

Table 3. Comparison of mean stripe rust severities of RILs carrying Sentinel and Nyb 3 alleles for each QTL.

QTL	Experiment	Chromosome	Marker	Severity (%)		Severity Reduction (%)
				Sentinel Allele	Nyb3 Allele	
QYr.sun-1BL	2014 LDN	1BL	4406454	34.6	77.2	42.6
	2015 K	1BL	4406454	56.1	79.4	23.4
	2015 LDN	1BL	4406454	45.0	75.3	30.3
QYr.sun-2AS	2014 LDN	2AS	4539050	44.1	77.2	33.1
	2015 K	2AS	4539050	48.0	79.4	31.4
	2015 LDN	2AS	4539050	47.9	75.3	27.4
QYr.sun-3BS	2014 LDN	3BS	*YrSen*	26.7	77.2	50.6
	2015 K	3BS	*YrSen*	49.2	79.4	30.3
	2015 LDN	3BS	*YrSen*	50.0	75.3	25.3

2.2.5. Mendelization of *QYr.sun-2AS*

Based on the presence of the *QYr.sun-2AS* peak marker 4539050, moderately susceptible response (score 6), and the absence of peak markers for *QYr.sun-1BL* and *QYr.sun-3BS*, Sentinel/Nyb3 RIL#70 (SN#70) was selected as a carrier of *QYr.sun-2AS* singly. SN#70 was crossed with Nyb3 and was advanced to the F_5 generation through the single seed descent method. F_5 single plant progenies formed the F_6 RIL population. The entire RIL population (108 RILs) was phenotyped for stripe rust response variation in the field during the 2017 and 2018 crop seasons. SN#70 was scored '6', and Nyb3 was fully susceptible with a score of '9'. Stripe rust responses among the RILs varied from '6' to '9'. Monogenic segregation (61 HR:47 HS; $\chi^2_{1:1} = 1.81$, non-significant at $p = 0.05$ and 1 *d.f.*) for adult plant stripe rust response was observed among the SN#70/Nyb3 RIL population. Based on these results, *QYr.sun-2AS* was temporarily named *YrSen2*.

2.2.6. Interaction of QTL

A comparison of the mean percent stripe rust severities of individual QTL and different QTL combinations across sites and years is presented in Table 4. The critical difference (CD) was calculated for each experiment to see whether the stripe rust severities displayed on RILs with different QTL combinations were significantly different. RILs carrying single QTL exhibited significantly higher stripe rust severities compared to RILs with combinations of two or three QTL. In the 2014 crop season at the LDN site, RILs carrying *QYr.sun-3BS* exhibited significantly lower stripe rust severity (27%) compared to *QYr.sun-2AS* (44%), whereas *QYr.sun-2AS* and *QYr.sun-1BL* did not differ significantly (Table 4). The stripe rust severities of RILs with single QTL did not differ significantly in two other experiments. The RILs carrying *QYr.sun-1BL* + *QYr.sun-2AS* (11.7%) and *QYr.sun-1BL* + *QYr.sun-3BS* (17.1%) produced significantly lower stripe rust severities in 2014 than that exhibited by *QYr.sun-2AS* + *QYr.sun-3BS*-possessing RILs (28.5%). In the remaining two experiments, the severities exhibited by the three different dual combinations of QTL did not differ significantly. The mean rust severity of RILs carrying the Sentinel alleles for all three QTL produced significantly lower stripe rust severities than single- and dual-QTL-carrying RILs (Table 4).

Table 4. Mean stripe rust severities of Sentinel/Nyb3 RILs carrying different QTL combinations.

Loci	Rust Severity (%)		
	2014	2015	
	LDN	K	LDN
QYr.sun-1BL	34.58	56.07	45.00
QYr.sun-2AS	44.11	48.04	47.91
QYr.sun-3BS	26.67	49.17	50.00
QYr.sun-1BL + *QYr.sun-2AS*	11.67	34.50	28.75
QYr.sun-1BL + *QYr.sun-3BS*	17.06	34.38	30.74
QYr.sun-2AS + *QYr.sun-3BS*	28.50	28.75	26.25
QYr.sun-1BL + *QYr.sun-2AS* + *QYr.sun-3BS*	6.72	12.94	13.53
Critical Difference (CD)	9.68 ± 4.98	10.77 ± 5.54	9.88 ± 5.08

3. Discussion

The common wheat cultivar Sentinel has displayed a high level of resistance to stripe rust since its release in 2005. The continuous distribution of stripe rust response variation among the Sentinel/Nyb3 RIL population, when tested against the *Pst* pathotype 134 E16A+ *Yr17*+*Yr27*+, suggested quantitative inheritance of resistance. The involvement of two to three genes in conditioning APR to stripe rust in Sentinel was estimated. This estimation was confirmed with the detection of three consistent QTL, one each on chromosomes 1B, 2A and 3B of Sentinel.

The *QYr.sun-1BL* peaked at the marker 4406454 and mapped in the same genomic region as the APR gene *Yr29*. The presence of *Yr29* in Sentinel was confirmed using a closely linked SNP marker, *Lr46_SNP1G22*. The location of APR gene *Yr29* on the long arm of chromosome 1B has been reported in a number of studies [19]. In the present study, *QYr.sun-1BL* explained 13–26% of the stripe rust variation at an LOD score range of 6.2–10.5. The wide range of LOD scores (2.8–23) and phenotypic variation explanations (4.5–65%) attributed to the *Yr29* locus in several studies have been documented in a review by Rosewarne et al. [19]. The *QYr.sun-1BL* from this study was concluded to be *Yr29*.

The second QTL, *QYr.sun-2AS*, peaked at the marker 4539050 located on the short arm of chromosome 2A. Studies by Agenbag et al. [21], Bansal et al. [20], Bariana et al. [10], Boukhatem et al. [22], Chhuneja et al. [23], Dedryver et al. [24], Hao et al. [25] and Mallard et al. [26] reported QTL in chromosome 2A. Most of these QTL mapped in 2AL based on the location of associated markers, except for markers associated with the stripe rust resistance QTL reported by Bansal et al. [20], which mapped on chromosome 2AS in a durum wheat mapping population. The QTL reported by Bansal et al. [20] was later shown to express resistance at the four-leaf stage and was formally designated *Yr56* (https://wheat.pw.usda.gov/GG3/wgc; accessed on 25 October 2023). *QYr.sun-2AS* from this study does not express resistance at the four-leaf stage and, hence, is likely to be different from *Yr56* and represent a new locus.

The third QTL, *QYr.sun-3BS*, detected in this study peaked at the stripe rust resistance locus, *YrSen*, which was characterized at the four-leaf stage at 21 ± 2 °C and was flanked by DArTseq markers 4409093 and 1012045. Several studies have reported QTL on the short arm of chromosome 3B [19]. The QTL reported in these studies corresponded to the APR gene *Yr30* on chromosome 3BS. The APR gene *Yr30* does not express stripe rust resistance at the four-leaf stage at 21 ± 2 °C. These results indicated that *Yr30* and *YrSen* are different.

The short arm of chromosome 3B also carries the all-stage stripe rust resistance gene *Yr57*. This gene is located at the distal end of chromosome 3BS and is flanked by markers *gwm389* at 2.0 cM proximally and *BS00062676* at a genetic distance of 2.3 cM distally [27]. The post-seedling expression of *QYr.sun-3BS* differentiates it from *Yr57*, which produces near-immune responses at the 2-leaf stage. The stripe rust resistance gene *Yr58* was also mapped on the distal end of chromosome 3BS [28]. *Yr58* can be detected at the 4-leaf stage at 21 ± 2 °C. The detection of *Yr58* and *YrSen* under similar conditions in the greenhouse suggested that these genes may represent the same locus. However, *Yr58* (*QYr.sun-3BS*) explained 34 to 59% of the variation in stripe rust response in a W195/BT-Schomburgk RIL population and, in contrast, *QYr.sun-3BS* (*YrSen*) contributed 6 to 12% to phenotypic variation. In addition, *Yr58* is ineffective against pre-2002 *Pst* pathotypes [28]. Based on these facts, *QYr.sun-3BS* (*YrSen*) is likely to be different from *Yr58*. All these results demonstrated the uniqueness of *QYr.sun-3BS* (*YrSen*) and, hence, it deserves a permanent gene symbol. The expression of *QYr.sun-3BS* at the 4-leaf stage highlights the importance of this locus in the early-growth-stage management of stripe rust.

The combination of all three QTL (*QYr.sun-1BL*, *QYr.sun-2AS* and *QYr.sun-3BS*) detected in Sentinel/Nyb3 RILs exhibited significantly lower stripe rust severities across sites and years (Table 4) compared to RILs with two or single QTL. Likewise, the dual-QTL-carrying RILs displayed stripe rust severities significantly lower than the single-QTL RILs. Such results have been reported in earlier studies [29]. The maintenance of a high level of adult plant stripe rust resistance by Sentinel since 2005 demonstrated the durable nature of *QYr.sun-1BL* (*Yr29*), *QYr.sun-2AS* (*YrSen2*) and *QYr.sun-3BS* (*YrSen*).

4. Materials and Methods

4.1. Materials

4.1.1. Plant Material

Sentinel was crossed with a stripe-rust-susceptible genotype, Nyabing 3 (Nyb3), and a mapping population of 117 recombinant inbred line (RIL) F_6 individuals was developed. Susceptible genotypes Morocco and Nyb3 were included as stripe-rust-susceptible spreaders during field testing.

4.1.2. Pathogen Materials

Pst pathotype 134 E16A+17+27+ was used for creating stripe rust epidemics in the field and for greenhouse studies.

4.2. Stripe Rust Assessments

4.2.1. Field Assessment

The Sentinel/Nyb3 RIL population was grown in the field as 10 seed hill plots at the Plant Breeding Institute (PBI), Cobbitty, in the 2014 and 2015 crop seasons. The parents Sentinel and Nyb3 were included as controls. Each block of 70 hill plots was surrounded by a mixture of susceptible genotypes (Morocco and Nyb3) to create stripe rust epidemics. The RIL population was evaluated at one experimental site (Lansdowne = LDN) during the 2014 crop season and at two sites (Karalee = K and LDN) in the 2015 crop season. The 2014 experiment was sown in the first week of July, whereas the 2015 experiments were planted in the first week of June at the K site and the second week of June at the LDN site. The experimental area was irrigated to create favorable conditions for stripe rust development. The experiments were artificially inoculated using *Pst* pathotype 134 E16A+17+27+ suspended in light mineral oil (Isopar L®) and misted over the whole experiment using an ultra-low-volume Micron sprayer. Further inoculation was done by dropping rusty potted seedlings between susceptible spreader rows. Adult plant stripe rust response assessments were performed using a 1–9 scale (1 = very resistant (VR), 2 = resistant (R), 3 = resistant to moderately resistant (RMR) 4 = moderately resistant (MR), 5 = moderately resistant to moderately susceptible (MRMS), 6 = moderately susceptible (MS), 7 = moderately susceptible to susceptible (MSS), 8 = susceptible (S) and 9 = very susceptible) described in Bariana et al. [30]. The 1–9 scale-based scores were converted into percent severities for comparison of genotypes carrying different numbers of QTL [30].

4.2.2. Greenhouse Assessment

The Sentinel/Nyb3 RIL population and both parents were tested at the two-leaf, three-leaf and four-leaf stages with pathotype 134 E16A+17+r27+ under two temperatures regimes (18 ± 2 °C and 21 ± 2 °C) to observe the expression of resistance under greenhouse conditions. The sowing and inoculation details of the greenhouse experiments are described in Randhawa et al. [31]. Stripe rust assessments were made using a 0–4 scale described in McIntosh et al. [32].

4.3. Molecular Mapping

4.3.1. DNA Extraction

Leaf tissues of about 2.5 cm length, picked from at least eight seedlings from each of the RIL and parents, were put in well-labelled 2 mL Eppendorf tubes and dried on silica gel for 3 days. DNA was isolated using the modified CTAB method described in Bansal et al. [20]. DNA samples were quantified using a Nanodrop ND-1000 spectrophotometer (Thermo Fisher Scientific, Wilmington, NC, USA), and 30 ng/μL genomic DNA dilutions were made.

4.3.2. Linkage Map Construction and QTL Analysis

The DNA samples of 92 Sentinel/Nyb3 RILs and the two parents were sent to Diversity Arrays Technology, Canberra, Australia, for DArTseq genotyping. A total of 16,815 DArTseq markers were used to genotype the Sentinel/Nyb3 RIL population and scored '1' for presence and '0' for absence of the target marker. Chi-squared (χ^2) analysis was performed to check segregation distortion of markers from the expected ratio of 1:1, and markers with Chi-squared values greater than 4 were excluded from the linkage map. Monomorphic markers, redundant markers and markers with more than 10% missing data were excluded from the linkage map. Sentinel/Nyb3 linkage groupings were created using 'MapDisto' software version 1.7.5 for MS Windows 2007, and linkage groups were selected based on LOD score of 3. MapManager QTXb20 [33] was used for construction of the linkage map. Composite interval mapping (CIM) was performed using window QTL Cartographer V2.5_011.

4.4. Marker Screening for APR Genes Yr18 and Yr29

Yr18-linked marker *csLV34* [34] and *Yr18* gene-specific markers *cssfr1*, *cssfr2*, *cssfr3*, *cssfr4*, *cssfr5* and *cssfr6* [35] were used to detect the presence of *Yr18* in parents Sentinel and Nyb3. *Lr34*-carrying cultivar Janz was included as a positive control. PCR amplifications were performed in 10 μL reaction volumes containing 30 ng of genomic DNA, 0.2 mM dNTPs, 1× PCR buffer containing $MgCl_2$ (Bioline), 0.5 μM of each forward primer and reverse primer and 0.02 U Immolase Taq DNA polymerase (Bioline (Sydney, Australia) Pty Ltd.). PCR products were separated in 2% agarose gel.

Yr29-linked SNP marker *Lr46_SNP1G22* was employed to detect the presence of this gene in the parents. Lalbahadur + *Lr46* and Lalbahadur were included as positive and negative controls, respectively. PCR reactions were performed in 8 μL reaction volumes containing 3 μL of 30 ng/μL genomic DNA, 4 μL of 2× KASP mix (KBioscience LLC, London UK), 0.11 μL of assay mix (containing 12 μM of each allele-specific forward primer and 30 μM of reverse primer) and 0.89 μL of autoclaved double-distilled water. The PCR conditions used included 94 °C for 15 min; 10 touchdown cycles of 94 °C for 20 s and 65–57 °C for 60 s (dropping 0.8 °C per cycle); and 26–35 cycles of 94 °C for 20 s and 57 °C for 60 s. Readings were taken by fluorescence detection of the reactions at 40 °C for 30 s, and the data were analyzed using CFX Manager 3.1 software (Biorad, Hercules, CA, USA).

4.5. Statistical Analysis

The minimum number of genes segregating for stripe rust resistance among the Sentinel/Nyb3 RIL population in the field was estimated using Wright's method [36] adjusted for the level of inbreeding in the original formula [37]. The equation used is as below:

$$n = (GR)\ 2/4.27 \times \sigma^2 g$$

where n = minimum number of effective genes, GR = genotypic range and σ^2g = genetic variance of RILs.

Chi-squared (χ^2) analysis was performed to determine the goodness of fit of the observed segregation for stripe rust response at the 4-leaf stage with expected genetic ratios and to check for segregation distortion of markers. Critical differences (CDs) were calculated using the following formula: CD = Standard error × t value at $p = 0.05$.

5. Conclusions

The combination of three consistent QTL (*QYr.sun-1BL*, *QYr.sun-2AS* and *QYr.sun-3BS*) were demonstrated to condition high level of adult plant stripe rust resistance in wheat cultivar Sentinel. The additive nature of these QTL in conditioning a high level of stripe rust resistance was demonstrated through comparison of recombinant inbred lines (RILs) carrying these QTL in all different combinations. While *QYr.sun-1BL* corresponded to APR gene *Yr29*, *QYr.sun-3BS* and *QYr.sun-2AS* were temporarily named *YrSen* and *YrSen2*, respectively. Based on their uniqueness, *YrSen* and *YrSen2* deserved to be permanently designated. These loci are a valuable addition to a small number of permanently named APR genes. The durability of adult plant stripe rust resistance of Sentinel makes it an excellent donor source of stripe rust improvement in wheat breeding programs.

Author Contributions: H.B. and U.B. planned the project; B.C., H.B. and U.B. produced the recombinant inbred line population; B.C. and H.B. conducted stripe rust phenotyping; B.C. and U.B. performed QTL analysis; U.B., W.W. and H.B. supervised the project; B.C. wrote the initial draft; U.B., W.W. and H.B. edited the manuscript. All authors have read and agreed to the published version of the manuscript.

Funding: We thank National Agricultural Research Organisation (NARO), Uganda for the award of a scholarship to B.C. to pursue his Ph.D. program at the University of Sydney and the Grains Research and Development Corporation, Australia for financial support through the Australian Cereal Rust Control Program (Grant numbers: US00063 and 9176057).

Data Availability Statement: All data are presented in this manuscript.

Acknowledgments: We thanks Hanif Miah for assistance in development of RIL populations and Matthew Williams for creation of field epidemics.

Conflicts of Interest: The authors declare no conflicts of interest.

References

1. Wellings, C.R. Global status of stripe rust: A review of historical and current threats. *Euphytica* **2011**, *179*, 129–141. [CrossRef]
2. Chen, X.M. Epidemiology and control of stripe rust [*Puccinia striiformis* f. sp. *tritici] on wheat. Can. J. Plant Pathol.* **2005**, *27*, 314–337. [CrossRef]
3. Beddow, J.M.; Pardey, P.G.; Chai, Y.; Hurley, T.M.; Kriticos, D.J.; Braun, H.J.; Park, R.F.; Yonow, T. Research investment implications of shifts in the global geography of wheat stripe rust. *Nat. Plants* **2015**, *1*, 5132. [CrossRef] [PubMed]
4. O'Brien, L.; Brown, J.S.; Young, R.M.; Pascoe, I. Occurrence and distribution of wheat stripe rust in Victoria and susceptibility of commercial wheat cultivars. *Austr. Plant Pathol.* **1980**, *9*, 14. [CrossRef]
5. Wellings, C.R.; Wright, D.G.; Keiper, F.; Loughman, R. First detection of wheat stripe rust in Western Australia: Evidence for a foreign incursion. *Austr. Plant Pathol.* **2003**, *32*, 321–322. [CrossRef]
6. Wellings, C.R. Puccinia striiformis in Australia: A review of the incursion, evolution, and adaptation of stripe rust in the period 1979–2006. *Austr. J. Agric. Res.* **2007**, *58*, 567–575. [CrossRef]
7. Ali, S.; Rodriguez-Algaba, J.; Thach, T.; Sørensen, C.K.; Hansen, J.G.; Lassen, P.; Nazari, K.; Hodson, D.P.; Justesen, A.F.; Hovmøller, M.S. Yellow rust epidemics worldwide were caused by pathogen races from divergent genetic lineages. *Front. Plant Sci.* **2017**, *8*, 1057. [CrossRef]
8. Singh, R.P.; Singh, P.K.; Rutkoski, J.; Hodson, D.P.; He, X.; Jørgensen, L.N.; Hovmøller, M.S.; Huerta-Espino, J. Disease impact on wheat yield potential and prospects of genetic control. *Annu. Rev. Phytopathol.* **2016**, *54*, 303–322. [CrossRef]

9. Carmona, M.; Sautua, F.; Pérez-Hérnandez, O.; Reis, E.M. Role of fungicide applications on the integrated management of wheat stripe rust. *Front. Plant Sci.* **2020**, *11*, 733. [CrossRef]
10. Bariana, H.S.; Bansal, U.K.; Schmidt, A.; Lehmensiek, A.; Kaur, J.; Miah, H.; Howes, N.; McIntyre, C.L. Molecular mapping of adult plant stripe rust resistance in wheat and identification of pyramided QTL genotypes. *Euphytica* **2010**, *176*, 251–260. [CrossRef]
11. Chen, X.M. Challenges and solutions for stripe rust control in the United States. *Austr. J. Agric. Res.* **2007**, *58*, 648–655. [CrossRef]
12. Bariana, H.S. Breeding for disease resistance. In *Encyclopedia of Applied Plant Sciences*; Thomas, B., Murphy, D.J., Murray, B.G., Eds.; Academic Press: London, UK, 2003; pp. 244–253.
13. Lagudah, E.S. Molecular genetics of race non-specific rust resistance in wheat. *Euphytica* **2011**, *179*, 81–91. [CrossRef]
14. Bariana, H.S.; McIntosh, R.A. Genetics of adult plant stripe rust resistance in four Australian wheats and the French cultivar 'Hybride-de-Bersee'. *Plant Breed.* **1995**, *114*, 485–491. [CrossRef]
15. Singh, R.P.; Huerta-Espino, J.; Rajaram, S. Achieving near-immunity to leaf and stripe rusts in wheat by combining slow rusting resistance genes. *Acta Phytopathol. Entomol. Hung.* **2000**, *35*, 133–139.
16. Bariana, H.S.; Bansal, U.K. Breeding for disease resistance. In *Encyclopedia of Applied Plant Sciences*; Brian, T., Brian, G.M., Dennis, M., Eds.; Academic Press: Waltham, MA, USA, 2017; Volume 3, pp. 69–76.
17. Elshire, R.J.; Glaubitz, J.C.; Sun, Q.; Poland, J.A.; Kawamoto, K.; Buckler, E.S.; Mitchell, S.E. A robust, simple genotyping-by-sequencing (GBS) approach for high diversity species. *PLoS ONE* **2011**, *6*, e19379. [CrossRef]
18. Varshney, R.K.; Nayak, S.N.; May, G.D.; Jackson, S.A. Next-generation sequencing technologies and their implications for crop genetics and breeding. *Trends Biotechnol.* **2009**, *27*, 522–530. [CrossRef]
19. Rosewarne, G.M.; Herrera-Foessel, S.A.; Singh, R.P.; Huerta-Espino, J.; Lan, C.X.; He, Z.H. Quantitative trait loci of stripe rust resistance in wheat. *Theor. Appl. Genet.* **2013**, *126*, 2427–2449. [CrossRef]
20. Bansal, U.K.; Kazi, A.G.; Singh, B.; Hare, R.A.; Bariana, H.S. Mapping of durable stripe rust resistance in a durum wheat cultivar Wollaroi. *Mol. Breed.* **2014**, *33*, 51–59. [CrossRef]
21. Agenbag, G.M.; Pretorius, Z.A.; Boyd, L.A.; Bender, C.M.; Prins, R. Identification of adult plant resistance to stripe rust in the wheat cultivar Cappelle-Desprez. *Theor. Appl. Genet.* **2012**, *125*, 109–120. [CrossRef]
22. Boukhatem, N.; Baret, P.V.; Mingeot, D.; Jacquemin, J.M. Quantitative trait loci for resistance against yellow rust in two wheat-derived recombinant inbred line populations. *Theor. Appl. Genet.* **2002**, *104*, 111–118. [CrossRef]
23. Chhuneja, P.; Kaur, S.; Garg, T.; Ghai, M.; Kaur, S.; Prashar, M.; Bains, N.S.; Goel, R.K.; Keller, B.; Dhaliwal, H.S.; et al. Mapping of adult plant stripe rust resistance genes in diploid A genome wheat species and their transfer to bread wheat. *Theor. Appl. Genet.* **2008**, *116*, 313–324. [CrossRef] [PubMed]
24. Dedryver, F.; Paillard, S.; Mallard, S.; Robert, O.; Trottet, M.; Negre, S.; Verplancke, G.; Jahier, J. Characterization of genetic components involved in durable resistance to stripe rust in the bread wheat 'Renan'. *Phytopathology* **2009**, *99*, 968–973. [CrossRef] [PubMed]
25. Hao, Y.; Chen, Z.; Wang, Y.; Bland, D.; Buck, J.; Brown-Guedira, G.; Johnson, J. Characterization of a major QTL for adult plant resistance to stripe rust in US soft red winter wheat. *Theor. Appl. Genet.* **2011**, *123*, 1401–1411. [CrossRef] [PubMed]
26. Mallard, S.; Gaudet, D.; Aldeia, A.; Abelard, C.; Besnard, A.L.; Sourdille, P.; Dedryver, F. Genetic analysis of durable resistance to yellow rust in bread wheat. *Theor. Appl. Genet.* **2005**, *110*, 1401–1409. [CrossRef] [PubMed]
27. Randhawa, M.S.; Bariana, H.S.; Mago, R.; Bansal, U.K. Mapping of a new stripe rust resistance locus *Yr57* on chromosome 3BS of wheat. *Mol. Breed.* **2015**, *35*, 65. [CrossRef]
28. Chhetri, M.; Bariana, H.S.; Kandiah, P.; Bansal, U.K. *Yr58*: A new stripe rust resistance gene and its interaction with *Yr46* for enhanced resistance. *Phytopathology* **2016**, *106*, 1530–1534. [CrossRef] [PubMed]
29. Maree, G.J.; Prins, R.; Boyd, L.A.; Castelyn, H.D.; Bender, C.M.; Boshoff, W.H.P.; Pretorius, Z.A. Assessing the individual and combined effects of QTLfor adult plant stripe rust resistance derived from Cappelle-Desprez. *Agronomy* **2019**, *9*, 154. [CrossRef]
30. Bariana, H.S.; Miah, H.; Brown, G.N.; Willey, N.; Lehmensiek, A. Molecular mapping of durable rust resistance in wheat and its implication in breeding. In *Wheat production in stressed environments: Proceedings of the 7th International Wheat Conference, Mar del Plata, Argentina, 27 November–2 December 2005*; Buck, H.T., Nisi, J.E., Salomón, N., Eds.; Springer: Netherlands, Dordrecht, 2 December 2007; pp. 723–728.
31. Randhawa, M.; Bansal, U.; Valarik, M.; Klocova, B.; Dolezel, J.; Bariana, H. Molecular mapping of stripe rust resistance gene *Yr51* in chromosome 4AL of wheat. *Theor. Appl. Genet.* **2014**, *127*, 317–324. [CrossRef]
32. McIntosh, R.A.; Park, R.F.; Wellings, C.R. *Wheat Rusts—An Atlas of Resistance Genes*; CSIRO Publishing: Collingwood, Australia, 1995; pp. 9–12.
33. Manly, K.F.; Cudmore, R.H., Jr.; Meer, J.M. Map Manager QTX, cross-platform software for genetic mapping. *Mammal. Genom.* **2001**, *12*, 930–932. [CrossRef]
34. Lagudah, E.S.; McFadden, H.; Singh, R.P.; Huerta-Espino, J.; Bariana, H.S.; Spielmeyer, W. Molecular genetic characterization of the *Lr34/Yr18* slow rusting resistance gene region in wheat. *Theor. Appl. Genet.* **2006**, *114*, 21–30. [CrossRef]
35. Lagudah, E.S.; Krattinger, S.G.; Herrera-Foessel, S.; Singh, R.P.; Huerta-Espino, J.; Spielmeyer, W.; Brown-Guedira, G.; Selter, L.L.; Keller, B. Gene-specific markers for the wheat gene *Lr34/Yr18/Pm38* which confers resistance to multiple fungal pathogens. *Theor. Appl. Genet.* **2009**, *119*, 889–898. [CrossRef] [PubMed]

36. Wright, S. Volume 1. Genetic and Biometric Foundations. In *Evolution and the Genetics of Populations*; University of Chicago Press: Chicago, IL, USA, 1968.
37. Cockerham, C.C. Covariances of Relatives from Self-Fertilization 1. *Crop Sci.* **1983**, *23*, 1177–1180. [CrossRef]

plants

MDPI

Article

Exploitation of the Genetic Variability of Diverse Metric Traits of Durum Wheat (*Triticum turgidum* L. ssp. *durum* Desf.) Cultivars for Local Adaptation to Semi-Arid Regions of Algeria

Zine El Abidine Fellahi [1,2], Tahar Boubellouta [3,4], Abderrahmane Hannachi [5], Haroun Belguet [6], Nasreddine Louahdi [6], Amar Benmahammed [2,7], Aleksandra O. Utkina [8] and Nazih Y. Rebouh [8,*]

1 Department of Agronomic Sciences, Faculty of Natural, Life and Earth Sciences and the Universe, University of Mohamed El Bachir El Ibrahimi, Bordj Bou Arreridj 34030, Algeria
2 Valorization of Natural Biological Resources Laboratory (VNBR), University of Ferhat Abbas Setif-1, Setif 19000, Algeria; benmahammeda@yahoo.com
3 Department of Biological Sciences, Faculty of Natural, Life, Earth and Universe Sciences, University Mohamed El Bachir El Ibrahimi, Bordj Bou Arreridj 34030, Algeria; t.boubellouta@univ-bba.dz
4 Characterization and Valorization of Natural Resources Laboratory (CVNRN), University Mohamed El Bachir El Ibrahimi, Bordj Bou Arreridj 34030, Algeria
5 National Agronomic Research Institute of Algeria (INRAA), Setif Research Unit, Setif 19000, Algeria
6 Experimental Farm, Field Crop Institute (ITGC), Farm Road-BP03, Setif 19000, Algeria; harounbelguet@gmail.com (H.B.); n.louahdi@gmail.com (N.L.)
7 Department of Ecology and Plant Biology, Faculty of Natural and Life Science, University of Ferhat Abbas Setif-1, Setif 19000, Algeria
8 Department of Environmental Management, Institute of Environmental Engineering, RUDN University, 6 Miklukho-Maklaya St., 117198 Moscow, Russia; utkina-ao@rudn.ru
* Correspondence: rebukh_nya@rudn.ru; Tel.: +7-9774188644

Abstract: Abiotic stresses pose significant challenges to wheat farming, yet exploiting the genetic variability within germplasm collections offers an opportunity to effectively address these challenges. In this study, we investigated the genetic diversity of key agronomic traits among twenty durum wheat cultivars, with the intention to pinpoint those better suited to semi-arid conditions. Field trials were conducted at the ITGC-FDPS Institute, Setif, Algeria, during the winter season of 2021/22. A completely randomized design was used with three replicates. Statistical analyses revealed significant variation among the genotypes for most of the studied traits, with some cultivars exhibiting a superior performance in a stressful environment. Notably, traits like the number of grains per spike (NGS) and the grain yield (GY) displayed high genotypic coefficients of variation (CVg). Except for membrane thermostability (MT) and biological yield (BY), the majority of the assessed traits exhibited moderate-to-high heritability estimates. Genotypic and phenotypic correlation studies have confirmed the importance of many yield-related traits in the expression of GY. The harvest index (HI) underscored the highest genotypic direct effect on GY, followed closely by spike number (SN), serving as consistent pathways through which most of the measured traits indirectly influenced GY. The cluster analysis categorized the durum wheat cultivars into seven distinct clusters. The largest inter-cluster distance was observed between clusters G3 and G4 (D^2 = 6145.86), reflecting maximum dissimilarity between the individuals of these clusters. Hybridizing divergent clusters may benefit future breeding programs aiming to develop potential durum wheat varieties through cross combinations. This study's findings contribute to sustainable agriculture efforts by facilitating the selection of genotypes with enhanced resilience and productivity, particularly for cultivation in challenging semi-arid regions.

Keywords: *Triticum durum*; genotypic correlation; path analysis; Mahalanobis distance; adaptation; performance; semi-arid

1. Introduction

Durum wheat (*Triticum turgidum* L. ssp. *durum* Desf.) is an important cereal crop, particularly in regions with limited water availability [1,2]. Algeria, being a country with diverse climatic conditions, including harsh environments, faces challenges in durum wheat cultivation due to water scarcity and other biotic and abiotic stresses [3]. To address these challenges and enhance durum wheat production, researchers and breeders focus on exploiting the genetic variability of diverse quantitative and some qualitative traits within old and modern germplasm [4–7].

Durum wheat germplasm represents a diverse collection of genetic resources, which includes different varieties, landraces, and wild relatives [8,9]. This genetic diversity possesses a range of traits, including highly heritable characteristics like heading date, plant height, and kernel weight, but also a large number of quantitative traits, such as yield potential, drought tolerance, disease resistance, and nutrient use efficiency, which exhibit continuous variation [10–12].

Through the systematic evaluation of diverse durum wheat germplasm collections, breeders have identified cultivars that possess desirable traits. This can be based on phenotypic traits like plant height, flowering time, grain yield, and quality characteristics, as well as genotypic traits revealed through molecular markers [13–15]. After crossing cultivars carrying desirable traits, breeders—through traditional breeding methods—select superior progenies over multiple generations [16]. Molecular breeding approaches, such as marker-assisted selection and genomic selection, facilitate the identification and utilization of specific genes or genomic regions associated with target traits [17,18].

Genetic and non-genetic parameters, such as phenotypic (CV_p) and genotypic coefficient of variation (CV_g), along with heritability ($h^2{}_{bs}$), serve as valuable biometric indicators for assessing the genetic variation and adaptability within a germplasm collection. They provide insights into the relative contributions of genetic and environmental factors to observed variations in specific traits among individuals [19–21].

Because grain yield is a polygenic complex trait that is controlled by many factors, and is extremely influenced by environmental variation, selection for high-yielding genotypes should not be based on yield only, but other yield-contributing traits should be taken into consideration during the selection process [22,23]. Hence, knowledge of the degree of relationship between yield and different yield-related components can identify the traits that could be used as indirect selection criteria, increasing the efficiency of the selection process. The genotypic and phenotypic correlations are concepts used by breeders to describe the association between the genetic makeup (genotype) and the observable traits (phenotype) of individuals within a population [24]. Genotypic correlation focuses on the genetic basis of traits, while phenotypic correlation considers the observable expression of these traits in individuals, influenced by both genetic and environmental factors [25].

In fact, the correlation coefficient only quantifies the link between each pair of traits and does not truly indicate the relative importance of secondary traits on the yield. Path coefficient is a common approach used in the field of crop breeding, where researchers are interested in understanding the complex relationships between grain yield and its related traits in order to identify those with significant effects on yield materialization to be used as selection criteria [26]. This method has the advantage of portioning the correlation coefficient into direct and indirect relationships among a set of variables, helping to understand the complex interplay between the different factors in a theoretical model [26]. The direct effect refers to the influence of one trait on another without the mediation of any other traits, whereas the indirect effect considers the impact of one trait on another that is mediated through one or more intervening traits in the model [27]. According to Boulelouah et al. (2022) [28], both correlation and path coefficient techniques could achieve knowledge regarding the proper cause-and-effect relationship between yield and yield components.

By understanding the morpho-genetic dissimilarity among diverse cultivars within the available germplasm, breeders can strategically choose parents for crossing with the goal of maximizing the potential for genetic gain. Cluster analysis is a statistical method used in the context of genetic diversity studies to classify cultivars of germplasm collections in specific heterotic groups [29]. In this context, the generalized Mahalanobis (1936) distance method provides greater precision when selecting genotypes for future crosses. It was largely used as a consistent measure of dissimilarity in various crop species, including durum wheat [30,31]. This information is valuable for both understanding the genetic structure of the population and for guiding breeding strategies to harness the genetic diversity for crop improvement.

In view of this, the present investigation was conducted to explore the genetic divergence and variability present within a durum wheat germplasm collection through the phenotypic evaluation of various desirable characteristics of particular importance for local adaptation to semi-arid regions of Algeria. This evaluation allows breeders to identify individuals as potential parents for future breeding programs targeting the development of wheat varieties with enhanced productivity and resilience.

2. Results

2.1. Sources of Variation Analysis

The results showed highly significant differences ($p \leq 0.01$) among the tested genotypes for the majority of traits subjected to one-way analysis of variance (ANOVA), except for MT and BY, which were non-significant (Table 1).

Table 1. Summary of the analysis of variance for the measured traits.

SV	Blocks	Genotypes	Residual	CV (%)
df	2	19	38	
HD	0.02	20.85 **	0.02	0.10
CC	19.19	203.57 **	50.52	16.32
CT	2.73	10.73 **	1.16	3.28
RWC	16.07	59.65 **	12.78	5.06
MT	70.30	340.29 ^{ns}	183.79	16.81
FLA	10.51	38.54 **	11.09	11.00
PH	43.80	344.75 **	15.61	4.97
SN	5255.40	20,530.40 **	5001.5	18.68
SW	0.35	2.30 **	0.56	18.39
TKW	4.98	96.24 **	7.35	9.02
NGS	0.07	19.76 **	2.29	17.15
GY	0.03	1.23 **	0.23	24.91
BY	2.58	6.40 ^{ns}	3.73	17.70
HI	1.87	113.56 **	11.28	18.76

SV: Source of variation, df: degrees of freedom, HD: Heading date (days), CC: Chlorophyll content (CCI), CT: Canopy temperature (°C), RWC: Relative water content (%), MT: Membrane thermostability (%), FLA: Flag leaf area (cm^2), PH: Plant height (cm), SN: Spike number (No m^{-2}), SW: Spike weight (t ha^{-1}), TKW: Thousand kernel weight (g), NGS: Number of grains per spike (No spike^{-1}), GY: Grain yield (t ha^{-1}), BY: Biological yield (t ha^{-1}), HI: Harvest index (%). ns and **: non-significant and significant at 1% probability level by *F*-test.

2.2. Mean Performance of Genotypes

The mean performance of 20 durum wheat genotypes for 15 morpho-physiological and agronomic traits at the Setif location is presented in Table 2. The HD elapsed between 128.0 days in Ofanto and 138.0 days in Guemgoum Rkhem, with a mean of 132.4 days. The mean CC was 43.6 CCI. The minimum CC was 29.3 CCI, which was exhibited by Bousselam, whereas the maximum value of CC was recorded in Simeto, at 54.3 CCI. The CT varied between 29.7 °C in Megress and 36.9 °C in Bousselam, with an average value of

32.8 °C. The mean RWC was 70.6%. Belikh02 was the cultivar with the lowest RWC, i.e., 62.3%, whereas Mohamed Ben Bachir was the genotype with the highest water content in the leaves, at 81.7%. Despite the wide range observed for MT, the difference among the genotypes for this trait was not significant. The mean value of MT assessed in the wheat genotypes was 80.6%. Ofanto had the lowest cell MT, which was 51.4%, whereas Amar06 had the highest MT, which was 95.2%.

Table 2. Descriptive statistics for the measured traits in the wheat varieties.

Traits	Mean ± SE	Minimum	Maximum	HSD (5%)	HSD (1%)
HD	132.4 ± 0.07	128.0	138.0	0.40	0.47
CC	43.6 ± 4.10	29.3	54.3	22.07	25.61
CT	32.8 ± 0.62	29.7	36.9	3.34	3.88
RWC	70.6 ± 2.06	62.3	81.7	11.10	12.88
MT	80.6 ± 7.82	51.4	95.2	42.10	48.85
FLA	30.3 ± 1.92	21.9	36.3	10.34	12.00
PH	79.6 ± 2.28	69.3	108.7	12.27	14.24
SN	378.7 ± 40.83	225.0	496.7	219.62	254.85
SW	4.1 ± 0.43	2.3	5.2	23.15	26.87
TKW	30.1 ± 1.56	21.1	41.0	8.42	9.77
NGS	8.8 ± 0.87	4.3	15.8	4.70	5.45
GY	1.9 ± 0.28	1.1	3.2	1.49	1.73
BY	10.9 ± 1.11	7.2	12.3	6.00	6.96
HI	17.9 ± 1.93	9.6	32.1	10.43	12.10

HD: Heading date (days), CC: Chlorophyll content (CCI), CT: Canopy temperature (°C), RWC: Relative water content (%), MT: Membrane thermostability (%), FLA: Flag leaf area (cm^2), PH: Plant height (cm), SN: Spike number (No m^{-2}), SW: Spike weight (t ha^{-1}), TKW: Thousand kernel weight (g), NGS: Number of grains per spike (No spike^{-1}), GY: Grain yield (t ha^{-1}), BY: Biological yield (t ha^{-1}), HI: Harvest index (%).

The mean FLA was 30.3 cm^2. Belikh02 had the smallest leaves, at 21.9 cm^2, whereas Waha had the largest leaves, at 36.3 cm^2. The mean PH was 79.6 cm. Mexicali75 was the shortest cultivar, with a plant height of 69.3 cm, whereas Guemgoum Rkhem was the tallest genotype, with a PH of 108.7 cm. The mean value of all of the genotypes for SN was 378.7 spikes m^{-2}. The lowest mean value of SN (225.0 spikes m^{-2}) was obtained in the Guemgoum Rkhem cultivar, while the highest mean value of this trait (496.7 spikes m^{-2}) was exhibited by Waha. The overall average SW was 4.1 t ha^{-1}, with a range from 2.3 (Guemgoum Rkhem) to 5.2 t ha^{-1} (Wahbi). The mean value of TKW was 30.1 g. The lowest kernel weight (21.1 g) was exhibited by Bousselam, while the highest TKW (41.0 g) was recorded for Boutaleb. The minimum NGS recorded was 4.3 grain spike^{-1} for Vitron, whereas a maximum of 15.8 grains spike^{-1} was found for Manssourah, with an average value of 8.8 grains spike^{-1}. The mean GY was 1.9 t ha^{-1}. Zb/Fl had the lowest GY, at 1.1 t ha^{-1}, whereas Manssourah had the highest GY, at 3.2 t ha^{-1}. The minimum BY recorded was 7.2 t ha^{-1} for Belikh02, whereas a maximum of 12.3 t ha^{-1} was registered for Saoura, with a mean value of 10.9 t ha^{-1}. Mohamed Ben Bachir had the lowest HI, i.e., 9.6% t ha^{-1}, whereas Manssourah displayed the highest HI, i.e., 32.1%, with an average HI value of 17.9%.

2.3. Estimation of Components of Variation

The estimates of genetic and non-genetic parameters for the measured traits in the wheat cultivars are illustrated in Table 3. The phenotypic coefficient of variation (CV_p) values varied between 1.99 for HD and 38.95% for GY. The genotypic coefficient of variation (CV_g) ranged from 1.99 for HD to 32.62% for HI. The CV_p, as well as the CV_g, values were categorized as low (<10%), moderate (10 to 20%), and high (>20%) [32]. A high CV_p indicates a considerable variation for the trait within the population, which may be due to genetic differences, environmental influences, or their interactions. In this study, next to GY (37.63%), HI (37.63%), NGS (32.29%), SN (26.64%), SW (26.34%), CC (23.14%), and TKW (20.22%) also presented high CV_p estimates. By contrast, low values of CV_p were found in HD (1.99%), CT (5.76%), and RWC (6.31%), whereas the remaining recorded traits exhibited moderate values of CV_p.

Table 3. Estimates of genetic and non-genetic parameters for the measured traits in the wheat varieties.

Traits	$\sigma^2{}_p$	$\sigma^2{}_g$	$\sigma^2{}_e$	CV_p (%)	CV_g (%)	CV_g/CV_e	$h^2{}_{bs}$
HD	6.96	6.94	0.02	1.99	1.99	20.41	1.00
CC	101.54	51.02	50.52	23.14	16.40	1.00	0.50
CT	4.35	3.19	1.16	6.36	5.44	1.66	0.73
RWC	28.40	15.62	12.78	7.54	5.60	1.11	0.55
MT	235.96	52.17	183.79	19.05	8.96	0.53	0.22
FLA	20.24	9.15	11.09	14.86	9.99	0.91	0.45
PH	125.32	109.71	15.61	14.07	13.17	2.65	0.88
SN	10,177.76	5176.29	5001.47	26.64	19.00	1.02	0.51
SW	1.14	0.58	0.56	26.34	18.86	1.03	0.51
TKW	36.98	29.63	7.35	20.22	18.10	2.01	0.80
NGS	8.12	5.83	2.29	32.29	27.36	1.59	0.72
GY	0.57	0.33	0.23	38.95	29.92	1.20	0.59
BY	4.62	0.89	3.73	19.71	8.64	0.49	0.19
HI	45.37	34.09	11.28	37.63	32.62	1.74	0.75

HD: Heading date, CC: Chlorophyll content, CT: Canopy temperature, RWC: Relative water content, MT: Membrane thermostability, FLA: Flag leaf area, PH: Plant height, SN: Spike number, SW: Spike weight, TKW: Thousand kernel weight, NGS: Number of grains per spike, GY: Grain yield, BY: Biological yield, HI: Harvest index. $\sigma^2{}_p$: Phenotypic variance, $\sigma^2{}_g$: Genetic variance, $\sigma^2{}_e$: Environmental variance, CV_p: Phenotypic coefficient of variation, CV_g: Genotypic coefficient of variation, CV_e: Environmental coefficient of variation, $h^2{}_{bs}$: Broad-sense heritability.

2.4. Relationships among Measured Traits

Figure 1 represents the phenotypic (r_p) and genotypic (r_g) correlation coefficients among the 13 morpho-physiological and agronomic traits measured in the durum wheat cultivars. At both genotypic and phenotypic levels, GY showed significant to highly significant and positive relationships with CC (r_p = 0.332 ** and r_g = 0.343 **), SN (r_p = 0.443 ** and r_g = 0.270 *), SW (r_p = 0.497 ** and r_g = 0.270 *), TKW (r_p = 0.337 ** and r_g = 0.332 **), NGS (r_p = 0.558 ** and r_g = 0.693 **), and HI (r_p = 0.833 ** and r_g = 0.926**). BY, in addition to its relation to GY, it had other strong correlations, including positives with CC (r_p = 0.256 * and r_g = 0.628 **), FLA (r_p = 0.429 ** and r_g = 1.152 **), SN (r_p = 0.739 ** and r_g = 0.599 **), and SW (r_p = 0.651 ** and r_g = 0.332 **); and a negative correlation with NGS (r_p = −0.262 * and r_g = −0.326 *). As discussed above, the flag leaf area and the biological yield were largely influenced by similar environmental conditions (Table 3); therefore, the environmental factors that affected both traits may explain the overestimation of the genetic correlation coefficient value. On the other hand, HI demonstrated significant negative correlations with RWC (r_p = −0.270 * and r_g = −0.552 **) and PH (r_p = −0.287 * and r_g = −0.311 *), and positive associations with TKW (r_p = 0.336 ** and r_g = 0.357 **) and NGS (r_p = 0.762 ** and r_g = 0.792 **).

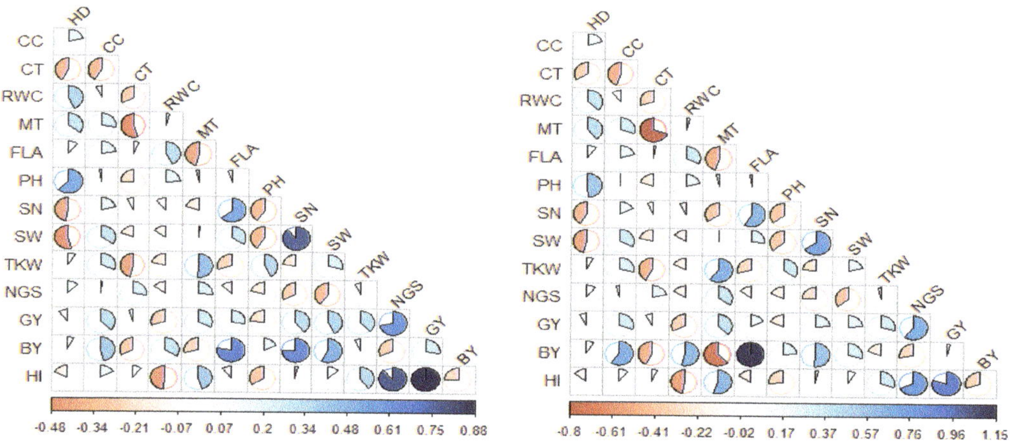

Figure 1. Correlogram representing phenotypic (**left**) and genotypic (**right**) correlation coefficients among the 13 measured traits in the durum wheat cultivars. HD: Heading date, CC: Chlorophyll content, CT: Canopy temperature, RWC: Relative water content, MT: Membrane thermostability, FLA: Flag leaf area, PH: Plant height, SN: Spike number, SW: Spike weight, TKW: Thousand kernel weight, NGS: Number of grains per spike, GY: Grain yield, BY: Biological yield, HI: Harvest index.

Worthy of note is the negative correlation of SN with HD ($r_p = -0.342$ ** and $r_g = -0.474$ **), PH ($r_p = -0.290$ * and $r_g = -0.416$ **), and NGS ($r_p = -0.292$ * and $r_g = -0.288$ *); and the positive correlations between SN and FLA ($r_p = 0.458$ ** and $r_g = 0.671$ **) and SN and SW ($r_p = 0.790$ ** and $r_g = 0.762$ **). The SW also showed negative relationships with HD ($r_p = -0.384$ ** and $r_g = -0.531$ **) and PH ($r_p = -0.293$ * and $r_g = -0.414$ **), and a positive correlation with FLA ($r_p = 0.307$ * and $r_g = 0.299$ *). The HD, in addition to its negative relations with SN and SW, showed highly significant correlations, including a negative correlation with CT ($r_p = -0.336$ ** and $r_g = -0.381$ **) and positive correlations with RWC ($r_p = 0.321$ * and $r_g = 0.438$ **) and PH ($r_p = 0.530$ ** and $r_g = 0.570$ **). The TKW exhibited a significant negative relationship with CT ($r_p = -0.490$ ** and $r_g = -0.286$ *) and significant positive correlations with MT ($r_p = 0.286$ * and $r_g = 0.699$ **) and PH ($r_p = 0.326$ * and $r_g = 0.390$ **). Furthermore, MT significantly correlated with CT ($r_p = -0.259$ * and $r_g = -0.804$ **) and FLA ($r_p = -0.361$ ** and $r_g = -0.512$ **). FLA and RWC were positively inter-correlated ($r_p = 0.358$ ** and $r_g = 0.382$ **) at both genotypic and phenotypic levels.

At the phenotypic level, GY presented significant to highly significant positive relationships with FLA ($r_p = 0.272$ * and $r_g = 0.210$ ^{ns}) and BY ($r_p = 0.390$ ** and $r_g = 0.045$ ^{ns}). Similarly, the data regarding the TKW measurements showed a significant positive correlation with SW ($r_p = 0.278$ * and $r_g = 0.238$ ^{ns}). At the genotypic level, the GY showed strong correlations, including a negative correlation with RWC ($r_p = -0.152$ ^{ns} and $r_g = -0.366$ **) and a positive correlation with MT ($r_p = 0.651$ ** and $r_g = 0.332$ ^{ns}). The BY significantly correlated with most of the physiological traits, including CT ($r_p = -0.176$ ^{ns} and $r_g = -0.497$ **), RWC ($r_p = 0.111$ ^{ns} and $r_g = 0.625$ **), and MT ($r_p = -0.059$ ^{ns} and $r_g = -0.721$ **). The BY also displayed a positive genotypic correlation with PH ($r_p = 0.085$ ^{ns} and $r_g = 0.281$ *) and a strong negative correlation with HI ($r_p = -0.152$ ^{ns} and $r_g = -0.354$ **). The latter also demonstrated a highly significant positive association with MT ($r_p = 0.241$ ^{ns} and $r_g = 0.629$ **). Other significant correlations were also observed among the measured traits. For example, negative correlations of SN with MT ($r_p = -0.033$ ^{ns} and $r_g = -0.409$ **) and TKW ($r_p = -0.113$ ^{ns} and $r_g = -0.269$ *) were found. The latter trait also demonstrated other important correlations, including a positive correlation with CC ($r_p = 0.222$ ^{ns} and $r_g = 0.322$ *) and a negative with correlation FLA ($r_p = -0.200$ ^{ns} and $r_g = -0.326$ *). The CC exhibited strong positive relationships with MT ($r_p = 0.186$ ^{ns} and $r_g = 0.353$**) and SW ($r_p = 0.241$ ^{ns} and $r_g = 0.378$ **) and a negative correlation with CT ($r_p = -0.186$ ^{ns}

and $r_g = -0.508$ **). The CT and RWC were negatively inter-related ($r_p = -0.221$ ^{ns} and $r_g = -0.349$ **). The data regarding the MT measurements showed highly significant positive correlations with HD ($r_p = 0.218$ ^{ns} and $r_g = 0.460$ **) and NGS ($r_p = 0.105$ ^{ns} and $r_g = 0.398$ **).

2.5. Estimate of Direct and Indirect Effects at a Genotypic Level

The results of the path coefficient analysis of 12 measured traits to GY are given in Table 4 and Figure 2. Dewey and Lu (1959) [26] classified path coefficient values as very high (>1.0), high (0.30–0.99), moderate (0.2–0.29), low (0.1–0.19), and negligible (0.00–0.09). Based on this, the harvest index (1.166) and RWC (0.381) exhibited very high and high positive direct effects on GY, respectively. Traits such as SN (0.503), CT (0.380), and CC (0.302) exerted moderate positive direct effects on GY, similar to their genotypic correlations with this primary trait (Table 4; Figure 1). The PH (0.215) and MT (0.169) expressed moderate-to-low direct effects on GY. Conversely, SW (−0.209), next to NGS (−0.183), yielded the first and second highest negative direct effects on GY, respectivley. While HD, FLA, and BY had non-significant associations with GY at a genotypic level, it has been stated in Table 4 and Figure 2 that the SW and NGS demonstrated positive and statistically significant correlations with the grain yield per hectare.

Table 4. Direct (diagonal) and indirect effects (off-diagonal) of 14 measured traits on grain yield in the wheat cultivars at a genotypic level.

Traits	HD	CC	CT	RWC	MT	FLA	PH	SN	SW	TKW	NGS	BY	HI
HD	**−0.047**	0.065	−0.145	0.167	0.078	−0.009	0.123	−0.239	0.111	0.004	−0.022	0.009	−0.200
CC	−0.010	**0.302**	−0.193	−0.054	0.060	−0.020	−0.002	0.104	−0.079	0.018	0.008	0.060	0.149
CT	0.018	−0.154	**0.380**	−0.133	−0.136	−0.002	−0.051	−0.042	0.051	−0.027	−0.045	−0.044	0.092
RWC	−0.021	−0.043	−0.133	**0.381**	0.007	−0.032	0.052	−0.049	0.038	−0.012	0.034	0.055	−0.644
MT	−0.022	0.107	−0.306	0.016	**0.169**	0.043	−0.008	−0.206	0.001	0.039	−0.073	−0.064	0.734
FLA	−0.005	0.072	0.008	0.146	−0.086	**−0.084**	−0.005	0.338	−0.063	−0.018	0.029	0.102	−0.221
PH	−0.027	−0.002	−0.090	0.092	−0.006	0.002	**0.215**	−0.209	0.087	0.022	0.037	0.025	−0.363
SN	0.022	0.062	−0.032	−0.037	−0.069	−0.056	−0.089	**0.503**	−0.159	−0.015	0.053	0.053	0.036
SW	0.025	0.114	−0.093	−0.069	−0.001	−0.025	−0.089	0.383	**−0.209**	0.013	0.075	0.029	0.116
TKW	−0.004	0.097	−0.186	−0.079	0.118	0.027	0.084	−0.135	−0.050	**0.055**	0.009	−0.019	0.415
NGS	−0.006	−0.013	0.094	−0.070	0.067	0.013	−0.043	−0.145	0.085	−0.003	**−0.183**	−0.029	0.925
BY	−0.005	0.206	−0.188	0.238	−0.122	−0.096	0.061	0.301	−0.070	−0.012	0.060	**0.088**	−0.413
HI	0.008	0.039	0.030	−0.210	0.106	0.016	−0.067	0.016	−0.021	0.020	−0.145	−0.031	**1.166**
r_g	−0.105 ns	0.343 **	−0.092 ns	−0.365 **	0.431 **	0.212 ns	−0.220 ns	0.272 *	0.270 *	0.331 **	0.693 **	0.048 ns	0.925 **

HD: Heading date, CC: Chlorophyll content, CT: Canopy temperature, RWC: Relative water content, MT: Membrane thermostability, FLA: Flag leaf area, PH: Plant height, SN: Spike number, SW: Spike weight, TKW: Thousand kernel weight, NGS: Number of grains per spike, GY: Grain yield, BY: Biological yield, HI: Harvest index, r_g: Genotypic correlation coefficient with grain yield. ns, * and **: non-significant and significant at 5% and 1% probability levels by *t*-test.

The NGS (0.925), MT (0.734), and TKW (0.415) depicted the highest positive indirect effects on GY via HI. Furthermore, SW (0.383), FLA (0.338), and BY (0.301) exhibited high positive indirect effects on GY through SN. The BY expressed moderate positive indirect effects on GY through RWC (0.238) and CC (0.206). On the other hand, the RWC (−0.644), followed by the BY (−0.413), PH (−0.363), FLA (−0.221), and HD (−0.200), demonstrated high-to-moderate negative indirect effects on GY via HI. Similarly, the HD (−0.239), next to the PH (−0.209) and MT (−0.206), had moderate negative indirect effects on GY through SN. MT, via CT (−0.306), and HI, via RWC (−0.210), also exerted high and moderate negative indirect effects on GY, respectively. The rest of the indirect effects on GY were low-to-negligible, as shown in Table 4 and Figure 2.

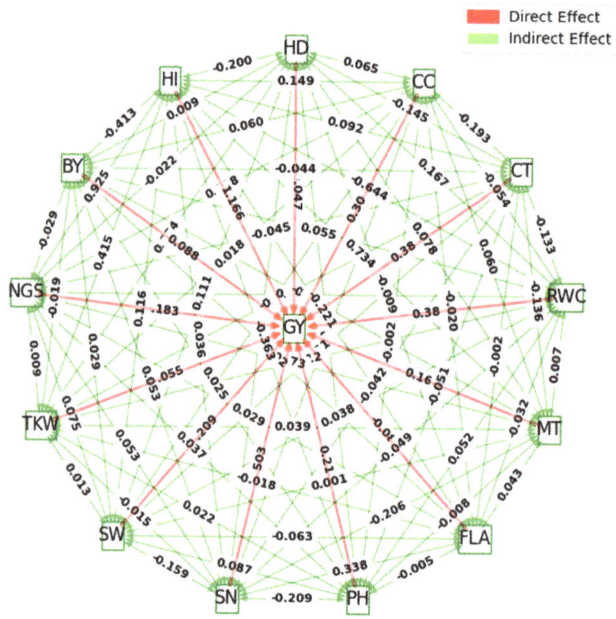

Figure 2. Genotypic path diagram representing the direct and indirect effects of 14 measured traits on grain yield in the durum wheat genotypes. HD: Heading date, CC: Chlorophyll content, CT: Canopy temperature, RWC: Relative water content, MT: Membrane thermostability, FLA: Flag leaf area, PH: Plant height, SN: Spike number, SW: Spike weight, TKW: Thousand kernel weight, NGS: Number of grains per spike, GY: Grain yield, BY: Biological yield, HI: Harvest index.

2.6. Divergence among Wheat Cultivars

The 20 durum wheat cultivars were classified into seven distinct clusters or groups, namely, G1, G2, G3, G4, G5, G6, and G7 (Table 5). Eight cultivars were found in the first cluster (G1), accounting for 40% of total genotypes, followed by four cultivars classified in the second cluster (G2), with a relative contribution of 20% to the genetic diversity of the durum wheat germplasm. The third cluster (G3) contained three genotypes, while the fourth cluster (G4) was formed by two cultivars only. Besides these four clusters, the remaining three clusters only had one cultivar (Table 5).

Figure 3 illustrates the average distances within and between the seven clusters formed by Tocher's method based on Mahalanobis Euclidean^2 distance. The average inter-cluster distance was observed at maximum (D^2 = 6145.86) between cluster G3 and cluster G4. The lowest inter-cluster distance (D^2 = 181.14) was recorded between clusters G2 and G7, which exhibited more genetic similarity. The maximum intra-cluster distance was obtained in cluster G4 (D^2 = 142.10), followed by cluster G1 (D^2 = 109.16), indicating that the cultivars belonging to these clusters were far diverged from those within clusters G2 (D^2 = 89.10) and G3 (D^2 = 93.75). Clusters G5, G6, and G7 did not show any intra-cluster distance. Among the fourteen traits studied, the greatest contribution to cluster divergence was made by HD (92.33%), followed by plant height (2.08%), thousand kernel weight (1.61%), and spike weight (1.52%).

Table 5. Clustering pattern of the durum wheat cultivars based on measured traits using Tocher's method.

Cluster No.	No. of Cultivars	Contribution (%)	Cultivars
G1	8	40	Waha, Simeto, Amar 06, Boutaleb, Saoura, Wahbi, Manssourah, Massinissa
G2	4	20	Megress, Mexicali75, Zb/Fl, Ain Lehma
G3	3	15	Oued El Bared, Belikh02, Ofanto
G4	2	10	Mohamed Ben Bachir, Guemgoum Rkhem
G5	1	5	Setifis
G6	1	5	Bousselam
G7	1	5	Vitron

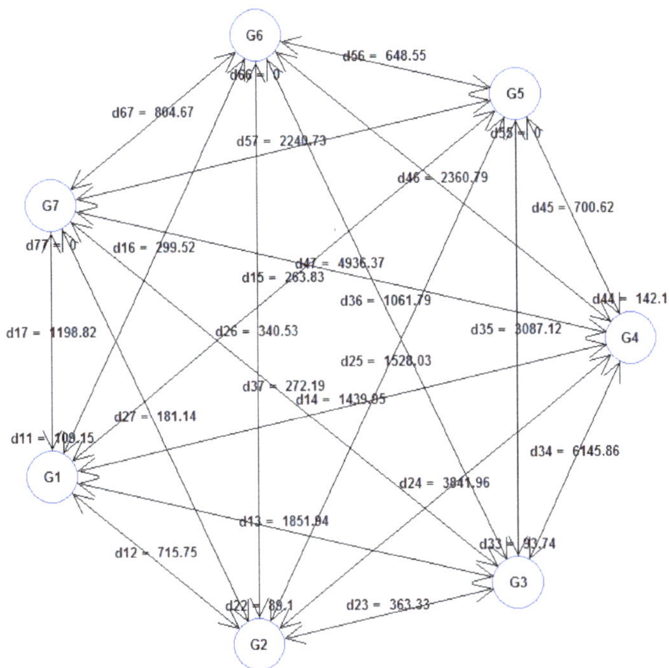

Figure 3. Clustering formed by Tocher's method representing the average intra- and inter-cluster distances for the studied traits in the durum wheat cultivars.

The average of the studied traits for each cluster has been represented in Table 6. Accordingly, cluster G4 was characterized by having the highest cluster mean for days to 50% heading (137.5 days), while cluster G3 had the lowest HD mean (128.7 days). G3 and G4 exhibited, respectively, the lowest (67.0) and the highest (75.4 CCI) chlorophyll content in the leaves. The cultivars grouped in cluster G1 showed the highest CT (47.2%), GY (2.4 t ha^{-1}), BY (11.6 t ha^{-1}), and HI (21.2%) values.

Cluster G2 showed the highest SN (411.3 spikes m^{-2}), whereas the lowest SN (240.8 spikes m^{-2}) was observed in cluster G4, which had also the longest plant stature (104.7 cm) and the lowest SW (2.4 t ha^{-1}), GY (1.1 t ha^{-1}), and HI (9.8%) values. Cluster G5 is characterized by the shortest plants (73.3 cm) and the highest RWC (91.2%) and NGS (10.1 grains spike^{-1}). Cluster G6 had the coolest canopy cover (29.3 °C), the largest leaves (33.8 cm^2), the highest MT (36.9%), and the smallest grain size (21.1 g). Cluster G7 had the lowest cluster mean for MT (31.5%), the smallest leaves (24.7 cm^2), the lowest spike fertility (4.3 grains spike^{-1}), and the lowest BY (9.2 t ha^{-1}). This cluster is characterized also by the highest SW (5.0 t ha^{-1}) and the largest grain size (40.1 g).

Table 6. Cluster mean values for 14 studied traits in the durum wheat cultivars.

Traits	Clusters						
	G1	G2	G3	G4	G5	G6	G7
HD	133.5	130.7	128.7	137.5	135.0	132.0	130.0
CT	47.2	46.9	39.1	45.2	30.0	29.3	38.9
MT	32.2	32.8	34.4	31.9	32.3	36.9	31.5
CC	71.0	71.0	67.0	75.4	69.8	71.7	67.5
RWC	84.7	81.3	69.7	79.3	91.2	64.8	86.3
FLA	32.2	30.6	28.1	28.8	25.1	33.8	24.7
PH	78.5	74.1	76.4	104.7	73.3	76.7	78.0
SN	408.8	411.3	373.9	240.8	288.3	408.3	358.3
SW	4.4	4.3	3.9	2.4	3.2	3.8	5.0
TKW	31.9	28.9	27.7	30.6	25.2	21.1	40.1
NGS	9.6	7.6	10.0	8.4	10.1	7.5	4.3
GY	2.4	1.8	1.9	1.1	1.5	1.3	1.2
BY	11.6	11.0	10.5	11.1	8.1	11.0	9.2
HI	21.2	16.6	19.7	9.8	18.2	11.6	13.6

HD: Heading date (days), CC: Chlorophyll content (CCI), CT: Canopy temperature (°C), RWC: Relative water content (%), MT: Membrane thermostability (%), FLA: Flag leaf area (cm^2), PH: Plant height (cm), SN: Spike number (No m^{-2}), SW: Spike weight (t ha^{-1}), TKW: Thousand kernel weight (g), NGS: Number of grains per spike (No spike^{-1}), GY: Grain yield (t ha^{-1}), BY: Biological yield (t ha^{-1}), HI: Harvest index (%).

3. Discussion

The results of the present study have revealed considerable variability in plant material, suggesting ample opportunities for wheat breeders to exploit this genetic diversity in breeding programs. Previous research in the Eastern Algerian High Plateaus supports these findings [33–35]. Researchers have noticed a wide variation in physiological, yield, and yield-related traits. The lack of significant differences in aboveground biomass may be due to water stress and increased temperatures at the post-anthesis growth stage, reducing cell division and elongation, leaf area, and grain filling period. The absence of significant differences in cell membrane thermostability could prompt further genetic studies by local breeders to enhance heat stress tolerance through hybridization and selection programs.

This study demonstrates that the mean trait values vary among the different genotypes of durum wheat, indicating distinct trait profiles and underlying genetic differences that influence morpho-physiological and agronomic features. This variability is crucial for selecting and breeding durum wheat varieties with desired characteristics, such as a higher yield, disease resistance, environmental stress tolerance, and enhanced nutritional content. By identifying the superior genotypes for specific traits, breeders can strategically design breeding programs to develop improved durum wheat varieties suited to various growing conditions and end-use requirements. [36].

This study has revealed varying levels of CV_g in durum wheat germplasm, with some traits exhibiting high-to-moderate CV_g values, offering opportunities for improvement through selective breeding. Conversely, the CV_g estimates were low for other traits, implying limited genetic variability that may hinder traditional selective breeding methods [37]. Additionally, the difference between CV_p and CV_g values has highlighted the influence of environmental factors on trait variability, particularly for traits like CC, MT, FLA, and yield components, excluding TKW. In such a situation, breeders should consider this information when developing strategies to enhance genotypic stability and environmental adaptability in wheat varieties [38,39]. Apart from MT and FLA, most of the measured traits in the durum wheat germplasm displayed high genetic control ($CV_g/CV_e > 1$), making them favorable targets for breeding programs, due to their responsiveness to selection efforts. On the other hand, those traits with a variation index below one ($CV_g/CV_e < 1$) are predominantly influenced by environmental factors, rendering them less predictable across various growing conditions [40]. Consequently, indirect selection, or the consideration of environmental factors, may be necessary to effectively improve these traits [22]. The broad-sense heritability ($h^2{}_{bs}$) has further supported these findings, indicating traits with high genetic determinism, like HD, CT, PH, TKW, NGS, and HI, which are desirable for breeding programs, due to their amenability to selection efforts. $h^2{}_{bs}$ was just intermediate for CC, RWC, FLA, SN, SW, and GY, suggesting that both genetic and environmental factors contribute to the observed variation in these traits. Conversely, less heritable traits pose challenges in breeding, due to their susceptibility to environmental influences.

Bendjama and Ramdani (2022) [20] also observed high heritability for most agro-morphological traits in local durum wheat varieties grown in Algeria. However, broad-sense heritability varies depending on the population and environmental conditions [41], and it does not reveal the specific genetic mechanisms underlying the traits [19]. It only quantifies the overall genetic contribution to trait variation. Understanding the balance between genetic and environmental factors is essential for assessing the risks in genotype selection and minimizing environmental impacts on breeding outcomes.

This study has uncovered significant relationships between various traits in durum wheat, at both genotypic and phenotypic levels. The genotypic coefficients of correlation were generally larger than their corresponding phenotypic coefficients. Some significant genotypic correlations were not statistically significant at the phenotypic level, possibly due to environmental factors or sample size limitations [42]. Conversely, instances where phenotypic correlation coefficients exceeded genotypic ones could indicate gene–environment interactions, where the environmental conditions modulate the impact of the genetic factors on traits, altering the observed correlations between genotypic and phenotypic levels [43].

The results indicate that all of the cultivars had a long growth cycle with fewer fertile tillers and light spikes, but they showed tolerance to water deficit with high above-ground biomass. Despite this, they had a low harvest index and productivity. The tall cultivars had a larger grain size, due to a longer grain-filling period, allowing for more nutrient and energy accumulation, leading to higher yields under favorable conditions [44]. Conversely, the short cultivars produced heavier spikes with better fertility, more spikes per unit surface, and efficient assimilate translocation, resulting in increased biological and grain yields. This suggests a link between dwarfing genes and genes controlling spike-related traits and the plant's capacity to allocate assimilates to grains (source–sink relationship).

Fellahi et al. (2023) [45] found that semi-dwarf wheat breeding lines carrying the Rht-D1b and Rht-B1b mutant alleles outperformed the tall lines in yield under supplemental irrigation and in spike number under rainfed conditions, along with other yield-related traits, such as harvest index and number of grains per spike, regardless of the environment. This observation is supported by the present investigation, which has revealed strong correlations between the yield and various yield attributes. Rabti et al. (2020) [5] similarly emphasized the significant influence of the spike number, spike fertility, and harvest index on the grain yield in both old and modern durum wheat varieties. These findings suggest that improving these agronomic components could substantially increase yield in the screened plant material.

According to Maeoka et al. (2020) [46], the improvement in yield of modern wheat varieties is strongly linked to reduced plant height, increased spike fertility, and higher harvest index, with no significant changes in biomass. Correlation analysis highlights the benefit of high above-ground biomass, particularly in adverse environmental conditions like drought. The cultivars with higher biological yield exhibited cooler canopy cover, better water status, and increased chlorophyll content, resulting in enhanced responses to post-anthesis water deficit and heat stress. Similar observations were made by Talebi (2011) [47], suggesting that wheat genotypes with a lower canopy temperature may have higher rates of transpiration and photosynthesis, leading to increased yield in water-stressed environments.

The genotypic path analysis results showed that the harvest index (HI) had a highly positive direct effect on grain yield (GY), consistent with findings in the existing literature [48,49]. Traits like spike number (SN), relative water content (RWC), canopy temperature (CT), and canopy cover (CC) had moderate positive direct effects on GY, reflecting their genotypic correlations with this primary trait. Despite RWC displaying a high positive direct effect on GY, it paradoxically showed a negative correlation with GY, suggesting a positive contribution to the overall crop yield. However, the negative genetic inter-correlation between RWC and GY indicates a tendency for genetic variations in RWC to be inversely related to variations in yield. This underscores the complexity of genetic interactions [50]. Plant height (PH) and maturity (MT) exhibited moderate-to-low direct effects on GY, with the effect of PH being rendered non-significant, due to the negative indirect effects of other yield components via this secondary morphological trait.

Our results suggest that more fertile tillers, along with a good plant capacity to allocate photosynthesis products (assimilates) from vegetative organs into the formed reproductive parts (grains), lead to incremented grain yield. Thus, the significant positive genotypic correlations of GY with SN and HI are predominately attributed to the direct effect of these two traits on yield per hectare. The results presented here align with the previous research findings documented in the literature [51–53]. TKW exerted a negligible direct effect on GY, implying the minimal impact of the grain size on the overall yield in wheat cultivars. However, the positive and significant genotypic inter-correlation between TKW and GY suggests that, while TKW may not directly influence yield, it could be associated with other factors or processes that indirectly contribute to the yield. This hidden contribution warrants further exploration for wheat improvement.

In the scope of this study, SW, followed by NGS, had the highest negative direct effects on GY. Despite this negative impact, they both demonstrated positive and statistically significant correlations with grain yield per hectare, which implies the true relationship between these two secondary traits with the main dependent trait, i.e., GY. This observation could be attributed to the influence of common genes with pleiotropic effects [54]. While the direct effect of the trait was negative, the other effects of these shared genes could be positive for grain yield. Moreover, traits like spike weight and spike fertility might negatively affect one aspect of the plant's physiology or morphology that directly contributes to the yield. However, compensatory mechanisms, possibly involving other traits or physiological processes, might be at play, mitigating their negative impacts. In addition, the relationship could be sensitive to environmental conditions. Under certain conditions, the negative direct effect might dominate, while under different conditions, the positive genetic correlation with yield becomes more evident. Thus, selection based on spike weight would be effective but should be practiced cautiously [38].

Path analysis has illustrated that NGS, MT, and TKW expressed the highest positive indirect effects on GY through HI. Additionally, SW, FLA, and BY depicted high positive indirect effects on GY via SN. BY also exhibited moderate positive indirect effects on GY via RWC and CC. These results confirm the importance of SN and HI and elucidate that the improvement of yield under drought stress during the grain filling stage could be more effective through selection.

Tocher's clustering partitioned the twenty durum wheat cultivars into seven distinct clusters. Notably, the maximum inter-cluster distance was found to be at maximum between clusters G3 and G4, indicating a wide dissimilarity between the cultivars in these clusters. The greatest contribution to cluster divergence was made by HD, followed by PH, TKW, and SW. This divergence may be due to historical breeding practices, geographic isolation, or other factors influencing the genetic makeup of these wheat cultivars [55].

Rabti et al. (2020) [5] have also stated that earliness had a strong influence on grain yield variability in both old and recent durum wheat germplasm.

Understanding the genetic relationships and distances between different accessions enables breeders to strategically select parents for hybridization, enhancing specific traits or overall performance [56]. Crossing genotypes from genetically divergent clusters can generate transgressive segregants, increasing the genetic variation in the offspring and facilitating the development of new potential varieties [57]. The genotypes in distant groups may possess unique or rare traits important for crop improvement, particularly in addressing challenges such as diseases, pests, or environmental stress [58]. However, the agronomic potential of those genotypes must be considered in order to develop potential varieties through cross combinations in wheat breeding programs [59,60]. The results have shown that cluster G1 could serve as a parent for developing high-yielding wheat varieties, while the presence of early heading cultivars in cluster G3 suggests the potential for developing early maturing varieties to mitigate heat waves during grain filling. The genotypes in other clusters may be selected based on their trait mean values, depending on the desired direction of selection (i.e., increase or decrease).

4. Materials and Methods

4.1. Plant Material

The plant material studied comprises a collection of 20 durum wheat (*Triticum turgidum* L. ssp. *durum* Desf.) cultivars (Table 7). These cultivars are selected based on their adaptability to semi-arid regions and their diverse genetic backgrounds. They represent a range of traits related to yield potential, drought tolerance, disease resistance, and quality characteristics [5,61].

Table 7. Plant material evaluated in the field experiment.

N°	Name	Origin	Pedigree
1	Mohamed Ben Bachir	Algeria	Local landrace
2	Bousselam	CIMMYT-ICARDA	Heider/Martes/Huevos de Oro. ICD-414
3	Oued El Bared	Algeria	Gta Dur/Ofanto
4	Waha	CIMMYT-ICARDA	Plc/Ruff//Gta/Rtte
5	Boutaleb	Algeria	Hedba03/Ofanto
6	Saoura	ACSAD	Belikh//Gediz/Bit
7	Megress	Algeria	Ofanto//Waha/Mohamed Ben Bachir
8	Manssourah	Algeria	Chinese spring/Mohamed Ben Bachir
9	Vitron	Spain	Turkey77/3/Jori/Anhinga//Flamingo
10	Amar06	ICARDA	Lgt3/Bicre// cham1//orlgt3/4/Bicre/3/Ch1//Gav/Starke
11	Wahbi	Algeria	Bidi17/Waha//Bidi17
12	Simeto	Italy	Capeiti8/Valnova
13	Ain Lehma	ICARDA	Bcr/Gro1//Mgn/1
14	Massinissa	Algeria	Ofanto/Bousselam
15	Setifis	Algeria	Bousselam/Ofanto
16	Belikh02	ICARDA	Crane/Stork
17	Mexicali75	CIMMYT	Gdovz469/3/Jo's'//61.130/Lds
18	Zb/Fl	CIMMYT	Zb/Fg's//Lk/3/Ko120/4/Ward
19	Ofanto	Italy	Appulo/Adamello
20	Guemgoum Rkhem	Algeria	Local landrace

ICARDA: International Center for Agricultural Research in the Dry Areas, CIMMYT: International Maize and Wheat Improvement Center, ACSAD: Arab Center for the Studies of Arid Zones and Dry Lands.

4.2. Site Description and Experimental Design

The field trials were conducted at the Technical Institute for Field Crops (ITGC), Demonstration and Seed Production Farm (FDPS) of Setif, during the 2021/22 crop season. The experimental farm is located at the geographical coordinates of 36°09′ N; 05°22′ E; 964 masl within the semi-arid region of Algeria, which represents the target ecological conditions. The experiment was set up under rainfed conditions in a randomized complete block design (RCBD) with three replicates. Each durum wheat variety was grown as a separate plot within the block to account for experimental variability. The plot consisted of 1.2 m wide by 5 m long plots, each containing 6 rows, with an inter-row distance of 20 cm.

4.3. Soil Characteristics

The characteristics of the soil at the experimental site are given in Table 8. Accordingly, the experimental field has a brown calcareous soil, belonging to clayey alluvial soils with an average cation exchange capacity (CEC) of 25 meq per 100 g of soil, a pH of 8.21, and 1.97% organic matter (OM) (Table 8). The percentage of soil nitrogen is 1.01‰, with 103.82 ppm of available phosphorus. The field electrical conductivity (EC) is 0.21 mmhos cm^{-1}, and the exchangeable potassium is 588.7 ppm.

Table 8. Soil characteristics at the Setif ITGC-FDPS experimental site.

Parameters	Values
Total $CaCO_3$ (%)	35.01
pH	8.21
CEC (meq 100 g^{-1})	25.0
EC (mmhos cm^{-1})	0.21
OM (%)	1.97
Total N (‰)	1.01
P_2O_5 (ppm)	103.82
K_2O (ppm)	588.7

The sowing was carried out on 29 December, using a Hege 80 experimental plot drill at 350 seeds m^{-2} seeding rate. Adequate care for wheat growth was carried out to ensure standard agronomic practices, such as appropriate soil preparation, fertilization, and weed control.

4.4. Climatic Conditions

The daily rainfall amount and the maximum and minimum temperature recorded during the cropping season are displayed in Figure 4. The accumulated rainfall from 1 September to 30 June was 330.46 mm. Compared to the two previous cropping seasons, in which the total precipitation amount was, respectively, 353.06 and 320.24 mm, the 2020/21 cropping season seems to be more suitable for the yield expression of the wheat cultivars under study. However, wheat productivity is not solely determined by the amount of rainfall received, but also by its distribution during the crop cycle [62]. As shown in Figure 4, the rainfall distribution of the growing season was extremely variable. Before sowing, a good quantity of rain was observed in November, and intense precipitation with a higher total hourly amount was registered at the end of April, corresponding to the wheat anthesis growth stage. Nonetheless, the grain filling period (May to June), a sensitive stage of the wheat plant's development, was highly affected by water stress. The rainfall amount received during the May–June period and the number of rainy days were only 6.34 mm and 7 days, respectively. The air temperature remained within the appropriate range for crop development, except for in June, in which the highest temperature was somewhat high, exceeding 35 °C. The daily mean temperatures were below 10 °C during most of the crop growing season. These climatic conditions of the field assessment cycle allowed us to screen the plant material for post-anthesis drought stress, which resulted in severe yield loss in most of the wheat cultivars tested.

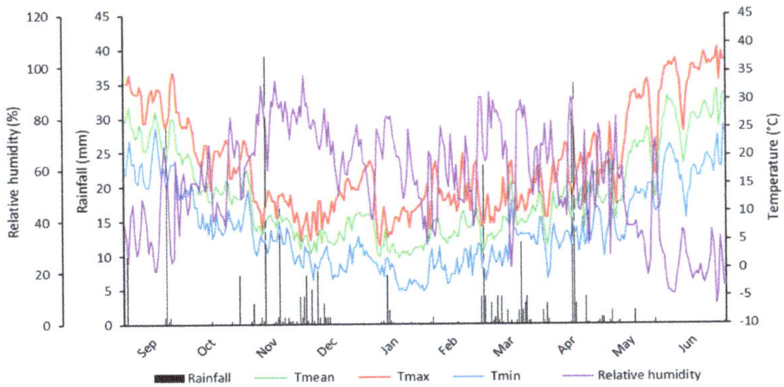

Figure 4. Meteorological data during the 2021/22 growth season recorded at the Setif ITGC-FDPS experimental site (https://www.tutiempo.net). accessed on 21 January 2024.

4.5. Data Collection

This study collected data on 15 phenotypic traits relevant to local adaptation in semi-arid regions. Data collection was carried out at appropriate growth stages, ensuring accurate and representative measurements for each trait. The heading date (HD, days) was recorded as the number of days from 1 January to the emergence of 50% of the spikes in a plot. Drought and heat tolerance were assessed based on the following physiological measurements: chlorophyll content (CC, CCI), canopy temperature (CT, °C), relative water content (RWC, %), and electrolyte leakage or leaf membrane thermostability (MT, %), which were determined at Zadoks growth stage (ZGS 55) [63] using appropriate laboratory techniques [64]. The morphological characteristics included flag leaf area (FLA, cm^2), which was obtained at the heading date (ZGS 55), according to the procedure described by Spagnoletti-Zeuli and Qualset (1990) [65]. Plant height (PH, cm) was measured from the base of the plant to the tip of the spike at the maturity growth stage (ZGS 95). Yield components, such as spike number (SN, No m^{-2}), spike weight (SW, t ha^{-1}), grain yield (GY, t ha^{-2}), biological yield (BY, t ha^{-1}), and harvest index (HI, %), were recorded in each plot at the ZGS 95 growth stage after the harvest of the wheat samples from 2-row segments, 1 m long, across all of the replicates. Thousand kernel weight (TKW, g) was determined from the count and weight of 250 kernels. The number of grains per spike (NGS, No. $spike^{-1}$) was derived from the grain number and spike number produced per square meter.

4.6. Data Analysis

The collected phenotypic data were subjected to one-way analysis of variance (ANOVA) to assess the significance of differences among the cultivars for each recorded trait. Tukey's honestly significant difference (HSD) at 5% and 1% levels of probability was employed to compare the means of the different cultivars. In parallel, the genetic and non-genetic parameters were calculated to explore the genetic variability within the plant material evaluated. This involved the computation of phenotypic (CV_p), genotypic (CV_g), and environmental (CV_e) coefficients of variation, as described by Falconer and Mackay (1996) [66]. The CV_p represents the total variation observed for a trait in the population, including both genetic and environmental factors. It was calculated as the standard deviation (σ_p) divided by the mean (μ) of the trait, expressed as a percentage. The CV_g estimates the proportion of total variation contributed by genetic factors alone, excluding environmental effects. It was calculated as the standard deviation of genotypic values (σ_g) divided by the mean value (μ) of the trait, expressed as a percentage. The CV_e represents the proportion of total variation attributed to environmental factors, excluding genetic effects. It was calculated as the standard deviation of the environmental values (σ_e) divided by the mean (μ) of the trait, expressed as a percentage. Additionally, the CV_g/CV_e ratio was computed to assess the relative importance of genetic and environmental factors contributing to the observed phenotypic variation for measured traits within the durum wheat germplasm collection [67]. A higher CV_g/CV_e ratio indicates a greater contribution of genetic factors to the observed variation, suggesting that the trait is predominantly influenced by the genetic differences among individuals rather than the environmental conditions. Thus, this ratio provides breeders with valuable information to optimize breeding programs. It guides decisions on trait selection, breeding methodologies, and the development of varieties that perform consistently across different conditions.

Broad-sense heritability ($h^2{}_{bs}$) provides an indication of the degree to which the observed variation in a trait is influenced by genetic factors, relative to environmental influences [66]. This statistical measure was obtained as the genetic variance ($\sigma^2{}_g$), divided by the phenotypic variance ($\sigma^2{}_e$) of the considered trait, expressed as a percentage. It is worth noting that these parameters are typically used in quantitative genetics and plant breeding research to understand the potential for genetic improvement and to make informed decisions in breeding programs [19,66].

Phenotypic (r_p) and genotypic (r_g) correlation coefficients were calculated based on the replicated data values to explore the relationships between different traits and to assess the potential for trait improvement through selection. Besides this, the r_g values were divided into the direct and indirect genotypic effects of the 13 measured traits (independent variables) on grain yield (dependent variable), using the path analysis method, as outlined by Dewey and Lu (1959) [26]. The durum wheat cultivars were classified into clusters using Tocher's optimization method [68]. The Mahalanobis (1936) distance was used to provide insights into the genetic structure and diversity within the durum wheat cultivars considering all recorded traits. To perform all of the mentioned statistical methods, 'variability' [69], 'ggplot2' [70], and 'corrplot' [71] statistical packages in R (Version 4.3.2) and Genes [67] software (Version 1990.2022.23) were used.

5. Conclusions

This study highlights the substantial genetic variation in durum wheat germplasm, offering promising prospects for effective breeding programs. Environmental influences contribute to observed trait variations alongside genetic effects. Traits such as canopy cover (CC), spike number (SN), spike weight (SW), thousand kernel weight (TKW), number of grains per spike (NGS), and harvest index (HI) exhibit significant positive correlations with grain yield at both genotypic and phenotypic levels, suggesting their importance in yield determination. The harvest index (HI) and spike number (SN) have the highest positive direct impact on grain yield, with substantial indirect effects through other yield components. Improvements in these traits could significantly increase yield per hectare. The identification of genetically diverse cultivars with desirable traits enables strategic crossbreeding to generate segregating progenies with maximum genetic variability, advancing durum wheat breeding programs. These findings contribute valuable insights into sustainable agriculture, particularly in semi-arid regions like Algeria, where environmental stresses pose significant challenges to durum wheat production.

Author Contributions: Conceptualization, Z.E.A.F. and A.H.; methodology, Z.E.A.F. and A.H.; software, A.O.U.; validation, T.B. and N.Y.R.; formal analysis, A.O.U., H.B. and N.Y.R.; investigation, A.O.U. and N.L.; resources, Z.E.A.F. and N.Y.R.; data curation, N.L., A.B. and H.B.; writing—original draft preparation, Z.E.A.F. and N.Y.R.; writing—review and editing, Z.E.A.F. and N.Y.R.; visualization, A.B. and A.H.; supervision, Z.E.A.F.; project administration, Z.E.A.F. All authors have read and agreed to the published version of the manuscript.

Funding: This paper has been supported by the RUDN University Strategic Academic Leadership Program.

Data Availability Statement: Data are contained within the article.

Conflicts of Interest: The authors declare no conflicts of interest.

References

1. Xynias, I.N.; Mylonas, I.; Korpetis, E.G.; Ninou, E.; Tsaballa, A.; Avdikos, I.D.; Mavromatis, A.G. Durum wheat breeding in the Mediterranean region: Current status and future prospects. *Agronomy* **2020**, *10*, 432. [CrossRef]
2. Martínez-Moreno, F.; Ammar, K.; Solís, I. Global changes in cultivated area and breeding activities of durum wheat from 1800 to date: A historical review. *Agronomy* **2022**, *12*, 1135. [CrossRef]
3. Hannachi, A.; Fellahi, Z. Efficiency of index-based selection for potential yield in durum wheat [*Triticum turgidum* (L.) ssp. *turgidum* convar. *durum* (Desf.) Mackey] lines. *Ital. J. Agron.* **2023**, *18*, 2182.
4. Lopes, M.S.; El-Basyoni, I.; Baenziger, P.S.; Singh, S.; Royo, C.; Ozbek, K.; Aktas, H.; Ozer, E.; Ozdemir, F.; Manickavelu, A.; et al. Exploiting genetic diversity from landraces in wheat breeding for adaptation to climate change. *J. Exp. Bot.* **2015**, *66*, 3477–3486. [CrossRef]
5. Rabti, A.; Mekaoussi, R.; Fellahi, Z.; Hannachi, A.; Benbelkacem, A.; Benmahammed, A.; Bouzerzour, H. Characterization of old and recent durum wheat [*Triticum turgidum* (L.) Tell. convar. *durum* (Desf.) Mackey] varieties assessed under South Mediterranean conditions. *Egypt. J. Agron.* **2020**, *42*, 307–320. [CrossRef]
6. Temirbekova, S.K.; Kulikov, I.M.; Afanasyeva, Y.V.; Beloshapkina, O.O.; Kalashnikova, E.A.; Kirakosyan, R.N.; Dokukin, P.A.; Kucher, D.E.; Latati, M.; Rebouh, N.Y. The Evaluation of Winter Wheat Adaptation to Climate Change in the Central Non-Black Region of Russia: Study of the Gene Pool Resistance of Wheat from the N.I. Vavilov Institute of Plant Industry (VIR) World Collection to Abiotic Stress Factors. *Plants* **2021**, *10*, 2337. [CrossRef]
7. Shewry, P.R.; Lovegrove, A.; Wingen, L.U.; Griffiths, S. Opinion Exploiting genomics to improve the benefits of wheat: Prospects and limitations. *J. Cereal Sci.* **2022**, *105*, 103444. [CrossRef]
8. Martínez-Moreno, F.; Solís, I.; Noguero, D.; Blanco, A.; Özberk, İ.; Nsarellah, N.; Elias, E.; Mylonas, I.; Soriano, J.M. Durum wheat in the Mediterranean Rim: Historical evolution and genetic resources. *Genet. Resour. Crop Evol.* **2020**, *67*, 1415–1436. [CrossRef]
9. Broccanello, C.; Bellin, D.; DalCorso, G.; Furini, A.; Taranto, F. Genetic approaches to exploit landraces for improvement of *Triticum turgidum* ssp. *durum* in the age of climate change. *Front. Plant Sci.* **2023**, *14*, 1101271. [CrossRef] [PubMed]
10. Maccaferri, M.; Sanguineti, M.C.; Demontis, A.; El-Ahmed, A.; Garcia del Moral, L.; Maalouf, F.; Nachit, M.; Nserallah, N.; Ouabbou, H.; Rhouma, S.; et al. Association mapping in durum wheat grown across a broad range of water regimes. *J. Exp. Bot.* **2011**, *62*, 409–438. [CrossRef]
11. El Haddad, N.; Kabbaj, H.; Zaïm, M.; El Hassouni, K.; Tidiane Sall, A.; Azouz, M.; Ortiz, R.; Baum, M.; Amri, A.; Gamba, F.; et al. Crop wild relatives in durum wheat breeding: Drift or thrift? *Crop Sci.* **2021**, *61*, 37–54. [CrossRef]
12. Temirbekova, S.K.; Kulikv, I.M.; Ashirbekov, M.Z.; Afanasyeva, Y.V.; Beloshapkina, O.O.; Tyryshkin, L.G.; Zuev, E.V.; Kirakosyan, R.N.; Glinushkin, A.P.; Potapova, E.S.; et al. Evaluation of Wheat Resistance to Snow Mold Caused by Microdochium nivale (Fr) Samuels and I.C. Hallett under Abiotic Stress Inflfluence in the Central Non-Black Earth Region of Russia. *Plants* **2022**, *11*, 699. [CrossRef] [PubMed]
13. Bassi, F.M.; Bentley, A.R.; Charmet, G.; Ortiz, R.; Crossa, J. Breeding schemes for the implementation of genomic selection in wheat (*Triticum* spp.). *Plant Sci.* **2016**, *242*, 23–36. [CrossRef] [PubMed]
14. Sukumaran, S.; Reynolds, M.P.; Sansaloni, C.; Morgounov, A. Association of grain attributes with yield and yield components under drought in spring wheat. *J. Agric. Sci.* **2018**, *156*, 28–46.
15. Crespo-Herrera, L.A.; Crossa, J.; Vargas, M.; Braun, H.J. Defining target wheat breeding environments. In *Wheat Improvement*; Springer: Cham, Switzerland, 2022; pp. 31–45.
16. Salmi, M.; Benmahammed, A.; Benderradji, L.; Fellahi, Z.; Bouzerzour, H.; Oulmi, A.; Benbelkacem, A. Generation means analysis of phyiological and agronomical targeted traits in durum wheat (*Triticum durum* Desf.) cross. *Rev. Fac. Nac. Agron. Medellín* **2019**, *72*, 8971–8981. [CrossRef]
17. Bernardo, R. Genomic selection in plant breeding: Current status and future perspectives. *Crop Sci.* **2016**, *56*, 1–12.
18. Crossa, J.; Pérez-Rodríguez, P.; Cuevas, J.; Montesinos-López, O.; Jarquín, D.; de los Campos, G.; Burgueño, J.; Gonzalez-Camacho, J.M.; Perez-Elizalde, S.; Beyene, Y.; et al. Genomic selection in plant breeding: Methods, models, and perspectives. *Trends Plant Sci.* **2017**, *22*, 961–975. [CrossRef] [PubMed]
19. Acquaah, G. *Principles of Plant Genetics and Breeding*, 2nd ed.; John Wiley and Sons, Ltd.: Chichester, UK, 2012; 740p.
20. Bendjama, A.; Ramdani, S. Genetic variability of some agronomic traits in a collection of wheat (*Triticum turgidum* L. sp. pl.) genotypes under South Mediterranean growth conditions. *Ital. J. Agron.* **2022**, *17*, 1976.
21. Lamara, A.; Fellahi, Z.; Hannachi, A.; Benniou, R. Assessing the phenotypic variation, heritability and genetic advance in bread wheat (*Triticum aestivum* L.) candidate lines grown under rainfed semi-arid region of Algeria. *Rev. Fac. Nac. Agron. Medellín* **2022**, *75*, 10107–10118. [CrossRef]
22. Fellahi, Z.; Hannachi, A.; Bouzerzour, H. Analysis of Direct and Indirect Selection and Indices in Bread Wheat (*Triticum aestivum* L.) Segregating Progeny. *Int. J. Agron.* **2018**, *2018*, 8312857. [CrossRef]
23. Casagrande, C.R.; Mezzomo, H.C.; Silva, C.M.; Lima, G.W.; Souza, D.J.P.; Borém, A.; Nardino, M. Selection indexes based on genotypic values applied to Brazilian tropical wheat breeding. *Agron. Sci. Biotechnol.* **2022**, *8*, 1–16. [CrossRef]

24. Miller, P.A.; Williams, J.C.; Robinson, H.F.; Comstock, R.E. Estimates of genotypic and environmental variances and covariances in upland cotton and their implications in selection 1. *Agron. J.* **1958**, *50*, 126–131. [CrossRef]
25. Baye, A.; Berihun, B.; Bantayehu, M.; Derebe, B. Genotypic and phenotypic correlation and path coefficient analysis for yield and yield-related traits in advanced bread wheat (*Triticum aestivum* L.) lines. *Cogent Food Agric.* **2020**, *6*, 1752603. [CrossRef]
26. Dewey, D.R.; Lu, K.A. Correlation and path–coefficient analysis of components of crested wheatgrass seed production 1. *Agron. J.* **1959**, *51*, 515–518. [CrossRef]
27. Kozak, M.; Kang, M.S. Note on modern path analysis in application to crop science. *Commun. Biometry Crop Sci.* **2006**, *1*, 32–34.
28. Boulelouah, N.; Berbache, M.R.; Bedjaoui, H.; Selama, N.; Rebouh, N.Y. Influence of Nitrogen Fertilizer Rate on Yield, Grain Quality and Nitrogen Use Efficiency of Durum Wheat (*Triticum durum* Desf) under Algerian Semiarid Conditions. *Agriculture* **2022**, *12*, 1937. [CrossRef]
29. Adilova, S.S.; Qulmamatova, D.E.; Baboev, S.K.; Bozorov, T.A.; Morgunov, A.I. Multivariate cluster and principle component analyses of selected yield traits in uzbek bread wheat cultivars. *Am. J. Plant Sci.* **2020**, *11*, 903. [CrossRef]
30. Lee, J.; Kaltsikes, P.J. The application of Mahalanobis's generalized distance to measure genetic divergence in durum wheat. *Euphytica* **1973**, *22*, 124–131. [CrossRef]
31. Cruz, C.D.; Regazzi, A.J.; Carneiro, P.C.S. *Modelos Biométricos Aplicados ao Melhoramento Genético*, 4th ed.; Ed. UFV: Viçosa, Brazil, 2012; 514p.
32. Burton, G.N.; Devane, E.M. Estimating Heritability in Fall Fescue (*Festuca arundiancea* L.) from Replicated Clonal Materials. *Agron. J.* **1953**, *45*, 478–481. [CrossRef]
33. Mansouri, A.; Oudjehih, B.; Benbelkacem, A.; Fellahi, Z.; Bouzerzour, H. Variation and relationships among agronomic traits in durum wheat [*Triticum turgidum* (L.) Thell. Ssp. *Turgidum* conv. *Durum* (Desf.) Mackey] under south Mediterranean growth conditions: Stepwise and path analyses. *Int. J. Agron.* **2018**, *2018*, 8191749. [CrossRef]
34. Kirouani, A.; Ould Kiar, R.; Boukhalfoun, L.; Fellahi, Z. Caractérisation de quelques variétés Algériennes de blé dur (*Triticum turgidum* L. var. *durum*) par le biais des marqueurs phénotypiques. *J. Appl. Biosci.* **2019**, *142*, 14464–14477.
35. Mekaoussi, R.; Rabti, A.; Fellahi, Z.; Hannachi, A.; Benmahammed, A.; Bouzerzour, H. Assessment of durum wheat (*Triticum durum* Desf.) genotypes based on their agro-physiological characteristics and stress tolerance indices. *Acta Agric. Slov.* **2021**, *117*, 1–16. [CrossRef]
36. Benbelkacem, A. The history of wheat breeding in Algeria. In *International Symposium on Genetics and Breeding of Durum Wheat*; Porceddu, E., Damania, A.B., Qualset, C.O., Eds.; Options Méditerranéennes: Série A. Séminaires Méditerranéens; CIHEAM: Bari, Italy, 2014; Volume 110, pp. 363–370.
37. Alambo, M.M.; Gessese, M.K.; Wachamo, E.W.; Melo, B.Y.; Lakore, Z.S.; Wassie, A.S.; Haile, W.T.; Kassie, F.C. Performance Evaluation of Ethiopian Bread Wheat (*Triticum aestivum* L.) Genotypes in Southern Ethiopia. *Adv. Agric.* **2022**, *2022*, 1338082. [CrossRef]
38. Matković Stojšin, M.; Petrović, S.; Banjac, B.; Roljević Nikolić, S.; Zečević, V.; Bačić, J.; Đorđević, R.; Knežević, D. Development of selection criteria for improving grain yield in wheat grown in different agro-ecological environments. *Acta Agric. Serbica* **2022**, *27*, 79–87. [CrossRef]
39. Regmi, S.; Poudel, B.; Ojha, B.R.; Kharel, R.; Joshi, P.; Khanal, S.; Kandel, B.P. Estimation of genetic parameters of different wheat genotype traits in Chitwan, Nepal. *Int. J. Agron.* **2021**, *2021*, 6651325. [CrossRef]
40. Fellahi, Z.; Hannachi, A.; Bouzerzour, H. Expected genetic gains from mono trait and indices based selection in advanced bread wheat (*Triticum aestivum* L.) populations. *Rev. Fac. Nac. Agron. Medellín* **2020**, *73*, 9131–9141. [CrossRef]
41. Schmidt, P.; Hartung, J.; Bennewitz, J.; Piepho, H.P. Heritability in plant breeding on a genotype-difference basis. *Genetics* **2019**, *212*, 991–1008. [CrossRef] [PubMed]
42. Colombo, M.; Roumet, P.; Salon, C.; Jeudy, C.; Lamboeuf, M.; Lafarge, S.; Dumas, A.-V.; Dubreuil, P.; Ngo, W.; Derepas, B.; et al. Genetic analysis of platform-phenotyped root system architecture of bread and durum wheat in relation to agronomic traits. *Front. Plant Sci.* **2022**, *13*, 853601. [CrossRef] [PubMed]
43. Fellahi, Z.; Hannachi, A.; Guendouz, A.; Rabti, A.; Bouzerzour, H. Héritabilité, corrélations et gain de sélection précoce en F2 de blé tendre (*Triticum aestivum* L.) sous conditions semi-arides. *J. Algér. Rég. Arid.* **2019**, *13*, 37–49.
44. Bogale, A.; Tesfaye, K. Relationship between grain yield and yield components of the Ethiopian durum wheat genotypes at various growth stages. *Trop. Subtrop. Agroecosyst.* **2016**, *19*, 81–91. [CrossRef]
45. Fellahi, Z.E.A.; Hannachi, A.; Dreisigacker, S.; Sehgal, D.; Yahyaoui, A.H.; Bouzerzour, H. Genetic effects of Rht-B1b and Rht-D1b dwarfing alleles on plant height and agronomic traits of bread wheat (*Triticum aestivum* L.) breeding lines. *Genetika* **2023**, *55*, 565–581. [CrossRef]
46. Maeoka, R.E.; Sadras, V.O.; Ciampitti, I.A.; Diaz, D.R.; Fritz, A.K.; Lollato, R.P. Changes in the phenotype of winter wheat varieties released between 1920 and 2016 in response to in-furrow fertilizer: Biomass allocation, yield, and grain protein concentration. *Front. Plant Sci.* **2020**, *10*, 1786. [CrossRef]
47. Talebi, R. Evaluation of chlorophyll content and canopy temperature as indicators for drought tolerance in durum wheat (*Triticum durum* Desf.). *Aust. J. Basic Appl. Sci.* **2011**, *5*, 1457–1462.
48. Hannachi, A.; Fellahi, Z.; Bouzerzour, H.; Boutekrabt, A. Diallel-cross analysis of grain yield and stress tolerance-related traits under semi-arid conditions in Durum wheat (*Triticum durum* Desf.). *Electron. J. Plant Breed.* **2013**, *4*, 1027–1033.

49. Meles, B.; Mohammed, W.; Tsehaye, Y. Genetic variability, correlation and path analysis of yield and grain quality traits in bread wheat (*Tritium aestivum* L.) genotypes at Axum, Northern Ethiopia. *J. Plant Breed. Crop Sci.* **2017**, *9*, 175–185.
50. Saeidnia, F.; Taherian, M.; Nazeri, S.M. Graphical analysis of multi-environmental trials for wheat grain yield based on GGE-biplot analysis under diverse sowing dates. *BMC Plant Biol.* **2023**, *23*, 198. [CrossRef]
51. Singh, T. Character association analysis in durum wheat (*Triticum durum* Desf.). *Agric. Sci. Dig. Res. J.* **2016**, *36*, 24–29.
52. Laala, Z.; Oulmi, A.; Fellahi, Z.; Benmahammed, A. Studies on the nature of relationships between grain yield and yield-related traits in durum wheat (*Triticum durum* Desf.) populations. *Rev. Fac. Nac. Agron. Medellín* **2021**, *74*, 9631–9642. [CrossRef]
53. Rebouh, N.Y.; Aliat, T.; Polityko, P.M.; Kherchouche, D.; Boulelouah, N.; Temirbekova, S.K.; Afanasyeva, Y.V.; Kucher, D.E.; Plushikov, V.G.; Parakhina, E.A.; et al. Environmentally Friendly Wheat Farming: Biological and Economic Efficiency of Three Treatments to Control Fungal Diseases in Winter Wheat (*Triticum aestivum* L.) under Field Conditions. *Plants* **2022**, *11*, 1566. [CrossRef] [PubMed]
54. Chebib, J.; Guillaume, F. Pleiotropy or linkage? Their relative contributions to the genetic correlation of quantitative traits and detection by multitrait GWA studies. *Genetics* **2021**, *219*, iyab159. [CrossRef]
55. Lehnert, H.; Berner, T.; Lang, D.; Beier, S.; Stein, N.; Himmelbach, A.; Kilian, B.; Keilwagen, J. Insights into breeding history, hotspot regions of selection, and untapped allelic diversity for bread wheat breeding. *Plant J.* **2022**, *112*, 897–918. [CrossRef] [PubMed]
56. Bertan, I.; de Carvalho, F.I.; Oliveira, A.C.D. Parental selection strategies in plant breeding programs. *J. Crop Sci. Biotechnol.* **2007**, *10*, 211–222.
57. Pandey, P.; Anurag, P.J.; Tiwari, D.K.; Yadav, S.K.; Kumar, B. Genetic variability, diversity and association of quantitative traits with grain yield in rice (*Oryza sativa* L.). *J. Biosci.* **2009**, *17*, 77–82. [CrossRef]
58. Devate, N.B.; Manjunath, K.K.; Ghajghate, R.; Shashikumara, P.; Reddy, U.G.; Kumar, M.; Krishna, H.; Jain, N.; Singh, G.P. Strategies to Develop Heat and Drought–Tolerant Wheat Varieties Following Physiological Breeding. In *Translating Physiological Tools to Augment Crop Breeding*; Springer Nature: Singapore, 2023; pp. 19–52.
59. Oliveira, C.E.D.S.; Andrade, A.D.F.; Zoz, A.; Lustosa Sobrinho, R.; Zoz, T. Genetic divergence and path analysis in wheat cultivars under heat stress. *Pesqui. Agropecu. Trop.* **2020**, *50*, e65493. [CrossRef]
60. Reynolds, M.P.; Braun, H.J. Wheat improvement. In *Wheat Improvement Food Security in a Changing Climate*; Reynolds, M.P., Braun, H.-J., Eds.; Springer: Cham, Switzerland, 2022; pp. 3–15.
61. Haddad, L.; Bachir, A.; Benmahammed, A.; Bouzerzour, H. Durum wheat (*Triticum turgidum* ssp *durum*) improvement during the past 67-year in Algeria: Performance assessment of a set of local varieties under rainfed conditions of the eastern high plateaus. *Jordan J. Biol. Sci.* **2021**, *14*, 327–336.
62. Bahrami, M.; Shabani, A.; Mahmoudi, M.R.; Didari, S. Determination of effective weather parameters on rainfed wheat yield using backward multiple linear regressions based on relative importance metrics. *Complexity* **2020**, *2020*, 6168252. [CrossRef]
63. Zadoks, J.C.; Chang, T.T.; Konzak, C.F. A decimal code for the growth stages of cereals. *Weed Res.* **1974**, *14*, 415–421. [CrossRef]
64. Pask, A.J.D.; Pietragalla, J.; Mullan, D.M.; Reynolds, M.P. *Physiological Breeding II: A Field Guide to Wheat Phenotyping*; CIMMYT: Mexico City, Mexico, 2012; 132p.
65. Spagnoletti-Zeuli, P.L.; Qualset, C.O. Flag leaf variation and the analysis of diversity in durum wheat. *Plant Breed.* **1990**, *105*, 189–202. [CrossRef]
66. Falconer, D.S.; Mackay, T.F.C. *Introduction to Quantitative Genetics*, 4th ed.; Longman: New York, NY, USA, 1996; 464p.
67. Cruz, C.D. Genes: A software package for analysis in experimental statistics and quantitative genetics. *Acta Scientiarum. Agron.* **2013**, *35*, 271–276.
68. Rao, C.R. *Advanced Statistical Methods in Biometric Research*; John Wiley and Sons: New York, NY, USA, 1952; 590p.
69. Popat, R.; Patel, R.; Parmar, D. Variability: Genetic variability analysis for plant breeding research. In *R Package Version 0.1.0*; R Foundation: Vienna, Austria, 2020.
70. Wickham, H. *ggplot2-Elegant Graphics for Data Analysis*; Springer International Publishing: Cham, Switzerland, 2016.
71. Wei, T.; Simko, V.; Levy, M.; Xie, Y.; Jin, Y.; Zemla, J. Package *'corrplot'*. *Statistician* **2017**, *56*, e24.

plants

MDPI

Article

Optimized Phosphorus Application Enhances Wheat Stem Lodging Resistance Under Spring Low-Temperature Stress

Xiang Chen [1], Qianqian Liu [1], Baoqiang Zheng [1] and Jincai Li [1,2,*]

1 College of Agriculture, Anhui Agricultural University, Hefei 230036, China; cx2468@ahau.edu.cn (X.C.); 15855105367@stu.ahau.edu.cn (Q.L.); zhengbaoqiang@ahau.edu.cn (B.Z.)
2 Jiangsu Collaborative Innovation Centre for Modern Crop Production, Nanjing 210095, China
* Correspondence: ljc5122423@126.com

Abstract: Spring low-temperature stress (LTS) has become a major limiting factor for the development of high yield, quality and efficiency in wheat production. It not only affects the function of wheat leaves and the development of spikes but also impacts stem lodging resistance, and may experience elevated risk of stem lodging. This study conducted a field pot experiment to assess the effect of phosphorus fertilizer application mode on wheat stem lodging resistance under spring LTS. Two wheat varieties, Yannong19 (YN19, cold-tolerant variety) and Xinmai26 (XM26, cold-sensitive variety) used as the experiment material. Two phosphorus fertilizer application models including traditional phosphorus application (TPA) and optimized phosphorus application (OPA) were employed. Temperature treatment was conducted at 15 °C (CK) and −4 °C (LT) in a controlled phytotron. Our results showed that spring LTS decreased the stem wall thickness and internode fullness, and altered stem anatomical structure and chemical composition, resulting in a decrease in wheat stem mechanical strength and lodging resistant index. Compared with TPA, the OPA increased the stem wall thickness and internode fullness. The thickness of the stem mechanic tissue layer and parenchymatous tissue, and the area of the large vascular bundle and small vascular bundle were increased by the OPA, which alleviated the damage to stem cell walls caused by spring LTS. At the same time, the OPA also increased the contents of lignin, cellulose, and soluble sugar, improving the C/N ratio in wheat stem. Due to the improved stem morphological characteristics, anatomical structure, and chemical compositions, the wheat stem exhibited enhanced lodging resistance, which increased the lodging resistant index of the 2nd and 3rd internodes of YN19 and XM26 by 27.27%, 11.63% and 14.15%, 15.73% at the dough stage compared with TPA under spring LTS. Meanwhile, OPA could not only alleviate the yield loss caused by spring LTS, but also increase the grain yield without spring LTS. This study indicated that OPA enhances wheat stem lodging resistance under spring LTS, and would be meaningful and practical for improving wheat resistance to low-temperature stress.

Keywords: *Tritium aestivum* L.; cold stress; base stem characteristics; phosphorus fertilizer postpone; lodging resistance index; yield

1. Introduction

Wheat (*Tritium aestivum* L.) is not only one of the world's three major food crops, but also a key staple crop in China [1]. According to data from the National Bureau of Statistics (NBS) of China, in 2023, the cultivation area and yield of wheat in China were 23,627.2 thousand hectares and 136,590 thousand tons, accounting for approximately 19.86% and 19.64% of the total cultivation area and yield of grain crops [2]. Therefore, the safe production of wheat is of great significance for ensuring national food security, promoting socio-economic development, and improving people's quality of life. The Sixth Assessment Report of the Intergovernmental Panel on Climate Change (IPCC) reported that global surface temperatures have risen by approximately 0.8–1.3 °C from 1850 to 1900 compared to that in 2010 to 2019, and is expected to rise exceed 1.5 °C in the future [3]. The ongoing global climate warming poses major challenges for food production and adverse climate change-related events such as low temperatures, high temperatures, strong winds, and heavy rainfall are becoming increasingly frequent [4]. Among them, low-temperature stress (LTS) poses a great threat to the production of major wheat-producing countries and regions such as China [5], Australia [6], Europe [7], and the United States [8]. The Huang-Huai region is the largest wheat-producing area in China. Global warming has accelerated the growth of winter wheat, advancing the critical growth period (jointing and booting), and increasing its vulnerability and susceptibility to damage from spring LTS [9–11]. Therefore, spring LTS has become one of the important factors limiting the high yield, high quality, and efficient development in the Huang-Huai region [12].

Spring LTS in the Huang-Huai wheat region often occurs from March to April [13], which means that the temperature in the season of rebirth and growth rises faster, and in the late spring the temperature is lower than normal [14]. The jointing to booting stage is a critical period of simultaneous nutritional and reproductive growth of wheat. Meanwhile, the wheat young spike is in a key period of differentiation and development, and is highly sensitive to LTS [12]. In this stage, if encountering spring LTS, causing damage or death to the main stem, large tillers, and young spike, resulting in spikelets or whole spike infertile. Generally, mild to moderate spring LTS leads to a 10% to 30% reduction in wheat yield, while severe spring LTS can result in a reduction of over 50% in wheat yield and deterioration of grain quality [15–17].

Regarding the disaster mechanism of wheat spring LTS, previous studies have conducted extensive research from the perspective of source-sink organs. The studies by Zhang et al. [5] and Liu et al. [18] showed that spring LTS not only damages the integrity of the functional leaf mesophyll cell structure of wheat, but also leads to an imbalance in reactive oxygen species (ROS) metabolism in the leaves, and an increase in malondialdehyde (MDA) content leads to a decrease in its photosynthetic performance. Ke et al. [19] found that spring LTS can affect the accumulation and distribution of dry matter in wheat plants, especially decreasing the transport and distribution of photosynthetic compounds to spike. Currently, further research by Lin et al. [20] showed that the accumulation of assimilates in wheat spikes and the decrease in their allocation to different spike positions under spring LTS could restrain spike and floret development, affecting the grain number at different spike positions and resulting in a significant decrease in wheat yield. However, previous studies mostly focus on leaves or spikes, and there are few reports on the effects of spring LTS on wheat stem lodging resistance. Wheat stems have the functions of supporting aboveground plants, transporting and storing nutrients. Global climate change has led to frequent occurrences of extreme weather such as strong winds and heavy rainfall during the later stages of wheat growth, resulting in frequent lodging phenomena and yield losses of up to 10–30% [21]. For example, in 2024, some wheat fields experienced lodging due to the large plant density and strong winds in the middle and later stages of wheat growth, seriously affecting the yield and quality of wheat [22].

Phosphorus, as the most important nutrient element in crop nutrition after nitrogen, is a component of various compounds such as nucleic acids, phospholipids, adenosine triphosphate, and plays an important role in various metabolic activities [23,24]. However, in natural ecosystems, the phosphorus that is lost from the soil-plant cycling system has to be replaced by the slow process of rock weathering or added via fertilizer in human-managed systems, and approximately 40% of the global agricultural lands are deficient in phosphorus [25,26]. Therefore, improving the efficiency of phosphorus fertilizer utilization in agricultural production is a necessary measure for sustainable agricultural development. Many studies have shown that the rational application of phosphorus fertilizer was able to improve the lodging resistance of crop stems [27–29]. Deng et al. [27] showed that the stem diameter and stalk strength of Quinoa applied with 100 kg ha^{-1} phosphorus was increased by 9.5% and 23.2%, and the lodging rate was decreased by 44.5% in a 2-year field experiment, compared with CK. Liang et al. [28] found that the amount of phosphorus applied is closely related to the lodging resistance of maize and optimized phosphorus application (OPA) could promote the accumulation of dry matter between maize internodes, change the ear position coefficient, and improve lodging resistance. The study by Xiang et al. [29] also showed that OPA can improve the thickness, C/N ratio, cell wall cellulose and lignin content, mechanical strength, and lodging resistance index of intercropping soybean stem internodes, enhancing the lodging resistance of soybean. Previous studies have found that optimized phosphorus application could alleviate the damage of spring LTS by improving the root physiology, photosynthesis, spike development and setting [30–32]. However, there have been no reports on the effect of OPA on the lodging resistance of wheat stems under the spring LTS.

In this study, a controlled phytotron experiment was conducted with two contrasting wheat varieties with different cold sensitivities under spring LTS (−4 °C for 4 h) and different phosphorus applications during the anther differentiation period. The objective was to (1) quantify the effects of OPA on the morphological characteristics, anatomical structure, and chemical composition of wheat stems under spring LTS, (2) assess the effects of OPA on the wheat stem lodging resistance under spring LTS, (3) analyze the effects of OPA on wheat yield under spring LTS. The results of this study could provide a theoretical basis for research and development of stress-resistant cultivation measures for wheat, and also be used to guide disaster prevention and reduction in wheat production.

2. Results

2.1. Stem Wall Thickness and Internode Fullness

The stem wall thickness and internode fullness of the second and third internode showed a decreasing trend in each treatment of the two varieties with the growth process moving forward (Figure 1). Compared with CKR1, the thickness of the 2nd and 3rd internodes of YN19 in LTR1 and LTR2 significantly decreased by 35.29%, 32.68% and 39.44%, 28.33% at the dough stage, respectively; meanwhile the thickness of the 2nd and 3rd internodes of XM26 in LTR1 and LTR2 decreased by 40.57%, 17.14% and 30.68%, 0.57%, respectively. Compared with the TPA at the same temperature levels, the thickness of the 2nd and 3rd internodes of YN19 and XM26 in LTR2 under OPA increased by 4.04%, 18.35% and 39.42%, 43.44%, respectively. At the dough stage in LTR1 and LTR2 treatments, the fullness of the 2nd and 3rd internodes decreased by 7.69%~28.32% and 11.14%~28.13% in YN19, and 7.36%~22.50% and 10.73%~20.66% in XM26, respectively, compared with CKR1; whereas the fullness of the 2nd and 3rd internodes of YN19 and XM26 in LTR2 under OPA increased by 28.78%, 23.63% and 19.54%, 12.51%, compared with LTR1, respectively. Therefore, spring LTS decreased the thickness of stem internodes and internode fullness, while OPA was beneficial for increasing the thickness of stem internodes and internode fullness under spring LTS.

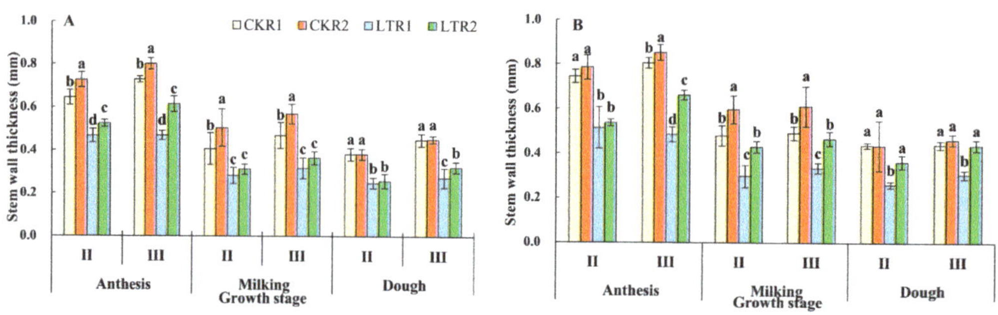

Figure 1. *Cont.*

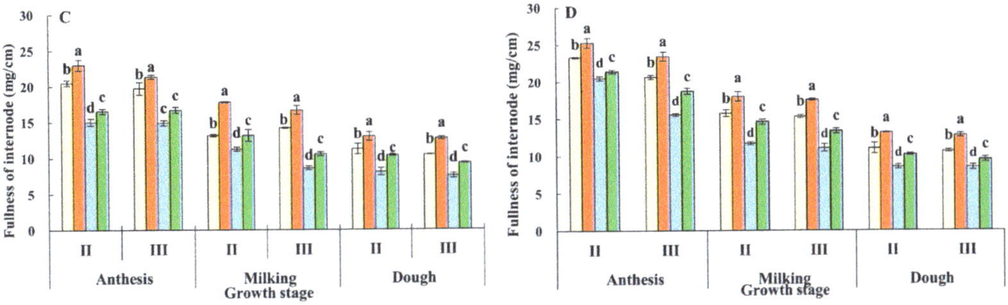

Figure 1. Effects of OPA on stem wall thickness and internode fullness of wheat under spring LTS. (**A**) Stem wall thickness of YN19. (**B**) Stem wall thickness of XM26. (**C**) Internode fullness of YN19. (**D**) Internode fullness of XM26. CKR1 and LTR1 represent the 15 °C and −4 °C under traditional phosphorus application, respectively. CKR2 and LTR2 represent the 15 °C and −4 °C under optimized phosphorus application, respectively. II and III indicate the second and third internode of the main stem. Values are mean ± standard deviation. Different lowercase letters above the bars indicate significant differences among treatments for the same cultivar at the 0.05 level.

2.2. Stem Anatomical Parameters

2.2.1. Mechanic Tissue Layer Thickness and Parenchymatous Tissue Thickness

With the advancement of the growth stage, the mechanic tissue layer thickness and parenchymatous tissue thickness of the 2nd and 3rd internodes of YN19 and XM26 showed a decreasing trend in each treatment from the anthesis to dough stage (Figure 2). Compared with CKR1, the mechanic tissue layer thickness and parenchymatous tissue thickness of the 3rd internodes of YN19 in the LTR1 and LTR2 treatment were decreased by 16.96%, 7.59% and 24.26%, 15.66% at the dough stage, respectively; whereas it increased by 11.28% and 11.35% in the treatments of LTR2, compared with LTR1. The mechanic tissue layer thickness and parenchymatous tissue thickness of the 3rd internodes of XM26 under TPA were decreased by 27.24% and 13.94% after spring LTS and increased by 14.93% and 6.36% under LTR2 treatments compared with LTR1. Additionally, the mechanic tissue layer thickness and parenchymatous tissue thickness of the 2nd internodes of the two varieties showed a similar trend of changes in the 3rd internode. Thus, spring LTS resulted in decreased mechanic tissue layer thickness and parenchymatous tissue thickness of the 2nd and 3rd in two varieties, and OPA treatment alleviated the damage of spring LTS.

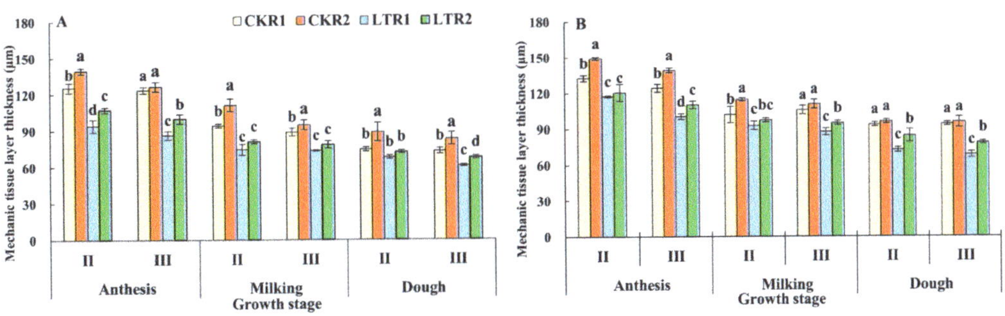

Figure 2. *Cont.*

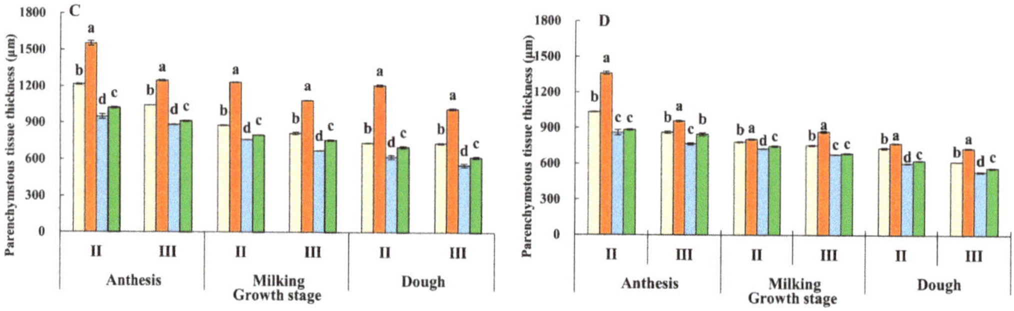

Figure 2. Effects of OPA on stem mechanic tissue layer thickness and parenchymatous tissue thickness of wheat under spring LTS. (**A**) Stem mechanic tissue layer thickness of YN19. (**B**) Stem mechanic tissue layer thickness of XM26. (**C**) Parenchymatous tissue thickness of YN19. (**D**) Parenchymatous tissue thickness of XM26. CKR1 and LTR1 represent the 15 °C and −4 °C under traditional phosphorus application, respectively. CKR2 and LTR2 represent the 15 °C and −4 °C under optimized phosphorus application, respectively. II and III indicate the second and third internode of the main stem. Values are mean ± standard deviation. Different lowercase letters above the bars indicate significant differences among treatments for the same cultivar at the 0.05 level.

2.2.2. Large Vascular Bundle Area and Small Vascular Bundle Area

The overall trend of large vascular bundle area and small vascular bundle area of the 2nd and 3rd internodes in each treatment of the two varieties were both decreased from the anthesis to dough stage (Figure 3). Compared with CKR1, large vascular bundle area of the 2nd and 3rd internodes of YN19 in LTR1 and LTR2 significantly decreased by 25.06%, 11.89% and 22.13%, 6.12% at the dough stage, respectively; meanwhile the large vascular bundle area of the 2nd and 3rd internodes of XM26 in LTR1 and LTR2 decreased by 28.28%, 19.54% and 17.56%, 7.19%, respectively. Compared with LTR1, the large vascular bundle area of the 2nd and 3rd internodes of YN19 and XM26 in LTR2 under OPA increased by 17.57%, 20.56% and 12.18%, 12.56%, respectively. At the dough stage in LTR1 and LTR2 treatments, except for the 2nd internode of YN19 and the 3rd internode of XM26, the small vascular bundle area of the 3rd internodes significantly decreased by 12.53% and 11.14% in YN19, and decreased by 20.26% and 6.26% in the 2nd internode of XM26, respectively, compared with CKR1; whereas the small vascular bundle area of the 2nd internodes of XM26 in LTR2 under OPA significantly increased by 17.56%, compared with LTR1. Therefore, spring LTS decreased the large vascular bundle area and small vascular bundle area, while OPA treatment alleviated the damage of spring LTS.

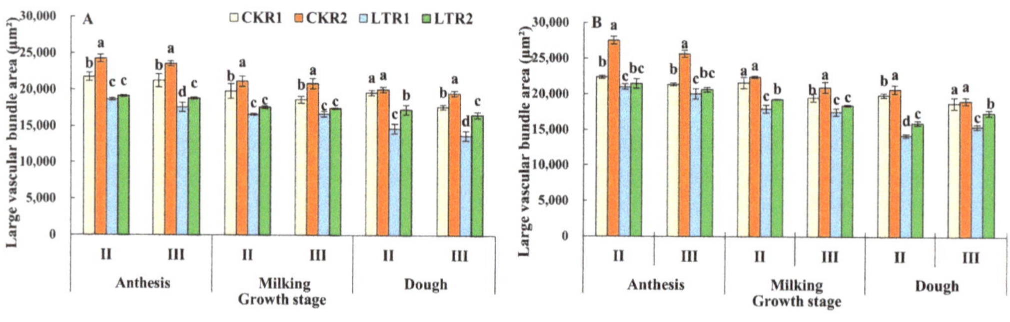

Figure 3. *Cont.*

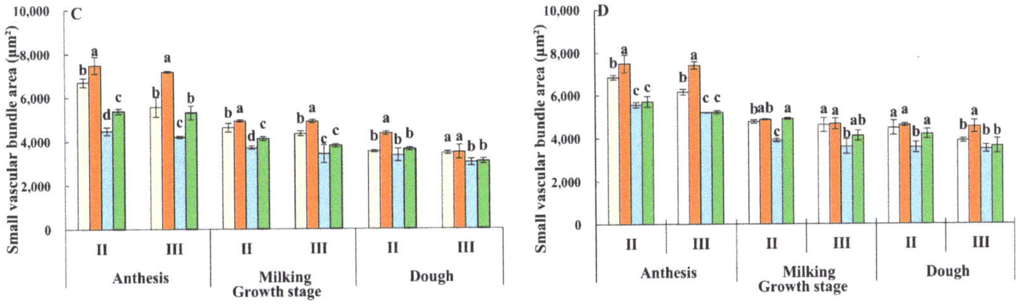

Figure 3. Effects of OPA on large vascular bundle area and small vascular bundle area of wheat under spring LTS. (**A**) Large vascular bundle area of YN19. (**B**) Large vascular bundle area of XM26. (**C**) Small vascular bundle area of YN19. (**D**) Small vascular bundle area of XM26. CKR1 and LTR1 represent the 15 °C and −4 °C under traditional phosphorus application, respectively. CKR2 and LTR2 represent the 15 °C and −4 °C under optimized phosphorus application, respectively. II and III indicate the second and third internode of the main stem. Values are mean ± standard deviation. Different lowercase letters above the bars indicate significant differences among treatments for the same cultivar at the 0.05 level.

2.3. *Stem Chemical Composition*

2.3.1. Lignin and Cellulose

Following the completion of anthesis, a gradual decline in the lignin and cellulose contents of the 2nd and 3rd internodes was observed in wheat (Figure 4), and this trend remained across different treatments of both varieties. The trend of lignin and cellulose contents at the 2nd and 3rd internodes of the two varieties during anthesis, milking and dough stages is CKR2 > CKR1 > LTR2 > LTR2. At the dough stage, the lignin content of 2nd and 3rd internodes of two varieties in LTR2 was significantly greater than that in the LTR1 treatment; meanwhile, except for the 2nd and 3rd internode of XM26, the cellulose content of 2nd and 3rd internodes of YN19 in LTR2 was significantly higher that of LTR1 treatment. Thus, spring LTS resulted in decreasing lignin and cellulose contents of the 2nd and 3rd internodes; whereas both the contents of lignin and cellulose increased in LTR2 under OPA, compared with the TPA at the same temperature levels, respectively.

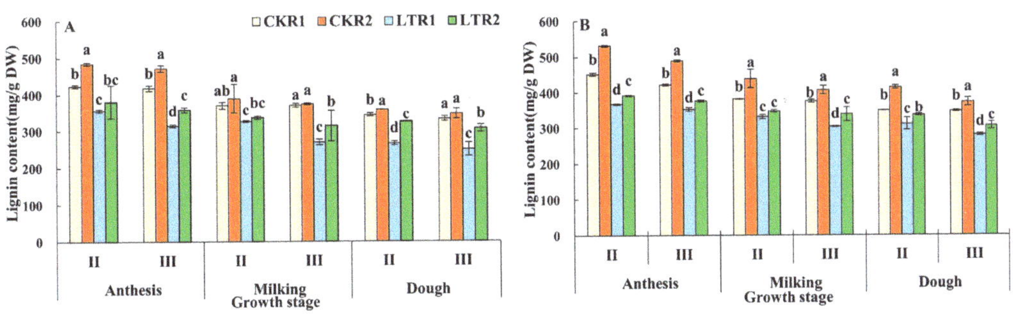

Figure 4. *Cont.*

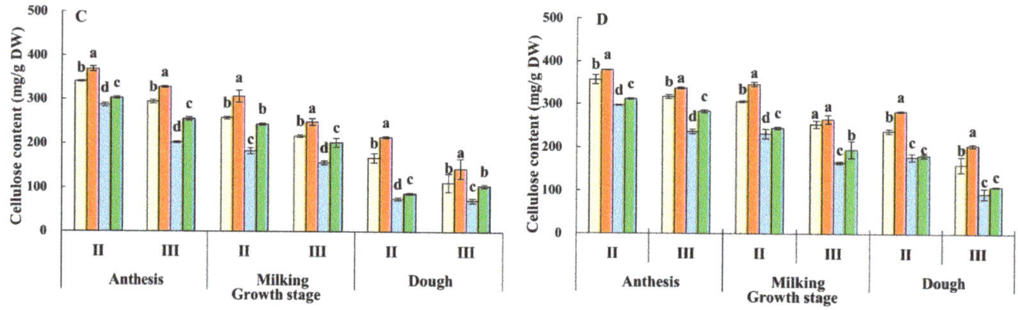

Figure 4. Effects of OPA on stem lignin and cellulose contents of wheat under spring LTS. (**A**) Lignin content of YN19. (**B**) Lignin content of XM26. (**C**) Cellulose content of YN19. (**D**) Cellulose content of XM26. CKR1 and LTR1 represent the 15 °C and −4 °C under traditional phosphorus application, respectively. CKR2 and LTR2 represent the 15 °C and −4 °C under optimized phosphorus application, respectively. II and III indicate the second and third internode of the main stem. Values are mean ± standard deviation. Different lowercase letters above the bars indicate significant differences among treatments for the same cultivar at the 0.05 level.

2.3.2. Soluble Sugar, Total Nitrogen and C/N Ratio

The soluble sugar content, total nitrogen content and the C/N ratio of the 2nd and 3rd internodes showed a decreasing trend in each treatment of the two varieties with the growth of plant (Figure 5). At the dough stage, except for the 2nd internode of XM26, the soluble sugar content of the 2nd and 3rd internodes of two varieties in LTR1 and LTR2 were significantly lower than that of the CKR1 treatment; meanwhile, the change of C/N ratio in all treatments had a similar trend. Additionally, both the soluble sugar content and C/N ratio of the 2nd and 3rd internodes of the two varieties increased in LTR2 under OPA at the dough stage, compared with LTR1, respectively. It is worth noting that spring LTS increased the total nitrogen content of the 2nd and 3rd internodes, and OPA treatment suppressed the increase in total nitrogen content under spring LTS. Thus, the trend of total nitrogen content at the 2nd and 3rd internodes of the two varieties during anthesis, milking and dough stages is LTR1 > LTR2 > CKR1 > CKR2. Therefore, spring LTS decreased the soluble sugar content and C/N ratio, while OPA was beneficial for increasing the soluble sugar content and C/N ratio.

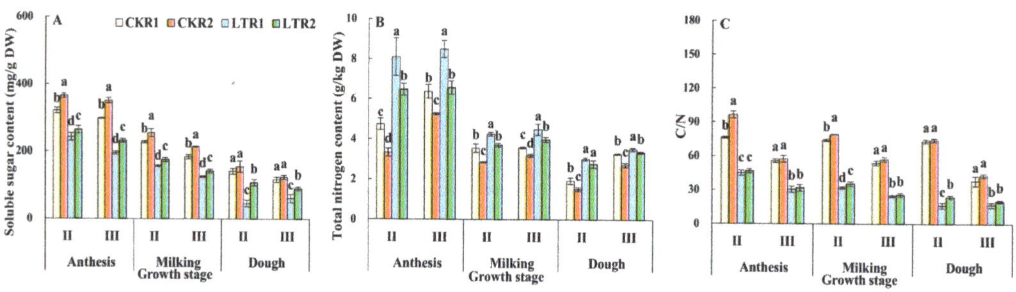

Figure 5. *Cont.*

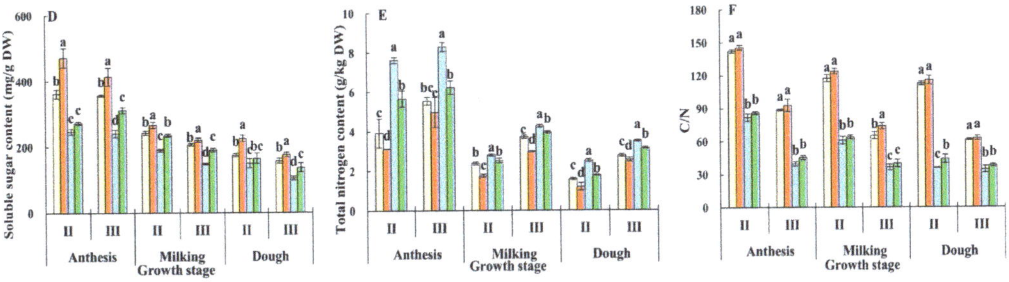

Figure 5. Effects of OPA on the contents of soluble sugar and total nitrogen, and C/N ratio of wheat stem under spring LTS. (**A**) The soluble sugar content of YN19. (**B**) Total nitrogen content of YN19. (**C**) The C/N ratio of YN19. (**D**) The soluble sugar content of XM26. (**E**) Total nitrogen content of XM26. (**F**) The C/N ratio of XM26. CKR1 and LTR1 represent the 15 °C and −4 °C under traditional phosphorus application, respectively. CKR2 and LTR2 represent the 15 °C and −4 °C under optimized phosphorus application, respectively. II and III indicate the second and third internode of the main stem. Values are mean ± standard deviation. Different lowercase letters above the bars indicate significant differences among treatments for the same cultivar at the 0.05 level.

2.4. Stem Mechanical Strength and Lodging Resistant Index

As shown in Figure 6, the stem mechanical strength and lodging resistant index of the 2nd and 3rd internodes showed a decreasing trend from the anthesis to dough stage in each treatment of two varieties. At the dough stage, the mechanical strength of 2nd and 3rd internodes of YN19 and XM26 in LTR2 under OPA significantly increased by 9.68%, 8.80% and 10.90%, 10.39% compared with LTR1, respectively; meanwhile, the lodging resistant index of the 2nd and 3rd internodes of YN19 and XM26 in LTR2 under OPA significantly increased by 27.27%, 11.63% and 14.15%, 15.73% compared with LTR1. Thus, spring LTS decreased the mechanical strength and lodging resistant index of the YN19 and XM26, while OPA could alleviate the decrease in stem mechanical strength and lodging resistant index caused by spring LTS.

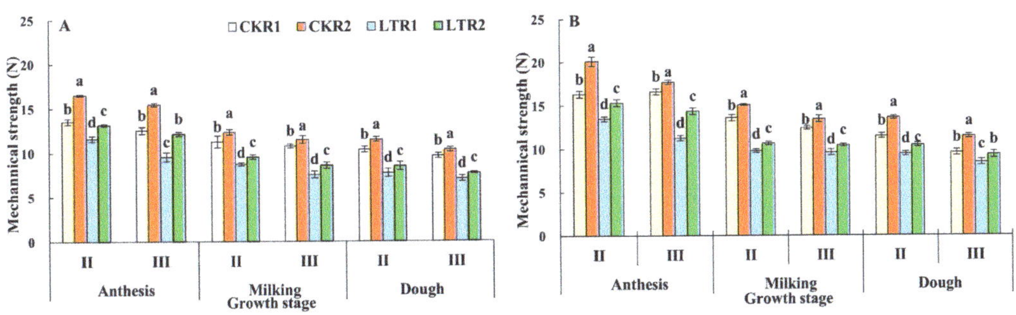

Figure 6. *Cont.*

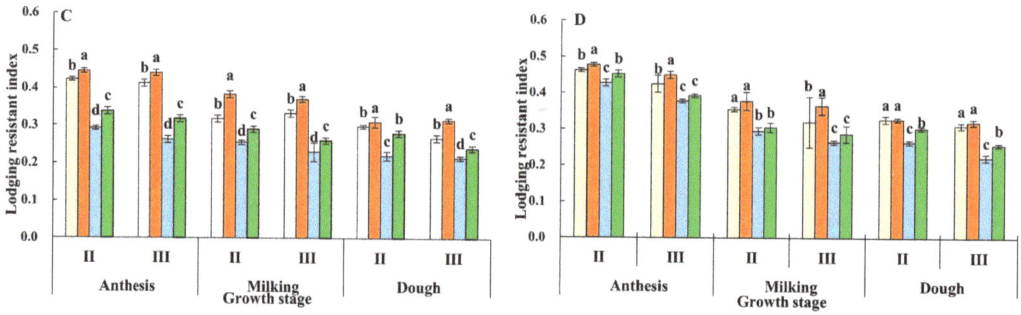

Figure 6. Effects of OPA on the mechanical strength and lodging resistant index of wheat stem under spring LTS. (**A**) Mechanical strength of YN19. (**B**) Mechanical strength of XM26. (**C**) Lodging resistant index of YN19. (**D**) Lodging resistant index of XM26. CKR1 and LTR1 represent the 15 °C and −4 °C under traditional phosphorus application, respectively. CKR2 and LTR2 represent the 15 °C and −4 °C under optimized phosphorus application, respectively. II and III indicate the second and third internode of the main stem. Values are mean ± standard deviation. Different lowercase letters above the bars indicate significant differences among treatments for the same cultivar at the 0.05 level.

2.5. Grain Yield and Yield Components

Spring LTS decreased the grains per spike, 1000-grain weight and grain yield of the YN19 and XM26, while OPA could alleviate the decrease in grains per spike, 1000-grain weight and grain yield caused by spring LTS (Table 1). Compared with CKR1, the grains per spike of YN19 and XM26 in LTR1 and LTR2 significantly decreased by 15.22%~18.04% and 12.64%~22.57%; whereas the grains per spike of YN19 and XM26 increased by 3.45% and 12.83% in LTR2 under OPA, compared with LTR1, respectively. Meanwhile, the grain yield of YN19 and XM26 in LTR1 and LTR2 treatments significantly decreased by 18.45%~25.73% and 16.75%~28.93%, compared with CKR1, respectively. Additionally, it is worth noting that the grain yield of two varieties under CKR2 treatment was the highest. Therefore, OPA was able to alleviate the yield loss caused by spring LTS, and increased the grain yield without spring LTS.

Table 1. OPA affects grain yield and its composition under spring LTS.

Variety	Treatment	Grains per Spike	1000-Grain Weight (g)	Grain Yield (g/stem)
YN19	CKR1	46.0 ± 2.0 a	44.80 ± 1.69 a	2.06 ± 0.07 a
	CKR2	48.0 ± 1.0 a	44.97 ± 0.54 a	2.16 ± 0.06 a
	LTR1	37.7 ± 1.5 b	40.56 ± 0.81 b	1.53 ± 0.01 c
	LTR2	39.0 ± 2.0 b	43.26 ± 2.11 ab	1.68 ± 0.01 b
XM26	CKR1	44.3 ± 2.1 b	44.34 ± 1.73 ab	1.97 ± 0.14 b
	CKR2	49.7 ± 2.9 a	46.56 ± 2.70 a	2.32 ± 0.26 a
	LTR1	34.3 ± 1.5 d	40.88 ± 2.48 b	1.40 ± 0.03 c
	LTR2	38.7 ± 1.5 c	42.62 ± 3.61 ab	1.64 ± 0.09 c

Note: CKR1 and LTR1 represent the 15 °C and −4 °C under traditional phosphorus application, respectively. CKR2 and LTR2 represent the 15 °C and −4 °C under optimized phosphorus application, respectively. Values are mean ± standard deviation. Different lowercase letters above the bars indicate significant differences among treatments for the same cultivar at the 0.05 level.

2.6. Correlation Analysis

The correlations among stem morphological traits, anatomical parameters, chemical composition, mechanical strength and lodging resistant index at the dough stage are shown in Figure 7. The results showed that the lodging resistant index was significantly and positively related to the SWT, IF, and anatomical parameters like MTLT, PTT, LVBA and SVBA, and the chemical compositions like lignin, cellulose, SS, and the C/N ratio, while it was significantly and negatively correlated with the TN. Stem mechanical strength had a similar correlation with other indexes. Moreover, IF was significantly and positively related to the MTLT, LVBA, SVBA, lignin, cellulose, SS, and the C/N ratio. The results indicated that the lodging resistant index is an integrated trait associated with morphological and anatomical traits, chemical constituents and stem strength.

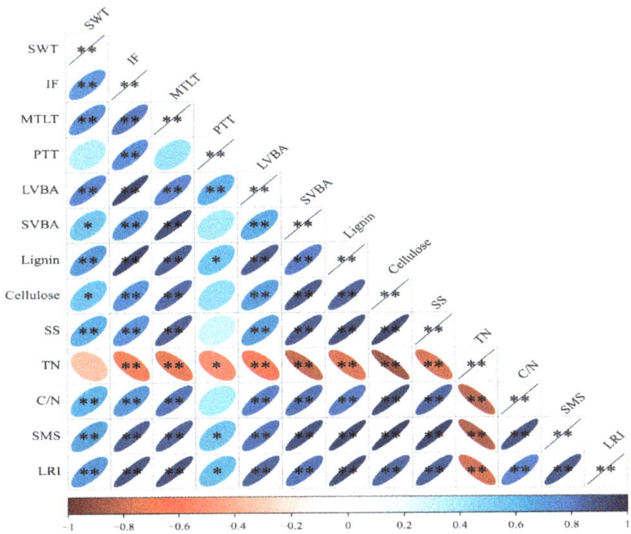

Figure 7. Correlation analysis of stem indexes. * and ** indicate significant differences at the $p < 0.05$ level and $p < 0.01$ level, respectively. SWT, stem wall thickness; IF, internode fullness; MTLT, mechanic tissue layer thickness; PTT, parenchymatous tissue thickness; LVBA, large vascular bundle area; SVBA, small vascular bundle area; SS, soluble sugar; TN, total nitrogen; SMS, stem mechanical strength; LRI, lodging resistant index.

3. Discussion

With the intensification of global climate warming, the growth of winter wheat was accelerated and was more sensitive to spring LTS [9–11]. Therefore, spring LTS has become an important limit for the development of high yield, quality and efficiency in wheat production. Many studies have implicated the effects of spring LTS on the function of wheat leaves and the development of spikes [5,14,18,20]. Spring LTS would affect the growth and development of the wheat stem, and may experience elevated risk of stem lodging, ultimately resulting in a significant yield loss in wheat production. However, previous research mostly focused on the source-sink organs, and less consideration of the effect of spring LTS on wheat stem lodging resistance. Wheat lodging resistance is closely related to stem morphology and structure, i.e., plant height, internode fullness, and internode fullness. Muhammad et al. [33] studied the wheat stem characteristics with lodging resistance and found that wall thickness is a key trait that determines stem lodging resistance and is positively correlated with lodging resistance. This study showed that spring LTS decreased the stem wall thickness and internode fullness. This may be due to spring LTS limiting the elongation and development of the internodes, resulting in decreased stem wall thickness and internode fullness. Phosphorus is a major nutrition element and plays a vital role in plant various metabolic activities [23,24]. Optimized phosphorus application could increase nutrition uptake by roots and fulfill the wheat nutrition demand, coordinate the development of source-sink organs and stem assimilate transport, and reduce the risk of stem lodging [31,34,35]. In the present study, the OPA not only increased the stem wall thickness but also increased the internode fullness.

From an anatomical perspective, the lateral growth and development of the stem form the epidermis, thick-walled tissue, vascular bundles, parenchymatous tissue, and medullary cavity from the outside to the inside, respectively [33]. The thickness of stem mechanical tissue could be used as an index to identify the anatomical structure related to lodging resistance. Kong et al. [36] compared the morphological, anatomical and chemical characteristics of four different genotypes of wheat, and found that the mechanical tissue thickness could explain 99% of the lodging resistance changes. The mechanical tissue thickness and vascular bundle area of the wheat stem had a significant correlation with stem strength, and when the mechanical tissue layer thickness is thicker and the vascular bundle area is larger, the stem strength is stronger [37]. In this study, stem mechanic tissue layer thickness, parenchymatous tissue thickness, large vascular bundle area and small vascular bundle area were decreased under spring LTS. However, the thickness of the stem mechanic tissue layer and parenchymatous tissue, and the area of the large vascular bundle and small vascular bundle were increased by the OPA. This may be due to the OPA could alleviate the damage to stem cell wall and promote the growth of stem wall under spring LTS.

Chemical substances in wheat stem participate in the construction of morphological skeleton and the entire process of growth and development, which is the foundation of stem robustness. The stem chemical composition included structural compounds such as cellulose, hemicellulose, lignin, as well as nonstructural compounds such as soluble sugars, all of which could affect stem strength and lodging resistance [38]. Lignin and cellulose are important components of secondary cell walls, which can enhance the mechanical strength of crop stems [39,40]. This study showed that spring LTS decreased the contents of lignin, cellulose, soluble sugar, and the C/N ratio. Xiang et al. [29] found that optimized phosphorus application could significantly increase the content of cellulose and lignin, promote carbohydrate transport in stems and increase stem C/N ratio, and ultimately enhance stem lodging resistance. Similarly, our study also found that OPA increases the content of lignin and cellulose, promotes the accumulation of soluble sugar, and improves the C/N ratio under spring LTS. The results indicate that OPA increases the accumulation of chemical components in wheat stem under spring LTS, especially improving stem C/N ratio.

Lodging is an integrated trait associated with many plant characteristics. In addition to morphological and anatomical traits, and chemical constituents, stem strength plays a significant role in stem lodging risk [41,42]. In this experiment, spring LTS decreased the mechanical strength and lodging resistant index of the two varieties, while OPA could alleviate the decrease in stem mechanical strength and lodging resistant index caused by spring LTS. Therefore, the OPA improved stem morphological characteristics, anatomical structure, and chemical compositions of wheat stem, resulting in enhanced mechanical strength and lodging resistance (Figure 8). However, the physiological and molecular mechanisms of OPA improving wheat stem lodging resistance still need to be further studied.

Many previous studies showed that spring LTS could lead to a decrease in wheat yield [5,14,15,20]. The results of this study also indicate that a yield loss in wheat was caused by spring LTS. The OPA could alleviate the decrease in grains per spike, 1000-grain weight and grain yield caused by spring LTS. This may be due to the OPA could promote the nutritional and reproductive growth of wheat, coordinate the competition for assimilates between the source and sink, and improve spikelet development and setting, leading to increased grains per spike and final yield [30,31]. Consequently, the OPA could not only improve the lodging resistance of wheat stems at low temperatures, but also alleviate yield loss. OPA will be a very promising and efficient cultivation measure for wheat stress resistance, which is conducive to the sustainable development of agriculture.

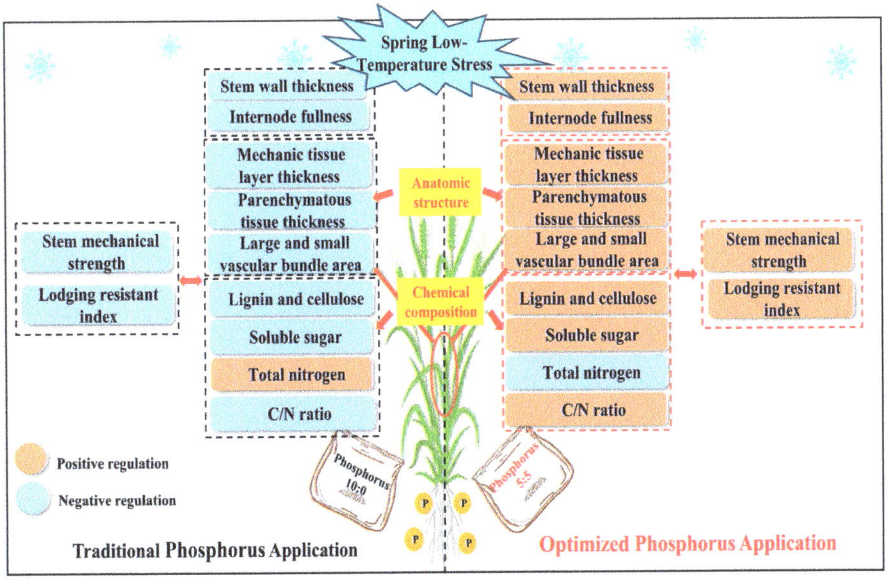

Figure 8. Schematic model in describing the mechanism of wheat stem characteristics under spring LTS by different phosphorus fertilizer treatments. Traditional phosphorus application represents one-off base application at pre-sowing. Optimized phosphorus application represents 50% each at the pre-sowing and jointing stage.

4. Materials and Methods

4.1. Experiment Design

The experiment was conducted from November 2022 to May 2023 at the Anhui Agricultural University Hefei High Technology Agricultural Garden, Anhui Province, China (31°93′ N, 117°21′ E) during the wheat growing season. Two wheat varieties, Yannong19 (YN19, cold-tolerant variety) and Xinmai26 (XM26, cold-sensitive variety), were used as the test varieties [5]. Wheat varieties were planted in plastic pots (30 cm diameter × 35 cm height; 3 drainage holes). The soil used in the pot experiment was brown soil collected from the 0–20 cm of the top layer. Physicochemical analyses revealed that the collected soil comprised 14.53 g kg^{-1} organic matter, 1.03 g kg^{-1} total nitrogen, 18.25 mg kg^{-1} available phosphorus, 256.02 mg kg^{-1} available potassium and a pH of 5.7. Each pot was filled with 10 kg of sieved soil [43].

Referring to the practice of field fertilization in the experiment region, phosphorus fertilizer was all used as base fertilizers before sowing [30,44]. In this experiment, the mode of phosphorus fertilizer application included traditional phosphorus application (TPA, R1, one-off base application at pre-sowing) and optimized phosphorus application (OPA, R2, 50% each at the pre-sowing and jointing stage) [30–32]. The method of temperature levels was divided into normal temperature control (CK, 15 °C) and low-temperature treatment (LT, −4 °C). Four treatments were performed in this experiment, i.e., TPA + CK (CKR1), OPA + CK (CKR2), TPA + LT (LTR1), and OPA + LT (LTR2). Thirty pots were planted in each treatment of two wheat varieties. During the whole wheat growing season, a total of 1.8 g urea (1.2 g of base fertilizer before sowing and 0.6 g of jointing fertilizer) and 1.7 g of potassium sulfate (one-off base application before sowing) were applied to each pot of wheat. The TPA treatment applied 5.0 g of superphosphate (active ingredient percentage ≥ 12.0%) per pot as base fertilizer, while the OPA treatment applied 2.5 g of superphosphate per pot as base fertilizer before sowing and jointing fertilizer, respectively. The top application during the jointing stage was on March 15 in 2023. The low-temperature treatment method was carried out in accordance with Xu et al. [30]. The other cultivation methods, such as pesticide application and irrigation, were conducted following the local optimum cultivation practices to ensure no other stress in these experiments.

4.2. Sampling and Measurement

4.2.1. Measurement of Stem Wall Thickness and Internode Fullness

Stem wall thickness and internode fullness of the second and third internode of the main stem were measured at the flowering stage, milking stage and dough stage, respectively. Stem wall thickness was measured using the methods of Zhou et al. [45]. Internode length (L) was measured with a steel rule of 100 cm, then the samples were placed at 105 °C for 30 min and dried to a constant weight at 75 °C, and the internode dry matter mass (M) was determined. Internode Fullness = M/L. Four samples were randomly chosen to measure stem wall thickness and internode fullness for each treatment.

4.2.2. Measurement of Stem Anatomical Parameters

At the flowering stage, milking stage and dough stage, three wheat plants of each treatment with similar size were sampled. The second and third basal internodes of the main stem were chosen to perform paraffin sectioning. The digital section browsing software (CaseViewer 2.4, 3DHISTECH, China) was used for the analysis of mechanic tissue layer thickness, parenchymatous tissue thickness, large vascular bundle area and small vascular bundle area.

4.2.3. Measurement of Stem Chemical Composition

Samples after the determination of internode dry matter mass were thoroughly ground into powder samples. The content of soluble sugar was determined by anthrone colorimetry [46]. Stem sample nitrogen contents were determined by using the H_2SO_4-H_2O_2 digestion method [47]. The contents of lignin and cellulose were measured using the lignin and cellulose assay kit (Beijing Boxbio Science & Technology Co., Ltd., Beijing, China). The above parameters were measured with three biological replicates. The ratio of C/N was calculated as the following formula:

$$C/N = \text{Soluble sugar content}/\text{Total nitrogen content} \quad (1)$$

4.2.4. Measurement of Stem Mechanical Parameters

Four wheat plants of each treatment with similar sizes were sampled at the anthesis stage, milking stage and dough stage, respectively. Stem mechanical strength of the second and third internode of the main stem was measured using a stem strength tester (YYD-1A, Zhejiang Topu Yunnong Technology Co., Ltd., Hangzhou, Chian) and the center of gravity height was measured with a steel rule of 100 cm according to Zhou et al. [45]. Lodging resistant index was calculated as the following formula:

$$\text{Lodging resistant index} = \text{internode mechanical strength}/\text{center of gravity height} \quad (2)$$

4.2.5. Yield Components and Grain Yield

At the maturity stage, three main stems of wheat plants were randomly selected from each treatment to investigate the grain number per spike, 1000-grain weight and grain yield per stem.

4.3. Statistical Analysis

Microsoft Excel (version 2019; Microsoft, Inc., Redmond, WA, USA) was used for data sorting and figure and table production. Statistical analyses were performed using SPSS 19.0 software (IBM SPSS 19.0, SPSS Inc., Chicago, IL, USA). The significant differences in measurement parameters among the treatments were identified by Duncan's multiple comparison tests ($p < 0.05$) via one-way ANOVA. The correlations analysis was plotted according to Pearson's method by using RStudio.

5. Conclusions

Spring LTS decreased the stem wall thickness and internode fullness, and altered stem anatomical structure and chemical composition, resulting in a decrease of wheat stem mechanical strength and lodging resistant index. Optimization of phosphorus application was able to the regulate stem characteristics. The results of our study showed that OPA could increase the stem wall thickness and internode fullness under spring LTS. OPA could increase stem mechanic tissue layer thickness, parenchymatous tissue thickness, large vascular bundle area and small vascular bundle area, and alleviate the damage to stem cell walls caused by spring LTS. Simultaneously, OPA increased the contents of lignin, cellulose, and soluble sugar, and improved the C/N ratio, thereby augmenting the stem mechanical strength and lodging resistant index under spring LTS. Notably, OPA was able to alleviate the yield loss caused by spring LTS, and increased the grain yield under normal temperature conditions. Therefore, optimized phosphorus application (50% each at the pre-sowing and jointing stage) would be meaningful and practical for improving crop resistance to stress, especially in the interest of sustainable agricultural development.

Author Contributions: Conceptualization, X.C. and J.L.; methodology, X.C. and Q.L.; software, X.C. and Q.L.; validation, X.C., B.Z. and J.L.; formal analysis, X.C. and Q.L.; investigation, X.C. and Q.L.; resources, J.L.; data curation, X.C.; writing—X.C. and Q.L.; writing—review and editing, X.C. and B.Z. All authors have read and agreed to the published version of the manuscript.

Funding: This research was supported by the Scientific Research Projects of Colleges and Universities in Anhui Province, China (No. 2023AH051016) and the National Key Research and Development Program of China (No. 2023YFD1901005).

Data Availability Statement: Data is contained within the article.

Conflicts of Interest: The authors declare no conflicts of interest.

References

1. Mao, H.; Jiang, C.; Tang, C.; Nie, X.; Du, L.; Liu, Y.; Cheng, P.; Wu, Y.; Liu, H.; Kang, Z.; et al. Wheat adaptation to environmental stresses under climate change: Molecular basis and genetic improvement. *Mol. Plant* **2023**, *16*, 1564–1589. [CrossRef] [PubMed]
2. Announcement by the National Bureau of Statistics on Grain Production Data for 2023. Available online: https://www.stats.gov.cn/sj/zxfb/202312/t20231211_1945417.html (accessed on 12 December 2023).
3. Ma, Z.Y.; Ren, J.X.; Chen, H.T.; Jiang, H.; Gao, Q.X.; Liu, S.L.; Yan, W.; Li, Z.M. Analysis and recommendations of IPCC working group I assessment report. *Res. Environ. Sci.* **2022**, *35*, 2550–2558.
4. Ye, L.; Wang, T.; Wu, R.; Zheng, C.; Zhan, L.; Chen, J.; Guo, S.; Chen, Y. Evaluation of cold resistance at seedling stage for 70 peanut genotypes based on photosynthetic fluorescence characteristics. *Agronomy* **2024**, *14*, 1699. [CrossRef]
5. Zhang, Y.; Cai, H.; Liu, L.; Xu, H.; Chen, X.; Li, J. Screening of varieties resistant to late-spring coldness in wheat and effects of late-spring coldness on the ultrastructure of wheat cells. *Agronomy* **2023**, *13*, 3011. [CrossRef]
6. Crimp, S.J.; Zheng, B.Y.; Khimashia, N.; Gobbett, D.L.; Scott Chapman, S.; Howden, M.; Nicholls, N. Recent changes in southern Australian frost occurrence: Implications for wheat production risk. *Crop Pasture Sci.* **2016**, *67*, 801–811. [CrossRef]
7. Trnka, M.; Rötter, R.; Ruiz-Ramos, M.; Kersebaum, K.C.; Olesen, J.E.; Zalud, Z.; Semenov, M.A. Adverse weather conditions for European wheat production will become more frequent with climate change. *Nat. Clim. Change* **2014**, *4*, 637–643. [CrossRef]
8. Holman, J.D.; Schlegel, A.J.; Thompson, C.R.; Lingenfelser, J.E. Influence of precipitation, temperature and 56 years on winter wheat yields in western Kansas. *Crop Manag.* **2011**, *10*, 11–21. [CrossRef]
9. Hao, Z.X.; Geng, X.; Wang, F.; Zheng, J.Y. Impacts of climate change on agrometeorological indices at winter wheat overwintering stage in northern China during 2021–2050. *Int. J. Climatol.* **2018**, *38*, 5576–5588. [CrossRef]
10. Jiang, G.; Hassan, M.A.; Muhammad, N.; Arshad, M.; Chen, X.; Xu, Y.H.; Xu, H.; Ni, Q.Q.; Liu, B.B.; Yang, W.K.; et al. Comparative physiology and transcriptome analysis of young spikes in response to late spring coldness in wheat (*Triticum aestivum* L.). *Front. Plant Sci.* **2022**, *13*, 811884. [CrossRef]
11. Li, X.; Jiang, D.; Liu, F. Winter soil warming exacerbates the impacts of spring low temperature stress on wheat. *J. Agron. Crop Sci.* **2016**, *202*, 554–563. [CrossRef]
12. Xiao, L.J.; Liu, L.L.; Senthold, A.; Xia, Y.M.; Tang, L.; Liu, B.; Cao, W.X.; Zhu, Y. Estimating spring frost and its impact on yield across winter wheat in China. *Agric. For. Meteorol.* **2018**, *260*, 154–164. [CrossRef]
13. Zhang, Y.; Ni, C.; Dong, Y.; Jiang, X.; Liu, C.; Wang, W.; Zhao, C.; Li, G.; Xu, K.; Huo, Z. The role of the ascorbic acid–glutathione cycle in young wheat ears' response to spring freezing stress. *Plants* **2023**, *12*, 4170. [CrossRef] [PubMed]
14. Zhang, Y.; Liu, L.Z.; Chen, X.; Li, J.C. Effects of low-temperature stress during the anther differentiation period on winter wheat photosynthetic performance and spike-setting characteristics. *Plants* **2022**, *11*, 389. [CrossRef] [PubMed]

15. Chen, X.; Liu, L.L.; Cai, H.M.; Zheng, B.Q.; Li, J.C. Effects of spring low-temperature stress on winter wheat seed-setting characteristics of spike. *Plant Soil. Environ.* **2024**, *7*, 84–92. [CrossRef]
16. Hu, X.Y.; Ma, J.F.; Qian, W.H.; Cao, Y.; Zhang, Y.; Liu, B.; Tang, L.; Cao, W.X.; Zhu, Y.; Liu, L.L. Effects of low temperature on the amino acid composition of wheat grains. *Agronomy* **2022**, *12*, 1171. [CrossRef]
17. Shi, K.J.; Yin, T.W.; Zhu, Y.; Liu, B.; Tang, L.; Cao, W.X.; Liu, L.L. Estimating the effect of low-temperature stress on the spatial distribution patterns of protein in wheat grains. *J. Cereal Sci.* **2022**, *105*, 103461. [CrossRef]
18. Liu, L.Z.; Zhang, Y.; Zhang, L.; Cai, H.M.; Yu, M.; Wei, F.Z.; Chen, X.; Li, J.C. Effects of low temperature stress during the anther differentiation period on leaf anatomical structure and photosynthetic characteristics of wheat. *Chin. J. Agrometeorol.* **2023**, *44*, 58–70.
19. Ke, Y.Y.; Chen, X.; Zhang, L.L.; Zhang, Y.; Xu, H.; Hassan, M.A.; Jiang, G.; Lin, F.F.; Wei, F.Z.; Li, J.C. Effects of low temperature stress at anther connective stage on dry matter accumulation, translocation and distribution and grain yield of wheat. *J. Anhui Agric. Univ.* **2021**, *48*, 701–706.
20. Lin, F.F.; Li, C.; Xu, B.; Chen, J.; Chen, A.H.; Hassan, M.A.; Liu, B.B.; Xu, H.; Chen, X.; Sun, J.Q.; et al. Late spring cold reduces grain number at various spike positions by regulating spike growth and assimilate distribution in winter wheat. *Crop J.* **2023**, *11*, 1272–1278. [CrossRef]
21. Zhang, L.; Chen, X.; Li, J.C. Bibliometric analysis of Chinese wheat lodging research literature based on CNKI. *J. Jianghan Univ. (Nat. Sci. Ed.)* **2022**, *50*, 80–89.
22. Technical Opinions on Promoting Recovery and Reducing Losses of Lodging Wheat. Available online: http://www.moa.gov.cn/gk/nszd_1/nszd_2/202405/t20240516_6455626.htm (accessed on 16 May 2023).
23. Poirier, Y.; Jaskolowski, A.; Clua, J. Phosphate acquisition and metabolism in plants. *Curr. Biol.* **2022**, *32*, 623–629. [CrossRef] [PubMed]
24. Yaakob, M.A.; Mohamed, R.M.S.R.; Al-Gheethi, A.; Ravishankar, G.A.; Ambati, R.R. Influence of nitrogen and phosphorus on microalgal growth, biomass, lipid, and fatty acid production: An overview. *Cells* **2021**, *10*, 393. [CrossRef] [PubMed]
25. Alewell, C.; Ringeval, B.; Ballabio, C.; Robinson, D.A.; Panagos, P.; Borrelli, P. Global phosphorus shortage will be aggravated by soil erosion. *Nat. Commun.* **2020**, *11*, 4546. [CrossRef]
26. Xu, H.; Hassan, M.A.; Sun, D.; Wu, Z.; Jiang, G.; Liu, B.; Ni, Q.; Yang, W.; Fang, H.; Li, J.C.; et al. Effects of low temperature stress on source–sink organs in wheat and phosphorus mitigation strategies. *Front. Plant Sci.* **2022**, *13*, 807844. [CrossRef]
27. Deng, Y.; Zhao, L.; Anwar, S.; Zhang, L.G.; Shafiq, F.; Guo, H.X.; Qin, L.X.; Wang, M.X.; Wang, C.Y. Phosphorus fertigation conferred lodging tolerance and improved grain quality in *Chenopodium quinoa* via enhanced root proliferation and stalk strength. *J. Soil. Sci. Plant Nutr.* **2022**, *22*, 5099–5110. [CrossRef]
28. Liang, Z.Y.; Xue, J.; Zhang, G.Q.; Min, B.; Shen, D.P.; Fang, L.; Zhou, L.L.; Zhang, Y.Q.; Yang, H.S.; Wang, K.R.; et al. Effects of phosphorus application rate on lodging resistance of maize under integrated water and fertilizer. *Crops* **2023**, *6*, 190–194.
29. Xiang, D.B.; Guo, K.; Lei, T.; Yu, X.B.; Luo, Q.M.; Yang, W.Y. Effect of phosphorus and potassium on stem characteristic and lodging resistance of relay cropping soybean. *Chin. J. Oil Crop Sci.* **2010**, *32*, 395–402.
30. Xu, H.; Hou, K.Y.; Fang, H.; Liu, Q.Q.; Wu, Q.; Lin, F.F.; Deng, R.; Zhang, L.J.; Chen, X.; Li, J.C. Twice-split phosphorus application alleviates low-temperature impacts on wheat by improved spikelet development and setting. *J. Integr. Agric.* **2023**, *22*, 3667–3680. [CrossRef]
31. Xu, H.; Wu, Z.; Xu, B.; Sun, D.; Hassan, M.A.; Cai, H.; Wu, Y.; Yu, M.; Chen, A.; Li, J.; et al. Optimized phosphorus application alleviated adverse effects of short-term low-temperature stress in winter wheat by enhancing photosynthesis and improved accumulation and partitioning of dry matter. *Agronomy* **2022**, *12*, 1700. [CrossRef]
32. Xu, H.; Hassan, M.A.; Li, J.C. Twice-split phosphorus application alleviates low temperature stress by improving root physiology and phosphorus accumulation, translocation, and partitioning in wheat. *Agronomy* **2023**, *13*, 2643. [CrossRef]
33. Muhammad, A.; Hao, H.H.; Xue, Y.L.; Alam, A.; Bai, S.M.; Hu, W.C.; Sajid, M.; Hu, Z.; Samad, R.A.; Li, Z.H.; et al. Survey of wheat straw stem characteristics for enhanced resistance to lodging. *Cellulose* **2020**, *27*, 2469–2484. [CrossRef]
34. Shen, J.B.; Yuan, L.X.; Zhang, J.L.; Li, H.G.; Bai, Z.H.; Chen, X.P.; Zhang, W.F.; Zhang, F.S. Phosphorus dynamics: From soil to plant. *Plant Physiol.* **2011**, *156*, 997–1005. [CrossRef] [PubMed]
35. Jahan, M.; Amiri, M.B. Optimizing application rate of nitrogen, phosphorus and cattle manure in wheat production: An approach to determine optimum scenario using response-surface methodology. *J. Soil. Sci. Plant Nutr.* **2018**, *18*, 13–26. [CrossRef]
36. Kong, E.Y.; Liu, D.C.; Guo, X.L.; Yang, W.L.; Sun, J.Z.; Li, X.; Zhan, K.H.; Cui, D.Q.; Lin, J.X.; Zhang, A.M. Anatomical and chemical characteristics associated with lodging resistance in wheat. *Crop J.* **2013**, *1*, 43–49. [CrossRef]
37. He, J.; Sun, S.G.; Ge, C.B.; Song, D.Y.; Qiao, J.L.; Li, S.P.; Su, Y.R.; Liao, P.G. Relationship between microstructure, biochemical composition and stem strength of different wheat varieties (lines). *Acta Agric. Boreali-Sin.* **2022**, *37*, 68–76.
38. Zhao, X.; Zhou, S.L. Research progress on traits and assessment methods of stalk lodging resistance in maize. *Acta Agron. Sin.* **2022**, *48*, 15–26. [CrossRef]
39. Ma, J.F.; Yamaji, N. Silicon uptake and accumulation in higher plants. *Trends Plant Sci.* **2006**, *11*, 392–397. [CrossRef] [PubMed]
40. Shah, A.N.; Tanveer, M.; Rehman, A.U.; Anjum, S.A.; Iqbal, J.; Ahmad, R. Lodging stress in cereal-effects and management: An overview. *Environ. Sci. Pollut. Res.* **2017**, *24*, 5222–5237. [CrossRef]
41. Berry, P.M.; Stering, M.; Spink, J.H.; Baker, C.J.; Sylvester-Bradley, R.; Mooney, S.J.; Tams, A.R.; Ennos, A.R. Understanding and reducing lodging in cereals. *Adv. Agron.* **2004**, *84*, 215–269.

42. Berry, P.M.; Sylvester-Bradley, R.; Berry, S. Ideotype design for lodging-resistant wheat. *Euphytica* **2007**, *154*, 164–179. [CrossRef]
43. Huang, W.X.; Dai, W.C.; Chen, T.T.; Cai, H.M.; Weng, Y.; Tang, Z.W.; Yin, C.; Wang, P.N.; Zheng, B.Q.; Li, J.C.; et al. Spraying KH_2PO_4 enhances the physiological activity of wheat flag leaves and roots to alleviate the damage caused by late spring coldness. *Plant Nutr. Fert. Sci.* **2024**, *30*, 1222–1233.
44. He, J.N.; Shi, Y.; Yu, Z.W. Subsoiling improves soil physical and microbial properties, and increases yield of winter wheat in the Huang-Huai-Hai Plain of China. *Soil. Tillage Res.* **2019**, *187*, 182–193. [CrossRef]
45. Zhou, J.; Wang, X.; Zhu, Y.L.; Liu, H.H.; Chen, X.; Wei, F.Z.; Sun, J.Q.; Song, Y.H.; Li, J. Effect of nitrogen fertilizer management on stem lodging resistance and yield of wheat. *J. Triticeae Crop.* **2019**, *39*, 979–987.
46. Wang, X.; Wu, Z.; Zhou, Q.; Wang, X.; Song, S.; Dong, S. Physiological response of soybean plants to water deficit. *Front. Plant Sci.* **2022**, *12*, 809692. [CrossRef]
47. Ni, Q.Q.; Cai, H.M.; Wu, H.; Jiang, G.; Xu, B.; Wei, F.Z.; Chen, X.; Li, J.C. Effect of straw returning in different seasons on dry matter accumulation, distribution and nitrogen use efficiency of winter wheat. *J. Triticeae Crop.* **2023**, *43*, 75–80.

plants

MDPI

Article

Induction of Salt Stress Tolerance in Wheat Seeds by Parental Treatment with Salicylic Acid

Lei Yan, Xue Jiang, Yuman Zhang, Yongwen Dong, Can Zhao, Ke Xu, Zhongyang Huo * and Weiling Wang *

Jiangsu Key Laboratory of Crop Cultivation and Physiology, Jiangsu Co-Innovation Center for Modern Production Technology of Grain Crops, Yangzhou University, No. 88 Daxue South Road, Yangzhou 225009, China; mz120231385@stu.yzu.edu.cn (L.Y.); mz120231361@stu.yzu.edu.cn (X.J.); mx120240739@stu.yzu.edu.cn (Y.Z.); mz120231414@stu.yzu.edu.cn (Y.D.); canzhao@yzu.edu.cn (C.Z.); xuke@yzu.edu.cn (K.X.)

* Correspondence: zyhuo@yzu.edu.cn (Z.H.); wangwl@yzu.edu.cn (W.W.)

Abstract: Soil salinization is an important factor that limits crop production. The effects of spraying salicylic acid (SA) during the grain-filling stage on the salt tolerance of progeny seeds in wheat (*Triticum aestivum* L.) were investigated in this study. The results showed that spraying SA during the grain-filling stage significantly increased the grain weight and yield of wheat plants. Meanwhile, the seeds from the SA-treated plants showed a higher germination rate, length and dry mass of the coleoptile and radicle, and a lower mean germination time compared to the seeds of water-treated plants under the salt germination condition, indicating that SA pretreatment during the grain-filling stage could effectively improve the salt tolerance of progeny seeds in wheat. SA pretreatment significantly increased the activities of amylases and the respiration rate, accompanied by a decrease in starch content, and a higher accumulation in the level of soluble sugars and adenosine triphosphate (ATP) in the germinated seedlings compared to the water pretreatment under salt stress. In addition, SA pretreatment obviously alleviated the increase in malondialdehyde (MDA) content and the reactive oxygen species (ROS) release rate in seedlings by activating antioxidant enzymes (superoxide dismutase (SOD) and peroxidase (POD)) under salt stress. Moreover, the seedlings of the SA-treated plants showed lower Na^+ and higher K^+ contents compared to the seeds of water-treated plants under salt stress. The results of this study indicate that spraying SA during the grain-filling stage improves the capacity of offspring seeds to maintain osmotic and ion balance and redox homeostasis under salt stress, thereby conferring salt tolerance to the wheat seeds.

Keywords: salicylic acid; salt tolerance; germination; offspring; physiological mechanism

1. Introduction

With the growth in population and the expansion of urbanization, global food demand is projected to increase by 35% to 56% by 2050 compared to 2010 [1]. Under the influence of global climate change, improper fertilizer application, unreasonable irrigation methods, and other factors have caused more than one-third of the world's soil to be affected by salinization and alkalinization [2]. Saline-alkaline soils impose multiple adverse effects on plants, mainly including osmotic stress and ion toxicity, severely inhibiting plant growth and development [3,4]. As a result, the efficiency of food production on saline-alkaline soils is significantly lower compared to normal soils [5]. Therefore, improving the salt tolerance of crops to enhance the productivity of saline-alkaline soils is an important strategy for addressing global food security challenges.

Seed germination is a critical stage in plant morphological establishment, but it is particularly sensitive to salt stress [6]. Salt stress has a direct and severe impact on seed germination. This is primarily manifested by the low water potential of saline soil solutions, which inhibits the ability of the seed to imbibe water [7]. This inhibition hampers the radical and plumule from breaking through the seed coat, thereby prolonging the germination time [8]. In addition, salt stress increases the content of abscisic acid (ABA) while decreasing the level of gibberellin (GA), which leads to the reduction in amylase activity in seeds, resulting in slow degradation of the endosperm starch and inhibited seed germination [9,10]. Moreover, the excessive accumulation of sodium (Na^+) leads to the buildup of reactive oxygen species (ROS) in seeds under salt stress, causing oxidative damage of the cell membrane, protein, and nucleic acid [11,12]. Therefore, enhancing the germination ability of crops under salt stress conditions can contribute to the growth of more robust seedlings, ultimately improving the overall crop yield and quality.

It has been documented that plants possess a transgenerational memory of environmental stresses, which is an important mechanism for plant offspring to acquire stress tolerance [13]. For example, the activity of antioxidant enzymes in plants under drought conditions can be effectively inherited by their offspring in wheat plants [14]. The progeny of plants subjected to waterlogging stress exhibited a higher net photosynthetic rate, antioxidant enzyme activity, and biomass than the progeny of plants that had not experienced waterlogging stress [15]. The seeds collected from plants cultivated under terminal drought stress showed a stronger salt tolerance compared to the seeds collected from plants under well-irrigated conditions [13]. Previous studies have found that epigenetic variations (such as DNA methylation, histone modifications, or non-coding RNAs) play a crucial role in transgenerational inheritance by mediating responses across generational layers to cope with environmental stresses [16]. It has been proven that salicylic acid (SA) is associated with epigenetic modifications to mediate bacterial infection-induced stress tolerance in the next generation.

Salicylic acid is an endogenous physiological regulator that belongs to the class of phenolic compounds, which plays an important role in the plants' responses to both biotic and abiotic stress including salt stress [17,18]. It has been reported that an appropriate dose of exogenous SA can alleviate the adverse effects of salt stress on seed germination in many crops [19]. For example, exogenous SA can alleviate ion toxicity by reducing the Na^+ content, thereby lowering the Na+/K+ ratio, which helps maintain ROS and hormone homeostasis. This process promotes starch hydrolysis, providing sufficient energy for seed germination, ultimately improving rice seed germination under salt stress [20]. Previous research has shown that exogenous SA alleviates salt stress damage in ryegrass by optimizing photosynthesis and antioxidant metabolism [21]. Exogenous SA also promoted the germination of *Limonium bicolor* (Kuntze) seeds by enhancing the activity of amylase [22]. However, there is limited research on the impact of exogenous SA application on the stress tolerance of subsequent generations, especially in food crops.

Wheat is a globally important staple crop, and is also known as the staple food for nearly 35 per cent of the world's population [23]. The grain-filling stage is a critical period for wheat grain development [24]. A previous study showed that spraying an appropriate concentration of SA solution during the grain-filling stage significantly increased the wheat yield and improved the grain quality [25]. However, the effects of spraying SA during this critical period on the salt tolerance of its progeny seeds remain unclear. In the present study, it was hypothesized that the SA-treated wheat plants could induce salt stress tolerance in their progeny. The effects of SA trans-generation priming were analyzed, and its underlying mechanisms were discussed in this study. The findings of this study are expected to contribute to the development of practical strategies for improving salt tolerance in wheat during the early stages of plant growth.

2. Materials and Methods

2.1. Chemical Treatments and Seed Collection

This experiment was conducted from 2021 to 2022 in Putou Town, Jiangdu District, Yangzhou City, Eastern China (geographical coordinates: 119.44° E, 32.43° N, with an elevation of approximately 3.5 m) (Figure 1). The location is situated on the lower reaches of the Yangtze River Plain, with an average annual temperature of about 15.7 °C and an annual precipitation of around 1000 mm. The wheat variety used in the experiment was Yangmai 23 (*Triticum aestivum* L.), a variety widely planted locally. The experiment was carried out using manual row sowing with a row spacing of 25 cm and a basic seedling density of 2.4 million plants per hectare. The sowing date was 9 November 2021. A single application of phosphorus and potassium fertilizers was used as the base application, and the application ratio of nitrogen to phosphorus and potassium as 2:1:1. Fertilization included the application of 240 kg of nitrogen (N) per hectare, and 120 kg each of phosphorus (P_2O_5) and potassium (K_2O) per hectare.

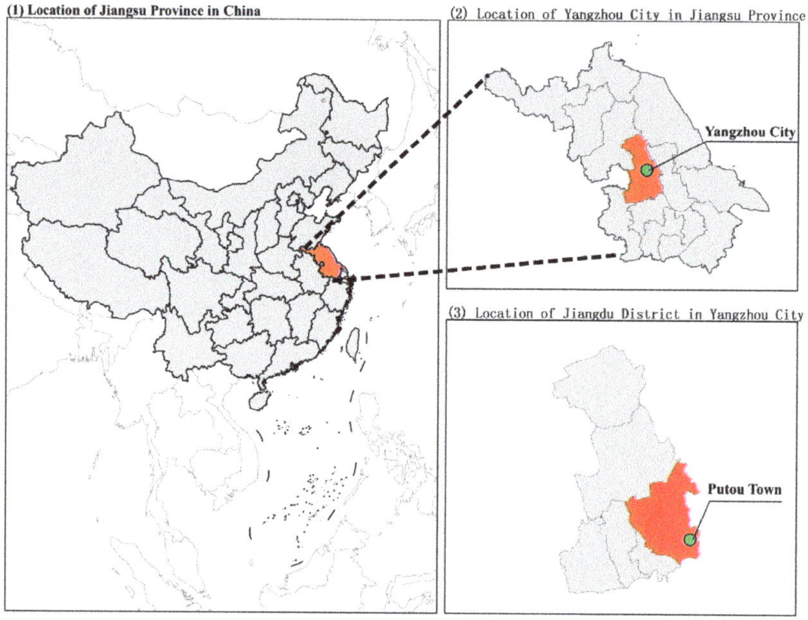

Figure 1. Location of Putou Town, Jiangdu District, Yangzhou City, Jiangsu Province, China.

Five days after the flowering of Yangmai 23 (in 20 April 2022), foliar spraying was applied to the wheat plants. Four different concentrations of SA solutions (prepared from Sigma-Aldrich Co., Ltd., St. Louis, MO, USA) were used in the experiment: 0, 50, 150, and 450 µmol/L. Spraying was conducted once every seven days for a total of three applications, with each spraying carried out at dusk. The volume of each application was 450 L per hectare, and the solution contained 0.1% Tween 20. Each treatment was repeated three times, with each replicate covering an area of 12 square meters. After the wheat matured, the grains were harvested (1 June 2022), and the thousand-grain weight and yield were measured.

2.2. Salt Stress Treatment During Seed Germination

The seeds collected from the pretreated wheat plants were surface sterilized with 10% hydrogen peroxide for 10 min and then thoroughly rinsed with distilled water. Fifty seeds pretreated with varying concentrations of salicylic acid (0, 50, 150, and 450 µmol/L) were placed evenly onto Petri dishes containing two layers of filter paper. To simulate salt stress, 10 mL of a 100 mmol/L sodium chloride (NaCl) solution was added to each dish. Seeds pretreated with water and germinated under normal conditions were used as the control group. Finally, the following five treatments were established: WW (water pretreatment + distilled water germination), WS (water pretreatment + salt treatment), $S_{50}S$ (50 µmol/L SA pretreatment + salt treatment), $S_{150}S$ (150 µmol/L pretreatment + salt treatment), and $S_{450}S$ (450 µmol/L SA pretreatment + salt treatment). All dishes were incubated at 20 ± 0.5 °C for 7 days in an incubator under darkness; each treatment was repeated 3 times. The germination rate, mean germination time, length of the radicle and coleoptile, and dry weight of the radicle, coleoptile, and seed residue in each treatment were investigated to evaluate the effects of the exogenous application of SA during the grain-filling stage on the salt tolerance of offspring in wheat.

To explore the mechanisms of the SA-induced transgenerational salt tolerance in wheat, we designed four treatments: WW (water pretreatment + distilled water germination), WS (water pretreatment + NaCl treatment), SW (150 μmol/L SA pretreatment + distilled water germination), and SS (150 μmol/L SA pretreatment + NaCl treatment), and the seed samples were collected at 24, 48, and 72 hours after salt treatment for the biochemical and physiological measurements; each treatment was replicated in 20 Petri dishes. The seed surface sterilization, NaCl concentration, and germination conditions were consistent with those above.

2.3. Grain Yield and 1000-Kernel Weight

At maturity, 1 m^2 wheat plants from each treatment were harvested, threshed, and weighed. Three random samples were then selected to determine the 1000-kernel weight. The grain moisture content was measured using a FOSS-370 near-infrared grain analyzer, and both the actual yield and 1000-kernel weight were adjusted to a standard moisture content of 13%.

2.4. Germination Rate and Mean Germination Time

Germinated seeds are defined as those in which, by the seventh day of germination, the radicle has grown to the full length of the seed and the coleoptile has reached half the length of the seed. The percentage of seeds that germinate each day is recorded. Mean germination time (MGT) is calculated using the formula

$$\mathrm{MGT} = \sum(\mathrm{f} \times \mathrm{i})/\sum \mathrm{f} \quad (1)$$

where f is the number of newly germinated seeds on day i, and i is the day of germination. A higher MGT value indicates slower germination [26].

2.5. Dry and Fresh Weight of Germinating Seeds

On the seventh day of germination, the coleoptile, radicle, and seed remnants were separated. These parts were then dried at 105 °C for 30 min to terminate the enzyme activity, followed by drying at 80 °C until a constant weight was achieved to determine the dry weight.

2.6. Water Content Determination

During seed germination, the initial weight of the 30 seeds was measured before the process began. At intervals of 3, 6, 12, 24, 48, and 72 h, seeds were removed from the Petri dishes and blotted dry with a paper towel, and then immediately weighed. The seeds were then returned to the Petri dishes. The seed moisture content was calculated based on the fresh mass using the formula [27]:

$$\text{Water uptake (\%)} = [(\text{Fresh mass} - \text{Initial mass})/\text{Initial mass}] \times 100\% \quad (2)$$

where 'Initial mass' refers to the initial sample weight; 'Fresh mass' refers to the final sample weight at different time periods.

2.7. Starch and Soluble Sugars Contents

To extract starch, 1 g of germinated seed powder was mixed with 10 mL of 0.33 mM HCl and heated at 100 °C for 10 min. After extraction, 0.5 mL of 30% (*w*/*v*) $ZnSO_4$ was added to the extract to precipitate proteins. Subsequently, 0.5 mL of 30% (*w*/*v*) K_3 [Fe $(CN)_6$] was added to the extract, and the mixture was shaken to ensure thorough mixing, then adjusted to a final volume of 20 mL. After vigorous shaking and filtration, the starch content was analyzed using a polarimeter (WZZ-2B, Shanghai, China) at 20–25 °C, following the method of Xie et al. [28]. For the determination of soluble sugars, 0.1 g of dried germinated seed powder was extracted twice with 80% (*v*/*v*) ethanol at 80 °C. The soluble sugar content was measured using the anthrone method as described by Fares [29].

2.8. α-Amylase and β-Amylase Activities

The activities of α-amylase and β-amylase were determined using assay kits (Item No. DFMB-1-Y) provided by Suzhou Keming Biotechnology Co., Ltd. (Suzhou, China). A colorimetric method was employed, where the generation of reducing sugars was detected using the 3,5-dinitrosalicylic acid method. The spectrophotometric analysis was performed at a wavelength of 540 nm. All measurements strictly adhered to the relevant regulations of the "Chinese Pharmacopoeia" and were performed in strict accordance with the manufacturer's instructions.

2.9. ATP Content and Respiration Rate in Seeds

The ATP content was determined using an assay kit (Item No. ATP-1-Y) provided by Suzhou Keming Biotechnology Co., Ltd. (Suzhou, China). The determination was based on the reaction catalyzed by creatine kinase, where creatine and ATP react to produce phosphocreatine. The phosphocreatine content was detected at 700 nm using the phosphomolybdic acid colorimetric method to infer the ATP content. All measurements strictly adhered to the relevant regulations of the "Chinese Pharmacopoeia" and were performed in strict accordance with the manufacturer's instructions.

Thirty seeds were placed in a small basket and enclosed within a 50 mL wide-mouth jar. To measure the CO_2 produced during seed respiration, a saturated solution of barium hydroxide [$Ba(OH)_2$] was used to absorb the CO_2. The unreacted barium hydroxide was then quantified by titration with a standard solution of oxalic acid. The formula for calculating the seed respiration rate is:

$$\text{Seed respiration rate (mg CO}_2\text{ g}^{-1}\text{ FW h}^{-1}) = (V_0 - V_1)/(mt) \quad (3)$$

where m is the fresh mass of the seeds (g), t is the measurement time (h), V_0 is the amount of oxalic acid removed for blank titration (mL), and V_1 is the amount of oxalic acid removed for sample titration (mL).

2.10. Antioxidant Capacity

The malondialdehyde (MDA) content was measured following the method described by Cui et al. [30]. Samples were thoroughly ground in a phosphate buffer solution at pH 7.8 and then centrifuged at 6000 rpm for 10 min. A total of 2 mL of the supernatant was transferred to a graduated test tube, and 1 mL of 0.5% thiobarbituric acid and 3 mL of 5% trichloroacetic acid solution were added. The mixture was heated in a boiling water bath for 10 min and then rapidly cooled. After centrifugation at 6000 rpm for 10 min, the absorbance at 532 nm and 600 nm was measured using a spectrophotometer (UV-2450, Shimadzu Corp., Kyoto, Japan) with distilled water as the blank and transmittance set to 100%. The difference in absorbance between 532 nm and 600 nm was used to calculate the MDA content.

The determination of the ROS release rates was performed using assay kits (Item No. ROS-1-Y) from Suzhou Keming Biotechnology Co., Ltd. (Suzhou, China). The method employs a specific fluorescent probe, 2′,7′-dichlorodihydrofluorescin diacetate. Fluorescence intensity was measured within 10 min at a constant temperature of 37 °C, with an excitation wavelength of 499 nm and an emission wavelength of 521 nm. The rate of change in fluorescence intensity (u) was used as an indicator of the ROS release rate. The activity of peroxidase (POD) was measured using assay kits (Item No. POD-1-Y) from Suzhou Keming Biotechnology Co., Ltd. (Suzhou, China). POD catalyzes the oxidation of H_2O_2 by specific substrates, resulting in a characteristic light absorption at

470 nm. The activity of superoxide dismutase (SOD) was determined using assay kits (Item No. SOD-1-Y) from the same company. The xanthine and xanthine oxidase reaction system generates superoxide anion (O_2^-), which reduces nitroblue tetrazolium to form a blue formazan with light absorption at 560 nm. SOD scavenges O_2^-, thereby inhibiting the formation of formazan. All measurements strictly adhered to the relevant regulations of the "Chinese Pharmacopoeia" and were performed in strict accordance with the manufacturer's instructions.

2.11. K^+ and Na^+ Contents

Around 0.1 g of dried germinated seed powder was fully digested at 350 °C with a few drops of 30% H_2O_2 (v/v) and 5 mL of 98% HNO_3. The resulting solution was then analyzed using inductively coupled plasma-mass spectrometry (ICP-MS; Perkin Elmer, Optima 7300DV, Waltham, MA, USA).

2.12. Statistical Analysis

All data were analyzed using one-way analysis of variance (ANOVA) with the SPSS 10.0 software package (SPSS Inc., Chicago, IL, USA). Differences were considered significant between treatments when $p < 0.05$.

3. Results

3.1. Effects of Exogenous SA Treatment During the Grain-Filling Stage on Grain Weight and Yield in Wheat

The effect of the foliar application of SA during the grain-filling stage on the grain weight was in a concentration dependent manner in wheat (Figure 2). The application of 150 μM SA solution significantly increased the grain weight by 6.98% compared to the water treatment, while concentrations of 50 μM and 450 μM resulted in minor changes, with 0.57% and 4.05%, respectively. The response of the wheat grain yield to the SA treatments exhibited a similar trend to grain weight, with the grain yield at 150 μM SA increasing more than at 50 μM and 450 μM (Figure 2). These results indicate that applying appropriate concentrations of SA during the grain-filling stage could significantly increase the grain weight and yield in wheat.

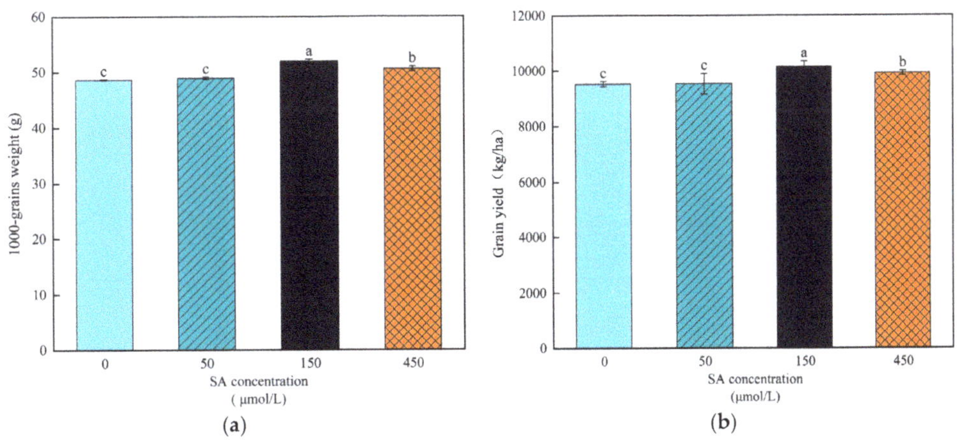

Figure 2. Effects of different SA concentrations on the wheat grain weight and yield. (**a**) Grain weight under different concentrations of SA. (**b**) Grain yield under different concentrations of SA. Each value is the mean ± SE of three biological replicates. The different lowercase letters indicate statistically significant differences at the $p < 0.05$ level.

3.2. Effects of Exogenous SA Treatment During the Grain Filling in Parental Wheat Plants on Seed Germination Under Salt Stress

As shown in Table 1, the WS treatment caused a significant decrease in the germination rate and an increase in the mean germination time compared to the WW treatment (Table 1). Exogenous SA treatments partially mitigated the salt stress caused the decrease in the germination rate and the increase in the mean germination time. Under the WS treatment, the length and dry mass of the radicle and coleoptile were markedly lower, and the dry mass of the seed residue was significantly higher than that under the WW treatment. However, the seeds of the SA-treated plants showed a higher length and dry mass of the radicle and coleoptile, and a lower dry mass of the seed residue compared to the seeds of the water-treated plants under the salt stress condition. These results suggest that the SA-treated wheat plants during the grain-filling stage could induce salt stress tolerance in their progeny. Among the different concentrations of SA treatment, 150 μmol /L SA ($S_{150}S$) was the most effective spraying concentration during the grain-filling stage to induce transgenerational salt tolerance (Table 1 and Figure 3). Therefore, the underlying mechanisms of SA-induced transgenerational salt tolerance were further explored under the concentration of 150 μmol/L SA.

Table 1. Effects of different SA concentrations on the germination and growth of the wheat progeny seeds under salt stress.

Treatment	Germination Rate (%)	Mean Germination Time (Day)	Length (cm)		Dry Weight (mg seed^{-1})		
			Coleoptile	Radicle	Seed Residue	Coleoptile	Radicle
WW	98.00 ± 0.02 a	3.27 ± 0.04 d	9.88 ± 0.13 a	7.76 ± 0.13 a	22.13 ± 0.46 d	7.67 ± 0.24 a	3.98 ± 0.19 a
WS	64.00 ± 0.03 d	3.72 ± 0.05 a	4.72 ± 0.09 e	4.37 ± 0.07 d	30.84 ± 0.41 a	4.37 ± 0.43 d	1.87 ± 0.28 c
S50S	73.33 ± 0.04 c	3.63 ± 0.07 ab	5.18 ± 0.12 d	4.59 ± 0.09 d	28.27 ± 0.83 b	5.28 ± 0.40 c	2.42 ± 0.07 b
S150S	80.00 ± 0.02 b	3.50 ± 0.08 c	5.78 ± 0.06 b	5.43 ± 0.12 b	24.60 ± 0.53 c	6.22 ± 0.10 b	2.56 ± 0.10 b
S450S	76.67 ± 0.03 bc	3.55 ± 0.06 bc	5.47 ± 0.14 c	5.03 ± 0.26 c	27.75 ± 0.59 b	5.58 ± 0.13 c	2.44 ± 0.12 b

Note: WW (water pretreatment + water germination), WS (water pretreatment + NaCl treatment), $S_{50}S$ (50 μmol/L SA pretreatment + NaCl treatment), $S_{150}S$ (150 μmol/L SA pretreatment + NaCl treatment), $S_{450}S$ (450 μmol/L SA pretreatment + NaCl treatment). Each value is the mean ± SE of three biological replicates. The different lowercase letters indicate statistically significant differences at the $p < 0.05$ level.

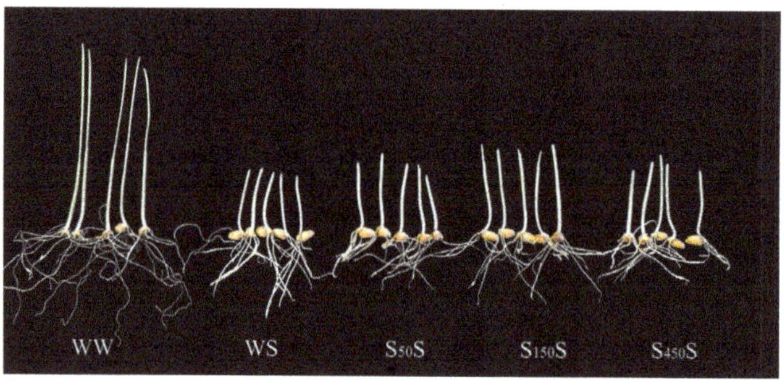

Figure 3. The phenotype of wheat plants at 7 d after germination. Note: WW (water pretreatment + water germination), WS (water pretreatment + NaCl treatment), $S_{50}S$ (50 μmol/L SA pretreatment + NaCl treatment), $S_{150}S$ (150 μmol/L SA pretreatment + NaCl treatment), $S_{450}S$ (450 μmol/L SA pretreatment + NaCl treatment).

3.3. Effects of Exogenous SA Treatment During the Grain Filling in Parental Wheat Plants on the Imbibition Rate of Progeny Seeds Under Salt Stress

As time progressed, the water content of the seeds showed an increasing trend. During the first 3, 6, 12, and 24 h of the experiment, there were no significant differences in the water uptake rates among the different treatment groups (Figure 4). This result indicates that the imbibition of the seeds was not affected by SA and the salt treatment. However, from the 48 h to the 72 h, the water uptake rate of the seeds under salt stress was significantly lower than that of the seeds germinated in water ($p < 0.05$). Notably, under both normal and saline germination conditions, seeds treated with SA showed varying degrees of increased water uptake compared to those treated with water.

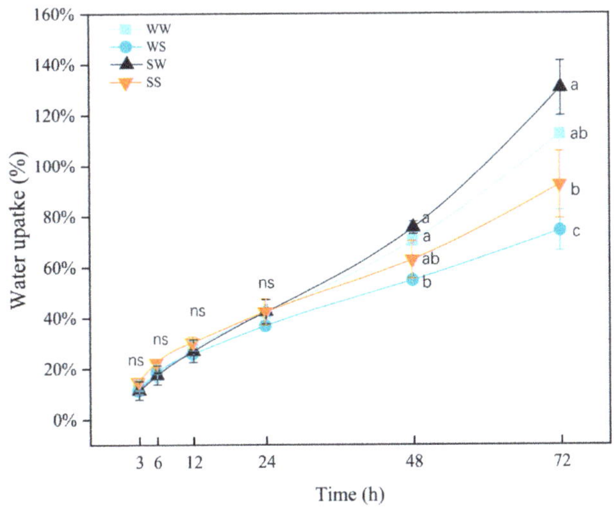

Figure 4. Effects of SA treatment on the water uptake rate of the wheat progeny seeds under salt stress. Note: WW (water pretreatment + water germination), WS (water pretreatment + NaCl treatment), SW (150 μmol/L SA pretreatment + water germination), SS (150 μmol/L SA pretreatment + NaCl treatment). Each value is the mean ± SE of three biological replicates. The different lowercase letters indicate statistically significant differences at the $p < 0.05$ level, "ns" represents no significant difference between the treatments.

3.4. *Effects of Exogenous SA Treatment During the Grain Filling in Parental Plants on the Degradation of Starch in Progeny Seeds Under Salt Stress*

Starch content gradually decreased while total soluble sugar content increased along with germination progress in wheat (Figure 5). Salt stress (WS) significantly increased the starch content and decreased the total soluble sugar content at 24 h, 48 h, and 72 h compared to the control treatment (WW). The seeds of plants treated with SA (SW and SS treatment) showed a lower starch content and higher total soluble sugar content than the seeds of the water-treated plants under normal (WW) and salt (WS) germination conditions, respectively.

The activity of α-amylase gradually increased, whereas the activity of β-amylase decreased during seed germination (Figure 5). The WS treatment significantly reduced the activities of both α-amylase and β-amylase in the seeds compared to the WW treatment, but the SS treatment significantly increased the activities of these enzymes related to the WS treatment. Under normal germination conditions, the seeds from the SA-treated plants showed a higher α-amylase activity at 48 h and 72 h than the seeds from the water-treated plants, but there was no significant difference in the activity of β-amylase (Figure 5).

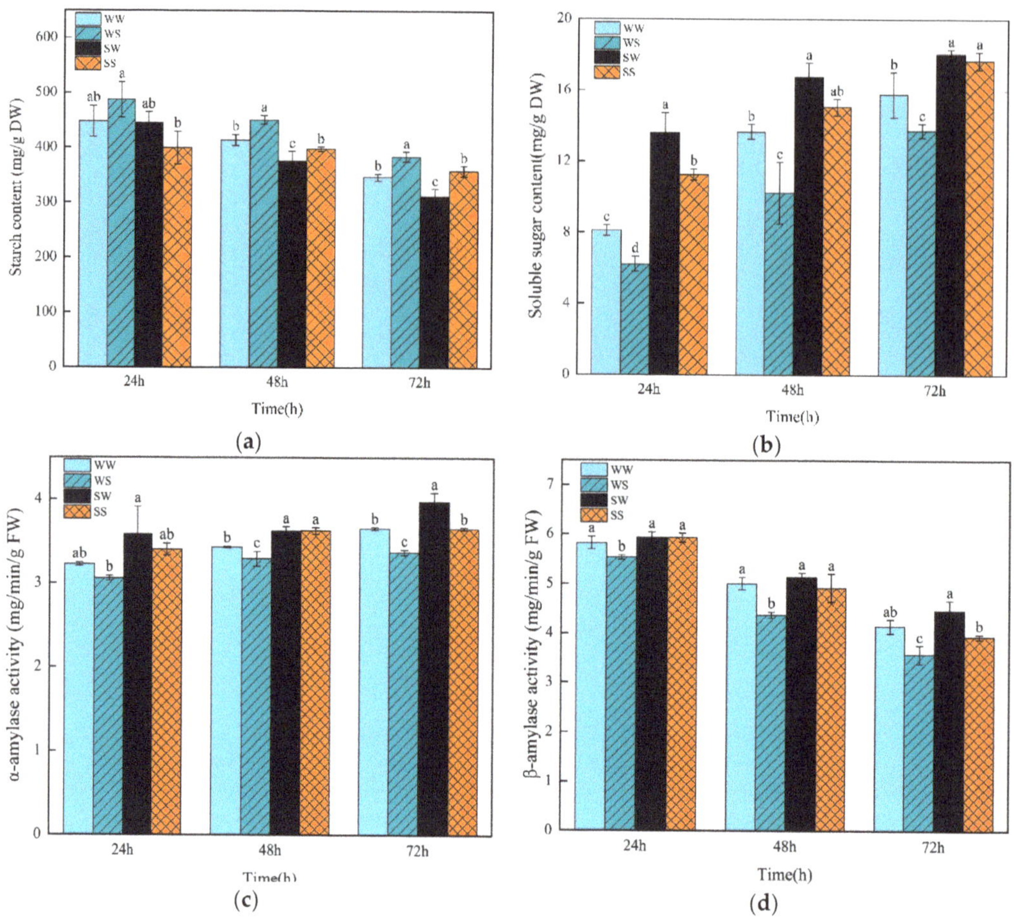

Figure 5. Effects of SA treatment on starch degradation in the wheat progeny seeds under salt stress. (**a**) Starch content; (**b**) soluble sugar content; (**c**) α-amylase activity; (**d**) β-amylase activity. WW (water pretreatment + water germination), WS (water pretreatment + NaCl treatment), SW (150 μmol/L SA pretreatment + water germination), SS (150 μmol/L SA pretreatment + NaCl treatment). Each value is the mean ± SE of three biological replicates. The different lowercase letters indicate statistically significant differences at the $p < 0.05$ level.

3.5. Effects of Exogenous SA Treatment During the Grain Filling in Parental Wheat Plants on ATP Content and Respiration Rate in Progeny Seeds Under Salt Stress

During the germination of wheat seeds, both the respiration rate and ATP content tended to increase over time (Figure 6). Salt stress significantly increased the respiration rate whereas decreased the ATP content compared to the control treatment. The SS treatment significantly increased the ATP content and respiration rate of the seeds in relation to the WS treatment. Specifically, the ATP content increased by 16.29%, 26.90%, and 19.07% at 24, 48, and 72 h, respectively, while the respiration rate increased by 14.17% and 9.71% at 48 and 72 h, respectively. There were no significant differences in respiration rate and ATP content between the treatment of WW and SW during germination (Figure 6).

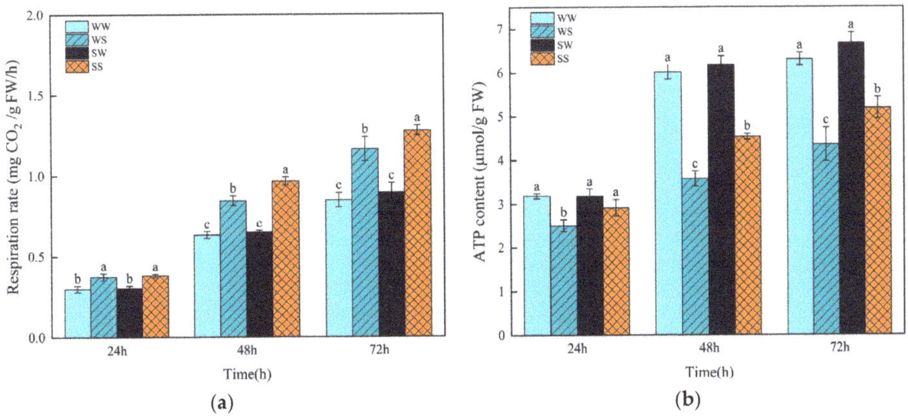

Figure 6. Effects of SA treatment on ATP content and respiration rate in the wheat progeny seeds under salt stress. (**a**) Respiration rate; (**b**) ATP content. WW (water pretreatment + water germination), WS (water pretreatment + NaCl treatment), SW (150 μmol/L SA pretreatment + water germination), SS (150 μmol/L SA pretreatment + NaCl treatment). Each value is the mean ± SE of three biological replicates. The different lowercase letters indicate statistically significant differences at the $p < 0.05$ level.

3.6. Effects of Exogenous SA Treatment During the Grain Filling in Parental Plants on MDA Content and ROS Release Rate in Progeny Seeds Under Salt Stress

The release rate of ROS and the content of MDA increased during the germination process of wheat seeds. There were no significant differences in the release rate of ROS and the content of MDA between the WW and SW treatment, except in the release rate of ROS at 72 h after germination, where the SW treatment showed significant lower (Figure 7). The WS treatment obviously increased the release rate of ROS and the content of MDA compared with the WW treatment. Compared with the WS treatment, SS treatment significantly reduced the release rate of ROS by 13.49%, 13.57%, and 14.25% at 24, 48, and 72 h after germination, respectively. There was no significant difference in the MDA content between the WS and SS treatment at 24 h after germination, but SS showed a much lower MDA content at 48 h (22.83%) and 72 h (18.85%) after germination.

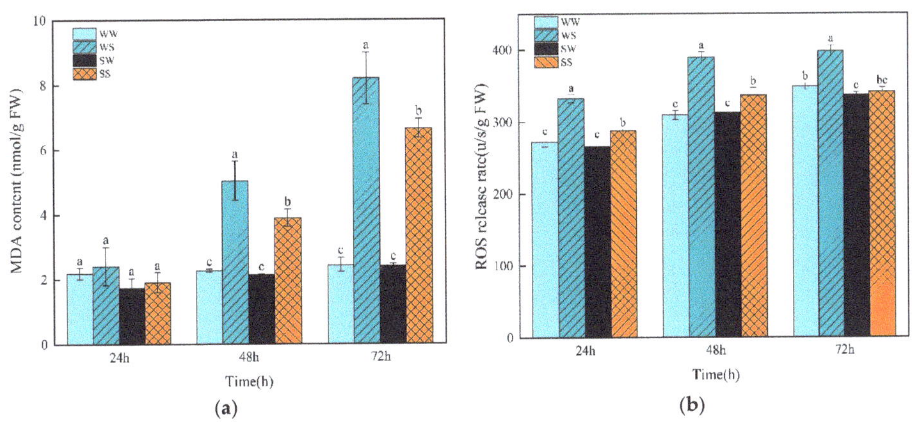

Figure 7. Effects of SA treatment on the MDA content and ROS release rate in the wheat progeny seeds under salt stress. (**a**) MDA content; (**b**) ROS release rate. WW (water pretreatment + water germination), WS (water pretreatment + NaCl treatment), SW (150 μmol/L SA pretreatment + water germination), SS (150 μmol/L SA pretreatment + NaCl treatment). Each value is the mean ± SE of three biological replicates. The different lowercase letters indicate statistically significant differences at the $p < 0.05$ level.

3.7. Effects of Exogenous SA Treatment During the Grain Filling in Parental Plants on the Activities of Antioxidant Enzymes in Progeny Seeds Under Salt Stress

The activities of SOD and POD increased during the germination of wheat seeds (Figure 8). Compared with the WW treatment, the activities of SOD and POD were significantly inhibited by the WS treatment at 24 h, 48 h, and 72 h (Figure 8). The SS treatment showed significantly higher SOD activity at 24 h (11.10%), 48 h (8.21%), and 72 h (26.59%), and a higher POD activity at 48 h (20.60%) than those in the WS treatment. The activity of SOD was much higher in the SW treatment at 48 h and 72 h after germination compared to the WW treatment. In addition, the SW treatment markedly increased the POD activity at 72 h after treatment relative to the WW treatment (Figure 8).

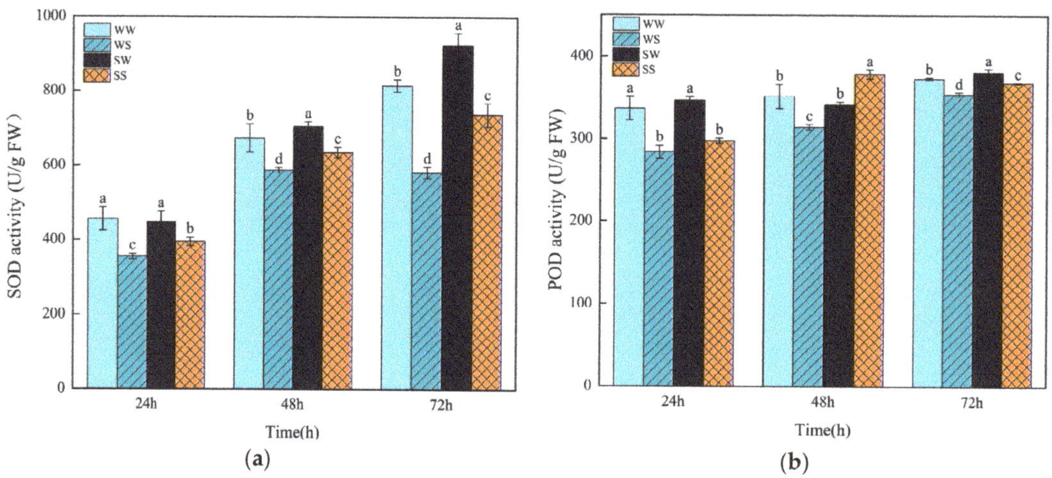

Figure 8. Effects of SA treatment on the SOD and POD activities in the wheat progeny seeds under salt stress. (**a**) SOD activity (**b**): POD activity. WW (water pretreatment + water germination), WS (water pretreatment + NaCl treatment), SW (150 μmol/L SA pretreatment + water germination), SS (150 μmol/L SA pretreatment + NaCl treatment). Each value is the mean ± SE of three biological replicates. The different lowercase letters indicate statistically significant differences at the $p < 0.05$ level.

3.8. Effects of Exogenous SA Treatment During the Grain Filling in Parental Plants on Sodium (Na^+) and Potassium (K^+) Ion Contents in Progeny Seeds Under Salt Stress

The WS treatment strikingly increased the Na^+ content while it decreased the K^+ content at 24 h, 48 h, and 72 h after germination relative to the WW treatment (Figure 9). Compared to the WS treatment, the SS treatment significantly reduced the Na^+ content by 27.13%, 35.65%, and 39.66% and increased the K^+ content by 5.58%, 11.00%, and 10.48% at 24, 48, and 72 h after germination, respectively. There were no significant differences in the contents of Na^+ and K^+ between the treatments of WW and SW.

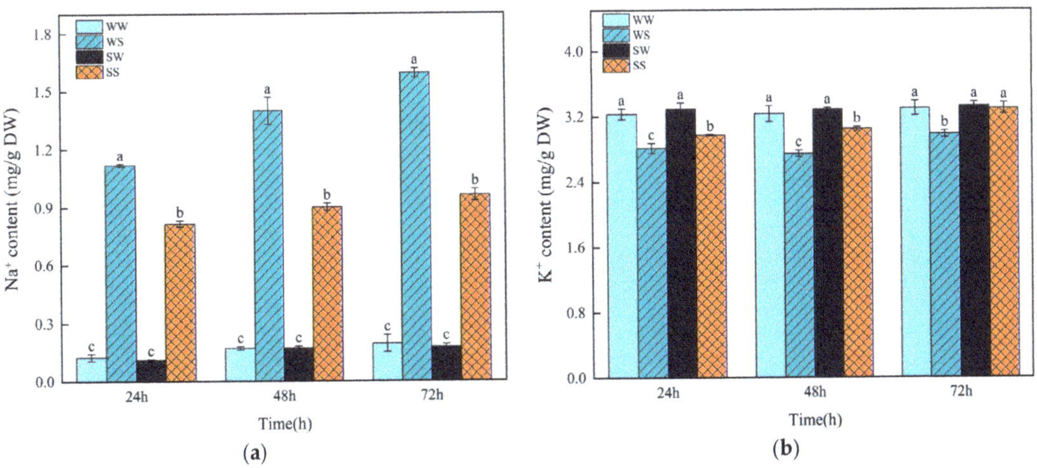

Figure 9. Effects of SA treatment on the Na^+ and K^+ contents in the wheat progeny seeds under salt stress. (**a**) Na^+ content; (**b**) K^+ content. WW (water pretreatment + water germination), WS (water pretreatment + NaCl treatment), SW (150 μmol/L SA pretreatment + water germination), SS (150 μmol/L SA pretreatment + NaCl treatment). Each value is the mean ± SE of three biological replicates. The different lowercase letters indicate statistically significant differences at the $p < 0.05$ level.

4. Discussion

Salicylic acid is a multifunctional signal substance in plants that plays an important role in regulating plant growth and development and stress responses [31]. Previous studies have shown that the foliar application of SA at early or later stages can significantly increase the grain yield of wheat [32,33]. In this study, the application of exogenous SA during the grain-filling stage significantly increased the grain yield of wheat, which was mainly due to the increase in grain weight (Figure 2). It has been documented that applying appropriate concentrations of SA could enhance the nitrogen use efficiency, leaf chlorophyll content, and photosynthetic efficiency in plants [34,35]. These results suggest that SA treatment increased the grain weight and yield of wheat by improving photosynthetic efficiency. In addition, the effects of SA on grain weight and yield were in a dose-dependent manner, and 150 μM SA was more effective than other concentrations in this study. This finding is consistent with the results of previous studies that reported that the application concentration is an important factor affecting the regulatory effects of SA on senescence and stress tolerance in plants [36].

Although the important role of endogenous SA in biotic stress induced trans-generational stress tolerance has been recorded [37,38], there are few studies on the effect of exogenous SA on offspring tolerance in plants. Numerous studies have shown that exogenous SA treatment could effectively increase the contemporary salt stress tolerance in many species including wheat [17,39,40]. In the present study, the exogenous application of SA during the grain-filling stage significantly enhanced the salt tolerance of the progeny seeds in wheat, as demonstrated by the higher germination rate, length and dry mass of the coleoptile and radicle, and a lower mean germination time compared to the seeds of the water-treated plants under the salt germination condition (Table 1). These results were consistent with findings that a SA soaking treatment promotes seed germination under salinity stress in crops including barely, rice, and wheat [20,41]. However, to our knowledge, this is the first report that the application of SA can induce trans-generational tolerance to abiotic stress in plants.

The germination process can be considered in terms of three sequential steps: imbibition, reserve mobilization, and radicle and coleoptile elongation [42]. The process of wheat seed imbibition is usually completed within 12 h [43]. In this study, SA and salt stress treatment did not affect the increase in fresh weight in the wheat seeds with 24 h after germination (Figure 4), indicating that the shorter germination time of seeds from the SA-treated plants under salt stress was not caused by enhanced seed imbibition. However, the seeds of SA-treated plants showed a greater increase in fresh weight at 48 h and 72 h after germination compared to the seeds of water-treated plants under salt stress (Figure 4). This result was associated with the higher soluble sugar content in the seeds of SA-treated plants than the seeds of water-treated plants under salt stress (Figures 4 and 5). Soluble sugar plays a crucial role in regulating osmotic balance in plants [44]. These results suggest that the seeds from SA-treated plants could accumulate more osmoregulatory substance to alleviate osmotic damage caused by salt stress.

Starch stored in the endosperm is hydrolyzed into soluble sugars, which can be further converted into energy and intermediate metabolites for seed germination. Previous studies have shown that salt stress inhibited the mobilizing of starch by reducing hydrolase activity, thereby delaying seed germination [10]. A recent study found that SA soaking treatment significantly increased the activity of α-amylase and the content of soluble sugar in rice seeds under salt stress [20]. In the present study, the seeds from the SA-treated plants showed higher hydrolase (α-amylase and β-amylase) activities and lower starch content compared to the seeds from the water-treated plants under salt stress, indicating that the seeds from SA-treated plants could better utilize starch in the endosperm, thus maintaining a higher soluble sugar content and water absorption rate under salt stress. However, high concentrations of SA inhibited the activity of amylase and seed germination [45]. From this perspective, the treatment method of spraying SA during the grain-filling stage to promote seed salt tolerance might be safer compared to SA soaking.

In this study, in addition to inhibiting starch hydrolysis, salt stress also significantly reduced the ATP content in wheat and increased the respiration rate of the wheat seeds. This result is consistent with previous reports that stated that salt stress increases seed respiration [46–48]. The increasing respiration rate was beneficial for producing more energy in response to stresses. However, the excessive energy consumption under salt stress might be the reason for the decrease in ATP content in the wheat seeds in the present study. Zapata [49] and others found that the respiration rates of lettuce, melon, and pepper seeds increased with rising salinity, but the salinity of spinach did not increase. This result suggests that the effect of salt on seed respiration rates may be related to the species' sensitivity to salt stress. Under salt stress, seeds from the SA-treated plants exhibited significantly higher respiration rates and ATP content compared to those from the water-treated plants (Figure 6), indicating that seeds from the SA-treated plants were able to maintain sufficient energy levels for seed germination under salt stress.

In addition to causing osmotic damage, salinity also induces an imbalance of the inner cellular ions, thus leading to ion toxicity in plant cells [50]. In this study, salt stress (NaCl treatment) dramatically increased the concentration of Na^+ while reducing the concentration of K^+ in the wheat seeds (Figure 9). These results are consistent with previous findings that plants absorb excessive Na^+ as the amount of Na^+ in the growth medium increases, resulting in an increase in K^+ efflux from cells [51]. However, SA treatment during the grain-filling stage significantly alleviated the changes in Na^+ and K^+ concentrations compared to the water treatment under the salt germination condition (Figure 9). This result suggests that SA could improve the wheat seed salt tolerance through maintaining ion homeostasis. Liu [20] et al. (2022) found that a SA soaking treatment upregulated the expression levels of *OsHKTs* in rice seeds, which promoted Na^+ efflux, thus maintaining ion balance under salt stress.

Reactive oxygen species are highly reactive and toxic; their overaccumulation can lead to the oxidative destruction of nucleic acids, proteins, lipids, etc. [52]. MDA is the product of the membrane lipid oxidation reaction and is often used to reflect the degree of cellular peroxidation damage under stresses [6]. In this study, salt stress significantly increased the rate of ROS production and MDA content in the wheat seeds, whereas SA pretreatment effectively relieved these changes (Figure 7). Similar observations have been recorded in rice, wheat, and *Arabidopsis thalianan*, where exogenous SA enhanced the salt tolerance by activating antioxidant systems under salt stress [53–56]. SOD and POD are two important enzymes in the plant antioxidant system [57]. Spraying SA during the grain-filling stage significantly increased the activities of SOD and POD in the progeny seeds under salt stress in this study (Figure 8), indicating that SOD and POD are involved in the enhancement of antioxidant capacity in the seeds of the SA-treated plants under salt stress. In addition, exogenous SA treatment increased the contents of glutathione and ascorbate in salt-stressed wheat seedlings [58]. These results indicate that enhancing the antioxidant capacity is an important mechanism by which SA enhances plant tolerance to salt stress.

Previous studies have found that exogenous SA treatment could improve the salt tolerance of seeds and plants in many crops in the current generation [19,20,22]. In this study, we found that the application of SA during the grain-filling stage in wheat plants could effectively improve the progeny seeds' salt tolerance by maintaining osmotic, ion, and redox homeostasis in cells under salt stress. Spraying SA during the grain-filling stage may have an impact on the formation of wheat seeds such as the content of endogenous SA in seeds, which in turn affects the salt tolerance of seeds. It has been documented that SA and its analogue induced long-term priming and transcriptional memory in plant immunity through the DNA methylation pathway [59]. For example, BTH (an SA analogue) significantly upregulated the expression level of *AMY1* (*amylase 1*) through DNA methylation in *Arabidopsis* [60]. Consistent with this result, in the present study, the seeds from the SA-treated plants exhibited higher activities of α-amylase and β-amylase under salt stress compared to the seeds from the water-treated plants (Figure 5). Thus, further investigation is needed to determine whether epigenetic modifications mediate SA-induced salt tolerance in wheat seeds across generations, as recorded in biotic stress [61]. Based on the results of this study, SA application may improve seed yield. The added expense of SA treatment for general agricultural production may be greater than the value of the yield increase. However, the value-added trait of seeds epigenetically primed with enhanced stress tolerance may be worth the expense of SA application in the field by growers producing seeds to sell to farmers for the next growing season.

Author Contributions: Methodology, C.Z.; Software, X.J. and Y.D.; Investigation, Y.Z.; Writing—original draft, L.Y.; Writing—review & editing, W.W.; Supervision, K.X. and Z.H. All authors have read and agreed to the published version of the manuscript.

Funding: This research was funded by Major core technology innovation projects of Jiangsu Province Independent Innovation Fund (CX(22)1001) and Lv Yang Jin Feng Research Plan and a project funded by the Priority Academic Program Development of Jiangsu Higher Education Institutions (PAPD).

Data Availability Statement: Data are contained within the article.

Conflicts of Interest: The authors declare no conflict of interest.

References

1. Kettlewell, P.; Byrne, R.; Jeffery, S. Wheat Area Expansion into Northern Higher Latitudes and Global Food Security. *Agric. Ecosyst. Environ.* **2023**, *351*, 108499. [CrossRef]
2. Zhao, C.; Zhang, H.; Song, C.; Zhu, J.-K.; Shabala, S. Mechanisms of Plant Responses and Adaptation to Soil Salinity. *Innovation* **2020**, *1*, 100017. [CrossRef]
3. Chen, D.; Yin, L.; Deng, X.; Wang, S. Silicon Increases Salt Tolerance by Influencing the Two-Phase Growth Response to Salinity in Wheat (*Triticum aestivum* L.). *Acta Physiol. Plant.* **2014**, *36*, 2531–2535. [CrossRef]
4. Zheng, L.; Ma, H.Y.; Jiao, Q.Q.; Ma, C.L.; Wang, P.P. Phytohormones: Important Participators in Plant Salt Tolerance. *Int. J. Agric. Biol.* **2020**, *24*, 319–332.
5. Jumaah, W.N.; Nabi, R.B.S.; Rolly, N.K.; Adamu, T.A.; Tayade, R.; Aye, N.C.; Hussain, A.; Yun, B.-W. Nitric Oxide Mediated Growth Enhancement of Tomato under Salinity Stress. *Plant Biotechnol. Rep.* **2024**, *18*, 437–446. [CrossRef]
6. Li, H.; Sun, H.; Ping, W.; Liu, L.; Zhang, Y.; Zhang, K.; Bai, Z.; Li, A.; Zhu, J.; Li, C. Exogenous Ethylene Promotes the Germination of Cotton Seeds under Salt Stress. *J. Plant Growth Regul.* **2023**, *42*, 3923–3933. [CrossRef]
7. Hameed, A.; Rasheed, A.; Gul, B.; Khan, M. Salinity Inhibits Seed Germination of Perennial Halophytes *Limonium stocksii* and *Suaeda fruticosa* by Reducing Water Uptake and Ascorbate Dependent Antioxidant System. *Environ. Exp. Bot.* **2014**, *107*, 32–38. [CrossRef]

8. Wang, J.; Lv, P.; Yan, D.; Zhang, Z.; Xu, X.; Wang, T.; Wang, Y.; Peng, Z.; Yu, C.; Gao, Y.; et al. Exogenous Melatonin Improves Seed Germination of Wheat (*Triticum aestivum* L.) under Salt Stress. *Int. J. Mol. Sci.* **2022**, *23*, 8436. [CrossRef]
9. Xia, K.; Liu, A.; Wang, Y.; Yang, W.; Jin, Y. Mechanism of Salt-Inhibited Early Seed Germination Analysed by Transcriptomic Sequencing. *Seed Sci. Res.* **2019**, *29*, 73–84. [CrossRef]
10. Hajihashemi, S.; Skalicky, M.; Brestic, M.; Pavla, V. Cross-Talk between Nitric Oxide, Hydrogen Peroxide and Calcium in Salt-Stressed *Chenopodium quinoa* Willd. At Seed Germination Stage. *Plant Physiol. Biochem.* **2020**, *154*, 657–664. [CrossRef]
11. Kumar, A.; Rodrigues, V.; Verma, S.; Singh, M.; Hiremath, C.; Shanker, K.; Shukla, A.K.; Sundaresan, V. Effect of Salt Stress on Seed Germination, Morphology, Biochemical Parameters, Genomic Template Stability, and Bioactive Constituents of *Andrographis paniculata* Nees. *Acta Physiol. Plant.* **2021**, *43*, 68. [CrossRef]
12. Zhu, J.K. Regulation of Ion Homeostasis under Salt Stress. *Curr. Opin. Plant Biol.* **2003**, *6*, 441–445. [CrossRef] [PubMed]
13. Molinier, J.; Ries, G.; Zipfel, C.; Hohn, B. Transgeneration Memory of Stress in Plants. *Nature* **2006**, *442*, 1046–1049. [CrossRef]
14. Robinson, G.I.; Robinson, J.W. Digest: Transgenerational Stress Memory Mechanisms in *Arabidopsis thaliana*. *Evolution* **2020**, *74*, 2423–2424. [CrossRef] [PubMed]
15. Feng, K.; Wang, X.; Zhou, Q.; Dai, T.; Cao, W.; Jiang, D.; Cai, J. Waterlogging Priming Enhances Hypoxia Stress Tolerance of Wheat Offspring Plants by Regulating Root Phenotypic and Physiological Adaption. *Plants* **2022**, *11*, 1969. [CrossRef] [PubMed]
16. Youngson, N.A.; Whitelaw, E. Transgenerational Epigenetic Effects. *Annu. Rev. Genomics Hum. Genet.* **2008**, *9*, 233–257. [CrossRef]
17. Shakirova, F.M.; Sakhabutdinova, A.R.; Bezrukova, M.V.; Fatkhutdinova, R.A.; Fatkhutdinova, D.R. Changes in the Hormonal Status of Wheat Seedlings Induced by Salicylic Acid and Salinity. *Plant Sci.* **2003**, *164*, 317–322. [CrossRef]
18. Song, W.; Shao, H.; Zheng, A.; Zhao, L.; Xu, Y. Advances in Roles of Salicylic Acid in Plant Tolerance Responses to Biotic and Abiotic Stresses. *Plants* **2023**, *12*, 3475. [CrossRef]
19. Jia, X.; Jiang, Z. Salicylic Acid Alleviates Salt Stress in Soybean by Improving Anti-Oxidant Capacity and Growth Performance. *Pak. J. Bot.* **2023**, *55*, 37–44. [CrossRef] [PubMed]
20. Liu, Z.; Ma, C.; Hou, L.; Wu, X.; Wang, D.; Zhang, L.; Liu, P. Exogenous SA Affects Rice Seed Germination under Salt Stress by Regulating Na+/K+ Balance and Endogenous GAs and ABA Homeostasis. *Int. J. Mol. Sci.* **2022**, *23*, 3293. [CrossRef]
21. Wang, Z.; Dong, S.; Teng, K.; Chang, Z.; Zhang, X. Exogenous Salicylic Acid Optimizes Photosynthesis, Antioxidant Metabolism, and Gene Expression in Perennial Ryegrass Subjected to Salt Stress. *Agronomy* **2022**, *12*, 1920. [CrossRef]
22. Liu, J.; Li, L.; Yuan, F.; Chen, M. Exogenous Salicylic Acid Improves the Germination of Limonium Bicolor Seeds under Salt Stress. *Plant Signal. Behav.* **2019**, *14*, e1644595. [CrossRef] [PubMed]
23. de Sousa, T.; Ribeiro, M.; Sabenca, C.; Igrejas, G. The 10,000-Year Success Story of Wheat! *Foods* **2021**, *10*, 2124. [CrossRef] [PubMed]
24. Li, S.Q.; Shao, M.A.; Li, Z.Y.; Wu, W.M.; Zhang, X.C. Research Progress on Grain Filling Characteristics and Influencing Factors of Wheat. *Acta Bot. Bor.-Occid. Sin.* **2003**, *23*, 2030–2038.
25. Liu, H.Y.; Guo, T.C.; Zhu, Y.J.; Wang, C.Y.; Ma, D.Y.; Liu, S.B.; Yue, C.F.; Yang, W.P.; Wang, H.H. Effects of salicylic acid spraying at flowering stage on grain starch and yield of different types of special wheat varieties. *J. Triticeae Crops* **2006**, *4*, 123–127.
26. Ellis, R.; Roberts, E.H. *Towards a Rational Basis for Testing Seed Quality*; Butterworths: London, UK, 1980.
27. Brzezinski, C.R.; Abati, J.; Zucareli, C.; Krzyzanowski, F.C.; França-Neto, J.d.B.; Henning, F.A. Water Uptake of Soybean Pods and Seeds with Different Lignin Contents. *Rev. Ceres* **2022**, *69*, 476–482. [CrossRef]
28. Xie, Z.; Jiang, D.; Cao, W.; Dai, T.; Jing, Q. Relationships of Endogenous Plant Hormones to Accumulation of Grain Protein and Starch in Winter Wheat under Different Post-Anthesis Soil Water Statusses. *Plant Growth Regul.* **2003**, *41*, 117–127. [CrossRef]
29. Fales, F.W. The assimilation and degradation of carbohydrates by yeast cells. *J. Biol. Chem.* **1951**, *193*, 113–124. [CrossRef]
30. Cui, G.; Zhao, X.; Liu, S.; Sun, F.; Zhang, C.; Xi, Y. Beneficial Effects of Melatonin in Overcoming Drought Stress in Wheat Seedlings. *Plant Physiol. Biochem.* **2017**, *118*, 138–149. [CrossRef]
31. Chen, S.; Zhao, C.-B.; Ren, R.-M.; Jiang, J.-H. Salicylic Acid Had the Potential to Enhance Tolerance in Horticultural Crops against Abiotic Stress. *Front. Plant Sci.* **2023**, *14*, 1141918. [CrossRef]
32. Matysiak, K.; Siatkowski, I.; Kierzek, R.; Kowalska, J.; Krawczyk, R. Effect of Foliar Applied Acetylsalicilic Acid on Wheat (*Triticum aestivum* L.) under Field Conditions. *Agron* **2020**, *10*, 1918. [CrossRef]
33. Ibrahim, O.M.; Bakry, B.A.; Thalooth, A.T.; El-Karamany, M.F. Influence of Nitrogen Fertilizer and Foliar Application of Salicylic Acid on Wheat. *Agric. Sci.* **2014**, *5*, 1316–1321.
34. Kumar, S.; Kumar, C.; Bose, B. Effects of Salicylic Acid on Seedling Growth and Nitrogen Metabolism in Cucumber (*Cucumis sativus* L.). *J. Stress Physiol. Biochem.* **2010**, *6*, 102–113.
35. Janda, T.; Kinga, B.; Yordanova, R.; Szalai, G.; Pál, M. Salicylic Acid and Photosynthesis: Signalling and Effects. *Acta Physiol. Plant.* **2014**, *36*, 2537–2546. [CrossRef]
36. Miura, K.; Tada, Y. Regulation of Water, Salinity, and Cold Stress Responses by Salicylic Acid. *Front. Plant Sci.* **2014**, *5*, 4. [CrossRef] [PubMed]
37. Singh, D.P.; Moore, C.A.; Gilliland, A.; Carr, J.P. Activation of Multiple Antiviral Defence Mechanisms by Salicylic Acid. *Mol. Plant Pathol.* **2004**, *5*, 57–63.
38. White, R.F. Acetylsalicylic Acid (Aspirin) Induces Resistance to Tobacco Mosaic Virus in Tobacco. *Virology* **1979**, *99*, 410–412. [PubMed]

39. Bastam, N.; Baninasab, B.; Ghobadi, C. Improving Salt Tolerance by Exogenous Application of Salicylic Acid in Seedlings of Pistachio. *Plant Growth Regul.* **2013**, *69*, 275–284.
40. Liu, S.; Dong, Y.; Xu, L.; Kong, J. Effects of Foliar Applications of Nitric Oxide and Salicylic Acid on Salt-Induced Changes in Photosynthesis and Antioxidative Metabolism of Cotton Seedlings. *Plant Growth Regul.* **2014**, *73*, 67–78.
41. Maqsood, M.F.; Shahbaz, M.; Zulfiqar, U.; Saman, R.U.; Rehman, A.; Naz, N.; Akram, M.; Haider, F.U. Enhancing Wheat Growth and Yield through Salicylic Acid-Mediated Regulation of Gas Exchange, Antioxidant Defense, and Osmoprotection under Salt Stress. *Stresses* **2023**, *3*, 372–386. [CrossRef]
42. Nautiyal, P.C.; Sivasubramaniam, K.; Dadlani, M. Seed Dormancy and Regulation of Germination. In *Seed Science and Technology, Biology, Production, Quality*; Springer: Berlin/Heidelberg, Germany, 2023; pp. 39–66.
43. Dong, K.; Zhen, S.; Cheng, Z.; Cao, H.; Ge, P.; Yan, Y. Proteomic Analysis Reveals Key Proteins and Phosphoproteins upon Seed Germination of Wheat (*Triticum aestivum* L.). *Front. Plant Sci.* **2015**, *6*, 1017. [CrossRef] [PubMed]
44. Sami, F.; Yusuf, M.; Faizan, M.; Faraz, A.; Hayat, S. Role of Sugars under Abiotic Stress. *Plant Physiol. Biochem.* **2016**, *109*, 54–61. [CrossRef]
45. Xie, Z.; Zhang, Z.-L.; Hanzlik, S.; Cook, E.; Shen, Q.J. Salicylic Acid Inhibits Gibberellin-Induced Alpha-Amylase Expression and Seed Germination via a Pathway Involving an Abscisic-Acid-Inducible *WRKY* Gene. *Plant Mol. Biol.* **2007**, *64*, 293–303. [CrossRef] [PubMed]
46. Che-Othman, M.H.; Jacoby, R.P.; Millar, A.H.; Taylor, N.L. Wheat Mitochondrial Respiration Shifts from the Tricarboxylic Acid Cycle to the GABA Shunt under Salt Stress. *New Phytol.* **2020**, *225*, 1166–1180. [CrossRef] [PubMed]
47. Livne, A.; Levin, N. Tissue Respiration and Mitochondrial Oxidative Phosphorylation of NaCl-Treated Pea Seedlings. *Plant Physiol.* **1967**, *42*, 407–414. [CrossRef] [PubMed]
48. Robertson, R.N.; Turner, J.S.; Wilkins, M.J. Studies in the Metabolism of Plant Cells; Salt Respiration and Accumulation in Red Beet Tissue. *Aust. J. Exp. Biol. Med. Sci.* **1947**, *25*, 1–8. [CrossRef] [PubMed]
49. Zapata, P.J.; Serrano, M.; Pretel, A.T.; Amorós, A.; Botella, M.A. Polyamines and Ethylene Changes during Germination of Different Plant Species under Salinity. *Plant Sci.* **2004**, *167*, 781–788. [CrossRef]
50. Balasubramaniam, T.; Shen, G.; Esmaeili, N.; Zhang, H. Plants' Response Mechanisms to Salinity Stress. *Plants* **2023**, *12*, 2253. [CrossRef] [PubMed]
51. Parvin, K.; Ahamed, K.; Mahbub Islam, M.; Haque, M. Modulation of Ion Uptake in Tomato (*Lycopersicon esculentum* L.) Plants with Exogenous Application of Calcium under Salt Stress Condition. *Poljoprivreda* **2016**, *22*, 40–49. [CrossRef]
52. Komatsu, S.; Diniyah, A.; Zhu, W.; Nakano, M.; Rehman, S.U.; Yamaguchi, H.; Hitachi, K.; Tsuchida, K. Metabolomic and Proteomic Analyses to Reveal the Role of Plant-Derived Smoke Solution on Wheat under Salt Stress. *Int. J. Mol. Sci.* **2024**, *25*, 8216. [CrossRef]
53. Lee, S.; Kim, S.-G.; Park, C.-M. Salicylic Acid Promotes Seed Germination under High Salinity by Modulating Antioxidant Activity in Arabidopsis. *New Phytol.* **2010**, *188*, 626–637. [CrossRef]
54. Singh, A.; Roychoudhury, A. Salicylic Acid-Mediated Alleviation of Fluoride Toxicity in Rice by Restricting Fluoride Bioaccumulation and Strengthening the Osmolyte, Antioxidant and Glyoxalase Systems. *Environ. Sci. Pollut. Res.* **2023**, *30*, 25024–25036. [CrossRef] [PubMed]
55. Talaat, N.B.; Mahmoud, A.W.M.; Hanafy, A.M.A. Co-Application of Salicylic Acid and Spermine Alleviates Salt Stress Toxicity in Wheat: Growth, Nutrient Acquisition, Osmolytes Accumulation, and Antioxidant Response. *Acta Physiol. Plant.* **2023**, *45*, 1. [CrossRef]
56. Yu, L.-L.; Liu, Y.; Zhu, F.; Geng, X.-X.; Yang, Y.; He, Z.-Q.; Xu, F. The Enhancement of Salt Stress Tolerance by Salicylic Acid Pretreatment in *Arabidopsis thaliana*. *Biol. Plant.* **2020**, *64*, 150–158. [CrossRef]
57. Abedini, M.; Hassani, B.D. Salicylic Acid Affects Wheat Cultivars Antioxidant System under Saline and Non-Saline Condition. *Russ. J. Plant Physiol.* **2015**, *62*, 604–610. [CrossRef]
58. Li, G.; Peng, X.; Wei, L.; Kang, G. Salicylic Acid Increases the Contents of Glutathione and Ascorbate and Temporally Regulates the Related Gene Expression in Salt-Stressed Wheat Seedlings. *Gene* **2013**, *529*, 321–325. [CrossRef] [PubMed]
59. Chen, J.; Clinton, M.; Qi, G.; Wang, D.; Liu, F.; Fu, Z.Q. Reprogramming and Remodeling: Transcriptional and Epigenetic Regulation of Salicylic Acid-Mediated Plant Defense. *J. Exp. Bot.* **2020**, *71*, 5256–5268. [CrossRef]
60. Gully, K.; Celton, J.-M.; Degrave, A.; Pelletier, S.; Brisset, M.-N.; Bucher, E. Biotic Stress-Induced Priming and de-Priming of Transcriptional Memory in Arabidopsis and Apple. *Epigenomes* **2019**, *3*, 3. [CrossRef]
61. Javed, T.; Wang, W.; Yang, B.; Shen, L.; Sun, T.; Gao, S.-J.; Zhang, S. Pathogenesis Related-1 Proteins in Plant Defense: Regulation and Functional Diversity. *Crit. Rev. Biotechnol.* **2024**. [CrossRef]

plants

MDPI

Article

PGPB Isolated from Drought-Tolerant Plants Help Wheat Plants to Overcome Osmotic Stress

Veronika N. Pishchik [1,*], Elena P. Chizhevskaya [1], Vladimir K. Chebotar [1,*], Galina V. Mirskaya [2], Yuriy V. Khomyakov [2], Vitaliy E. Vertebny [2], Pavel Y. Kononchuk [2], Dmitriy V. Kudryavtcev [1], Olga A. Bortsova [1], Nina G. Lapenko [3] and Igor A. Tikhonovich [1,4]

[1] All-Russia Research Institute for Agricultural Microbiology, Podbelskogo Hwy 3, Pushkin, 196608 Saint Petersburg, Russia; chizhevskaya@yandex.ru (E.P.C.); dv_kudryavtsev@arriam.ru (D.V.K.); bortsova17@yandex.ru (O.A.B.); i.tikhonovich@arriam.ru (I.A.T.)

[2] Agrophysical Scientific Research Institute, Grazhdanskiy pr. 14, 195220 Saint Petersburg, Russia; galinanm@gmail.com (G.V.M.); himlabafi@yandex.ru (Y.V.K.); verteb22@mail.ru (V.E.V.); 79117717774@yandex.ru (P.Y.K.)

[3] North Caucasus Federal Agrarian Research Centre, Zootechnical Lane, 15, 355017 Stavropol, Russia; lapenko31@yandex.ru

[4] Department of Genetics and Biotechnology, Faculty of Biology, Saint Petersburg State University, 7-9 Universitetskaya Embankment, 199034 Saint Petersburg, Russia

* Correspondence: veronika-bio@rambler.ru (V.N.P.); vladchebotar@arriam.ru (V.K.C.)

Abstract: The aim of this research was to study the effect of plant-growth-promoting bacteria (PGPB) isolated from the drought-tolerant plants camel thorn (*Alhagi pseudoalhagi* (M.Bieb.) Fisch) and white pigweed (*Chenopodium album* L.) on wheat (*Triticum aestivum* L.) plants cv. Lenigradskaya 6, growing under hydroponic conditions and osmotic stress (generated by 12% polyethylene glycol-6000 (PEG)). Based on the assumption that plants create a unique microbiome that helps them overcome various stresses, we hypothesized that bacteria isolated from drought-tolerant plants may assist cultivated wheat plants in coping with drought stress. PGPB were isolated from seeds and leaves of plants and identified as *Bacillus* spp. (strains Cap 07D, Cap 09D, and App 11D); *Paenibacillus* sp. (Cap 286); and *Arthrobacter* sp. (Cap 03D). All bacteria produced different phytohormones such as indole acetic acid (IAA), abscisic acid (ABA), and gibberellic acid (GAS_3) and were capable of stimulating wheat growth under normal and osmotic stress conditions. All PGPB reduced the malondialdehyde (MDA) content, increased the total chlorophyll content by increasing chlorophyll *a*, and modulated wheat hormone homeostasis and CAT and POX activities under osmotic conditions. Selected strains can be promising candidates for the mitigating of the drought stress of wheat plants.

Keywords: drought-tolerant plants; PGPB; wheat; phytohormones; osmotic stress

1. Introduction

Drought stress affects 45% of the total arable land, and the global aridity stress area consists of 27% of the world's agricultural areas [1–3].

Soft wheat (*T. aestivum*) is an important cereal crop cultivated in many regions of the world [4–6]. Wheat yield losses due to drought have been recorded in many countries and can be as high as 50% [7,8]. Drought damages various cellular compartments, inactivates enzymes, leads to protein degradation, reduces transpiration, and, eventually, leads to the wilting and desiccation of plants [9,10]. The following mechanisms are involved in drought: intracellular signal transduction (via serine/threonine-protein kinase); ABA (abscisic acid) signaling, together with the activation of the salicylic acid signaling pathway; the modulation of the antioxidant defense mechanism; the maintenance of the cellular energy balance; the closure of the stomata; and the regulation of the cell wall [11,12]. This stress-related process involves miRNAs which inhibit the translation of various target genes, mostly comprising transcription factors [13,14]. Candidate genes associated with adaptive gene networks in wheat to drought stress conditions have now been identified [15–18].

Plant hormonal signaling plays an important role in the process of plant adaptation to drought. Key genes have been identified that regulate ABA signaling, as well as those related to the signaling pathways of other hormones such as auxin, ethylene, jasmonic, salicylic, and gibberellic acids [18]. ABA-dependent genes associated with reactive oxygen species (ROS) inactivation and water use efficiency are involved in signal transduction and transcription pathways [19]. In general, an imbalance between ROS generation and antioxidant defense systems leads to an excessive ROS accumulation and induces oxidative stress in plants [20]. Under oxidative stress, lipid peroxidation occurs, nucleic acids and proteins are damaged, and the carbohydrate metabolism is altered [20,21]. Thus, ROS are signal transduction molecules that control various pathways in the process of plant acclimation under stress conditions [22].

Such antioxidant enzymes as peroxidase (POD) and ascorbate peroxidase (APX), catalase (CAT) and superoxide dismutase (SOD), and non-enzymatic components—proline and carotenoids—are involved in protecting plants from drought stress [23,24].

The intensity of drought is projected to increase consistently [25], hence the need for research to address this issue to ensure food security. One aspect of solving this problem, along with the use of drought-tolerant wheat genotypes [26], is the use of biologicals that help to increase the adaptive potential and yield of plants under changing environmental conditions [27].

It is known that the positive effect of PGPB on host plants under drought stress is realized due to (1) changes in enzyme activity and the induction of systemic stress resistance reactions in plants [28–30]; (2) the production of a wide range of biologically active compounds including hydrolytic enzymes [31], plant hormones [27], ACC deaminase [32], and exopolysaccharides [33]; (3) changes in the hormonal background of plants [34,35]; (4) changes in the root architecture and the increased availability of macro- and micronutrients [36]; and (5) the regulation of photosynthesis [12,34]. However, the efficacy of the same PGPB strain under drought stress can vary depending on the plant genotype and environment conditions [34,37,38].

It recently has been found that PGPB helped wheat plants to alleviate drought stress [39–44]. For example, the endophytic PGPB *Pantoea alhagi* isolated from the drought-tolerant plant *Alhagi sparsifolia* Shap. (*Leguminosae*) improved growth and drought tolerance in wheat [39].

Despite the current understanding of plant exposure mechanisms, some issues of wheat–microbial relationships under drought stress remain relevant for research. Different responses in the manifestation of wheat antioxidant enzyme activity in response to bacterial inoculation under drought conditions are described [28,44]. However, the bacterial mechanisms affecting endogenous plant hormone levels are not fully understood [34,45].

It is known that plants growing in different ecological niches form their own unique microbiome, which allows plants to tolerate various unfavorable conditions [46,47]. Drought-tolerant plants grow in an arid zone characterized by insufficient atmospheric moisture and high air temperatures that limit plant growth. Therefore, we hypothesized that bacteria isolated from the seeds and leaves of drought-tolerant plants could help other plants to overcome drought conditions.

This work aimed to study the effects of epiphytic bacteria isolated from the seeds and leaves of drought-tolerant plants on the growth, hormonal status, and antioxidant ferments activities of wheat plants under osmotic stress generated by 12% PEG6000.

2. Results

2.1. Bacterial Isolation

All the PGPB studied were isolated from the drought-resistant plants. The strains Cap 03 D, Cap 07D, and Cap 09D were isolated from the surface of the seeds of *C. album*,

Cap 286 was isolated from the surface of the leaves of *C. album*, and App 11D were isolated from the surface of the seeds of *A. pseudoalhagi*.

2.2. Bacterial Identification

Most of the isolates (Cap 07D, Cap 09D, and App 11D) were identified as *Bacillus* spp. The strain Cap 286 was identified as *Paenibacillus* sp., and the strain Cap 03D was identified as *Arthrobacter* sp. For each of the strains, representatives with the closest 16S rDNA sequences were identified in the GenBank database (Table S1). The phylogenetic analysis is presented in Figure 1.

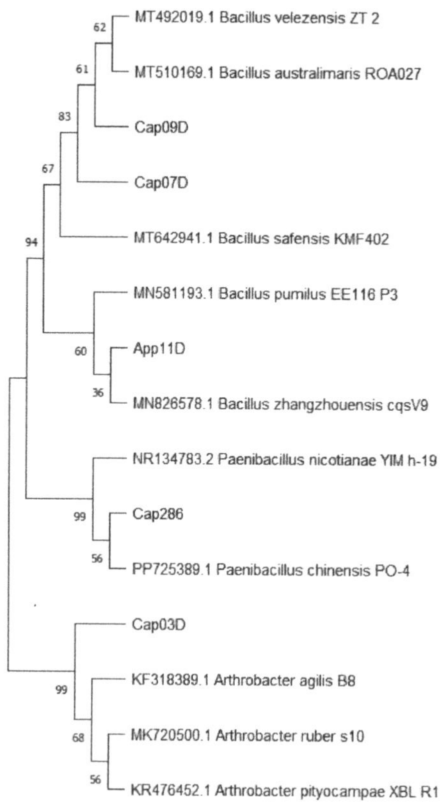

Figure 1. Phylogenetic analysis of 16S rRNA sequences of the epiphytic bacteria isolated from drought-tolerant plants. Notes: The Neighbor-Joining clustering method with bootstrap value (500 replicates), pairwise deletion algorithm, and p distance mathematical model were used.

Cap 03D, belonging to the genus *Arthrobacter*, was found to be in the same cluster with strains of *Arthrobacter agilis*, *A. ruber*, and *A. pityocampae* when compared with reference strains. Strains Cap 07D and Cap 09D (members of the genus *Bacillus)* are in the same cluster as strains of the *B. velezensis*, *B. safensis*, and *B. australimaris* species. Strain App 11D joined the cluster with *B. pumilus* and *B. zhangzhouensis*. Strain Cap 286, a member of the genus *Paenibacillus*, has merged into one cluster with *P. nicotianae* and *P. chinensis*.

2.3. The Effect of Temperature of the Growth of PGPB

The ability of PGPB to grow at different temperatures is presented in Table 1. Most strains were able to grow in the range of 10–40 °C, except the strain Cap 03D *Arthrobacter* sp. The maximal growth temperature of Cap 03D was 35 °C. The strain Cap 07D *Bacillus* sp.

was able to grow at 5 °C. The most heat-resistant strains were strains Cap 09D and App 11D *Bacillus* spp., which were able to grow at 50 °C. Moreover, the strain Cap 09D *Bacillus* sp. was found to have weak growth at a temperature of 55 °C.

Table 1. Growth temperature of the PGPB studied.

Strain	5 °C	10 °C	15 °C	28 °C	35 °C	37 °C	40 °C	45 °C	50 °C	55 °C
Arthrobacter sp. Cap 03D	−	++	++	++	+	−	−	−	−	−
Bacillus sp. Cap 07D	++	+	+++	++	++	++	++	−	−	
Bacillus sp. Cap 09D	−	+	++	+++	+++	+++	+++	+++	+++	+
Paenibacillus sp. Cap 286	+	+	++	+++	+++	++	+	−	−	−
Bacillus sp. App 11D	−	+	+++	+++	+++	+++	+++	+++	+++	−

Notes: +++ active growth, ++ medium growth, + weak growth, and − no growth.

2.4. The Effect of Different Concentrations of NaCl on Growth of PGPB

The ability of the isolates to grow at different concentrations of NaCl is present in Table 2. All strains were able to grow on a GMF medium with 1% NaCl. Strains Cap 07D, Cap 09D, and App 11D *Bacillus* spp. were able to grow in a medium with 7% NaCl. Only two strains, Cap 09D and App 11D, grew on a medium with 10% NaCl.

Table 2. Growth in presence of NaCl.

Bacteria	NaCl, 1%	NaCl, 5%	NaCl, 7%	NaCl, 10%	NaCl, 15%
Arthrobacter sp. Cap 03D	++	−	−	−	−
Bacillus sp. Cap 07D	++	++	+	−	−
Bacillus sp. Cap 09D	++	++	++	++	−
Bacillus sp. App 11D	++	++	++	++	−
Paenibacillus sp. Cap 286	+	−	−	−	−

Notes: ++ medium growth, + weak growth, and − no growth.

2.5. Metabolic Activities of PGPB

All strains were tested on different metabolic activities (Table 3). The strain Cap 03D *Arthrobacter* sp. and strains Cap 07D, Cap 09D, and Cap 286 *Bacillus* spp. possessed amylolytic activity. The highest amylolytic activity was detected in the strain Cap 286 (Figure 2).

Table 3. Metabolic activities of PGPB.

Strain	AA	LA	PM	PA	CA	NA	AP	IP	PP	SP
Arthrobacter sp. Cap 03D	+	−	−	−	−	−	−	+	−	+
Bacillus sp. Cap 07D	+	−	+	+	−	−	+	+	+	+
Bacillus sp. Cap 09D	−	+	−	++	−	−	+	+	+	+
Paenibacillus sp. Cap 286	++	−	−	+	+	+	−	+	+	+
Bacillus sp. App 11D	−	−	−	++	−	−	−	+	+	+

Notes: AA: amylolytic activity; LA: lypolytic activity; PM: phosphate-mobilizing activity; PA: proteolytic activity; CA: cellulolytic activity; NA: nitrogen-fixing activity; AP: (NH4) production; IP: indole production; PP: polysaccharide production; SP: siderophore production. ++ strong activity, + medium and weak activity, and − no activity.

Of all the strains studied, only strain Cap 09D *Bacillus* sp. possessed lipolytic activity. Phosphate-mobilizing activity was found only in the Cap 07D *Bacillus* sp. strain, whereas proteolytic activity was detected in almost all strains with the exception of Cap 03D *Arthrobacter* sp. Moreover, the strain Cap 286 *Bacillus* sp. had cellulolytic and nitrogen-fixing activities in contrast to the other strains studied (Figure 2). The strains Cap 07D and Cap 09D *Bacillus* spp. produced NH_4. All strains produced indoles and siderophores. Polysaccharides were revealed in all the studied strains with the exception of Cap 03D *Arthrobacter* sp. (Figure S1).

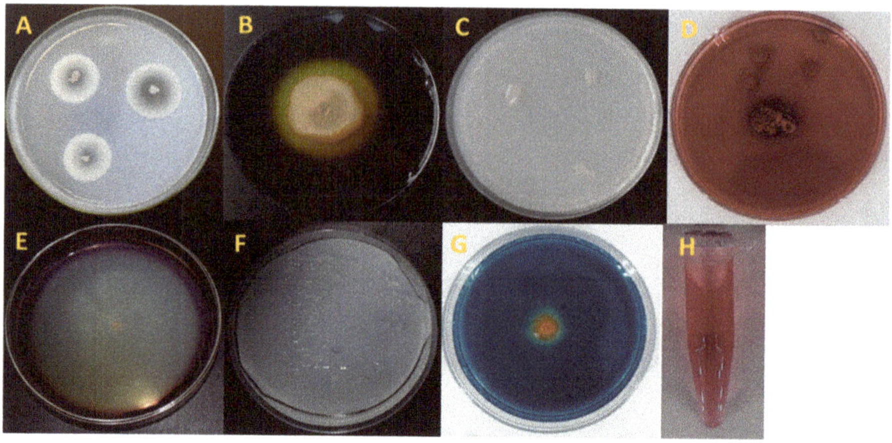

Figure 2. Metabolic activities of *Paenibacillus* sp. Cap 286. Notes: (**A**) Proteolytic activity; (**B**) amylolitic activity; (**C**) phosphate-mobilizing activity; (**D**) exopolysaccaride activity; (**E**) lypolitic activity; (**F**) nitrogen-fixing activity; (**G**) siderophore-producing activity; and (**H**): indole-producing activity.

2.6. Phytohormone Production Activity of PGPB

We analyzed the indole activity of the tested strains when grown on the R2A medium with tryptophan supplementation (Figure 3). All bacteria studied produced indoles. The highest indole activity was found for strains Cap 07D *Bacillus* sp. (58.8 mg L^{-1}) and Cap 286 *Paenibacillus* sp. (31.9 mg L^{-1}), and the lowest was found for strain Cap 03D *Arthrobacter* sp. (2.8 mg L^{-1}).

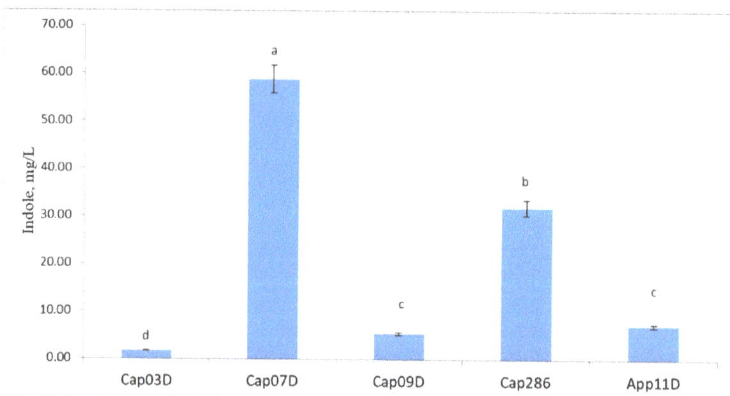

Figure 3. Indole production activities by PGPB studied with addition of tryptophan in the concentration 500 mg L^{-1}. Notes: Cap03D: *Arthrobacter* sp. Cap 03D; Cap07D: *Bacillus* sp. Cap 07D; Cap09D: *Bacillus* sp. Cap 09D; Cap286: *Paenibacillus* sp. Cap 286; and App11D: *Bacillus* sp App 11D. Bars show ±SEM. Values in columns followed by different letters (a–d) are significantly different at $p \leq 0.05$, as determined by Duncan's multiple-range test.

We have studied the ability of strains to produce such phytohormones as gibberellin (GAS_3), indole acetic acid (IAA), and abscisic acid (ABA) by using high-performance liquid chromatography–mass spectrometry (HPLC-MC) (Table 4). All strains studied were capable of producing phytohormones such as GAS_3 (13.7–29.6 ng mL^{-1}), IAA (0.37–5.27 µg mL^{-1}), and ABA (11.7–54.0 ng mL^{-1}). The highest level of GAS_3 was found for the strain Cap

03D *Arthrobacter* sp.—29.6 ng mL^{-1}. The strains App 11D *Paenibacillus* sp. and Cap 07D *Bacillus* sp. produced the maximum amount of abscisic acid (54.0 and 37.0 ng mL^{-1}), accordingly. The strains Cap 07 and Cap 286 *Bacillus* spp. were the most active producers of IAA, which is consistent with the results presented in Figure 3.

Table 4. Phytohormones production activity of PGPB.

Phytohormone Production	Cap 03D *Arthrobacter* sp.	Cap 07D *Bacillus* sp.	Cap 09D *Bacillus* sp.	App 11D *Paenibacillus* sp.	Cap 286 *Bacillus* sp.
GAS_3, ng/mL	29.6 ± 2.19 ^{a}	19.8 ± 1.06 ^{b}	14.1 ± 2.05 ^{c}	17.0 ± 3.05 ^{bc}	13.7 ± 2.6 ^{c}
IAA, μg/mL	0.37 ± 0.02 ^{e}	5.27 ± 1.06 ^{a}	1.38 ± 0.06 ^{d}	1.82 ± 0.32 ^{c}	3.82 ± 0.08 ^{b}
ABA, ng/mL	15.9 ± 3.62 ^{cd}	37.0 ± 2.80 ^{b}	19.3 ± 1.27 ^{c}	54.0 ± 4.30 ^{a}	11.7 ± 1.25 ^{d}

Notes: GAS_3—gibberellin, IAA—indole acetic acid (content was measured without tryptophan addition), and ABA—abscisic acid. Bars show ±SEM. Values in rows followed by different letters (a–d) are significantly different at $p \leq 0.05$, as determined by Duncan's multiple-range test.

2.7. Plant Growth Promotion Activity of PGPB

The growth-stimulating activity was tested after growing wheat cv. Leningradskaya 6 for 4 days (Figure 4). All strains have shown the growth-stimulating activity increasing the root growth (by 8.3–19.4%) and shoot growth (by 0.6–8.4%) as compared to the control variant. The highest root growth-stimulating activity was detected in the Cap 07D, Cap 09D, and Cap 286 *Bacillus* spp. strains.

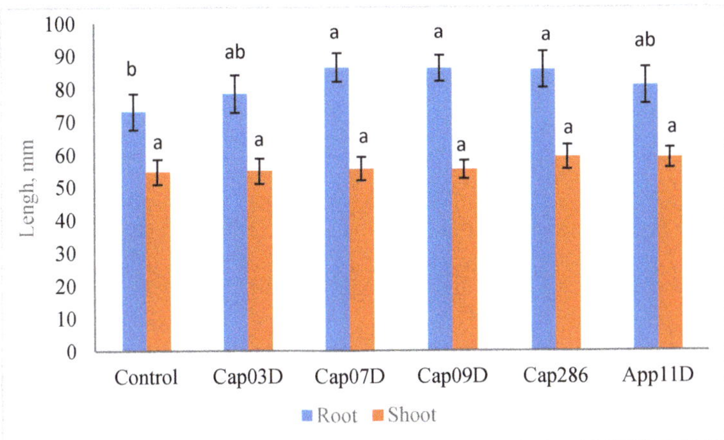

Figure 4. Effect of inoculation with PGPB on root and shoot lengths. Notes: The seedlings of wheat cv. Leningradskaya 6 were grown for 4 days. Control: non-inoculated wheat plants. Cap 03D, Cap 07D, Cap 09D, App 11D, and Cap 286: wheat plants inoculated with appropriate strains. Bars show ±SEM. Values in columns followed by different letters (a,b) are significantly different at $p \leq 0.05$, as determined by Duncan's multiple-range test.

2.8. Hydroponic Experiments

We studied the effect of PGPB strains on wheat growth under osmotic stress. In order to create osmotic stress, we used PEG6000 because it is widely used in similar studies [34,44,48,49]. Previously, we tested the effect of PEG6000 on the studied strains. It was found that the addition of 12% PEG6000 to the LB medium during cultivation does not significantly affect the growth of the strains (Figure S2).

2.8.1. Effect of PGPB on Wheat Morphological Characteristics Under Normal Conditions and Osmotic Stress

In the hydroponic experiments, the morphological characteristics of wheat plants under osmotic stress (generated by the addition of 12% PEG6000) were studied (Table 5, Figure 5).

Table 5. Effect of PGPB on wheat cv. Leningradskaya 6 grown under normal and osmotic stress conditions.

Treatment	Length, mm		FW, mg		DW, mg	
	Root	Shoot	Root	Shoot	Root	Shoot
Control	182 ± 6.9 ^{c}	360 ± 4.9 ^{c}	310 ± 17.2 ^{c}	381 ± 19.8 ^{c}	22.0 ± 0.6 ^{d}	52.2 ± 1.7 ^{c}
12PEG	160 ± 6.1 ^{d}	305 ± 11.8 ^{f}	153 ± 1.45 ^{f}	232 ± 14.7 ^{e}	10.7 ± 0.8 ^{g}	35.3 ± 1.7 ^{f}
Cap 03D	209 ± 2.2 ^{ab}	373 ±3.9 ^{ab}	359 ± 11.3 ^{a}	443 ± 23.8 ^{a}	26.6 ± 0.7 ^{c}	62.0. ± 2.8 ^{bc}
12PEG + Cap 03D	175 ± 4.7 ^{c}	316 ± 5.7 ^{ef}	157 ± 0.64 ^{ef}	251 ± 11.9 ^{de}	11.9 ± 0.6 ^{g}	39.7 ± 1.9 ^{e}
Cap 07D	199 ± 4.3 ^{b}	375 ± 7.2 ^{ab}	342 ± 27.7 ^{ab}	403 ± 26.8 ^{bc}	29.1 ± 1.1 ^{b}	65.3 ± 2.8 ^{b}
12PEG + Cap 07D	175 ±8.3 ^{c}	331 ± 9.9 ^{d}	183 ± 1.76 ^{e}	269 ± 14.2 ^{d}	16.5 ± 0.8 ^{f}	43.1. ± 2.1 ^{d}
Cap 09D	211 ± 1.2 ^{a}	364 ± 10.0 ^{bc}	353± 22.9 ^{a}	427 ± 22.6 ^{ab}	33.2 ± 1.7 ^{a}	77.3 ± 3.4 ^{a}
12PEG + Cap 09D	179 ± 5.0 ^{c}	327 ± 9.8 d^{e}	203 ± 1.02 ^{d}	266 ± 13.4 ^{d}	19.7 ± 1.0 ^{e}	48.9 ± 2.4 ^{cd}
App 11D	210 ± 9.7 ^{ab}	378 ± 2.7 ^{a}	347 ± 22.9 ^{ab}	429 ±18.9 ^{ab}	31.9 ± 1.1 ^{ab}	76.4 ± 3.1 ^{a}
12PEG + App 1D	180 ± 9.2 ^{c}	362 ± 5.6 ^{bc}	196 ± 2.07 ^{de}	279 ± 23.0 ^{d}	18.8 ± 1.1 ^{e}	51.9 ± 2.4 ^{c}
Cap 286	209 ± 10.8 ^{ab}	368 ± 9.3 ^{abc}	360± 26.1 ^{a}	419 ±24.5 ^{ab}	32.6 ± 2.1 ^{a}	76.7 ± 2.0 ^{a}
12 PEG + Cap 286	179 ± 2.8 ^{c}	327 ± 10.7 ^{de}	187 ± 1.02 ^{de}	275 ± 20.2 ^{d}	18.3 ± 1.0 ^{ef}	47.7 ± 1.8 ^{cd}

Notes: Control: non-inoculated wheat plants. 12PEG: the addition of 12% PEG6000. Cap 03D, Cap 07D, Cap 09D, App 11D, and Cap 286: wheat plants inoculated with appropriate strains. FW—fresh weight of shoots and roots. Bars show ±SEM. Values in columns followed by different letters (a–f) are significantly different at $p \leq 0.05$, as determined by Duncan's multiple-range test. Wheat plants cv. Leningradskaya 6 were grown under hydroponic conditions for 14 days (7 days of exposure to osmotic stress).

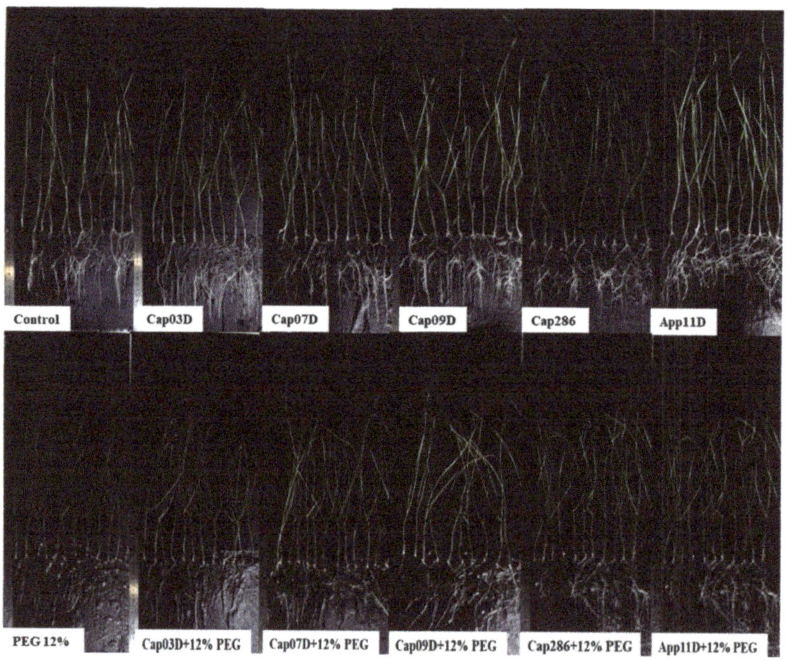

Figure 5. Effect of inoculation with PGPB on wheat plants cv. Leningradskaya 6 grown under normal and osmotic stress conditions. Notes: Control: non-inoculated wheat plants. PEG 12%: the addition of 12% PEG 6000; Cap03D, Cap07D, Cap09D, App11D, and Cap286: wheat plants inoculated with appropriate strains. Wheat plants cv. Leningradskaya 6 were grown under hydroponic conditions for 14 days (7 days of exposure to osmotic stress).

The inoculation with all PGPB studied under control conditions (without osmotic stress) significantly increased the wheat root length by 9.3–15.9%. The root fresh weight (FW) also increased significantly (by 9.6–15.8%) in all control variants when inoculated with the studied bacteria.

In general, the increase in the root FW of Leningradskaya 6 was associated with intensive root formation after the inoculation with PGPB (Figure 5). This is also confirmed by the significant increase in the dry weight (DW) of roots (by 20.9–50.9%) after the inoculation with all PGPB studied in the control variants. The dry matter (DM) in roots is also increased by 4–35%. The strains Cap 03D *Arthrobacter sp.*, Cap 07D, and App 11D *Bacillus* spp. significantly increased the shoot length (by 3.9–5.2%) in control variants, whereas Cap 09 *Bacillus* sp. and Cap 286 *Paenibacillus* sp. had no significant effect on the shoot length.

Osmotic stress had a negative effect on wheat plants. The addition of 12% PEG6000 significantly decreased the length of roots by 12.1% and leaves by 15.3%. Moreover, the FW of the roots and shoots decreased by 50.6% and 39.1% (Table 5), and the DW of the roots and shoots also decreased (by 51.4% and 32.4%), respectively.

Inoculation with the studied PGPB significantly reduced the effect of osmotic stress. PGPB significantly increased the length of the roots and shoots (except for strain Cap 03D *Arthrobacter* sp., which unreliably increased the shoot length) in all variants with the addition of 12% PEG600. No differences in the effect of the studied strains on the wheat root length were detected (Table 5); the efficiency of PGPB was 9–12.1%. PGPB under osmotic stress also increased the shoot length (by 3.8–11.9%) and root length (by 9.4–12.5%). The strain App 11D *Paenibacillus* sp. Showed the highest efficiency. PGPB also increased the DW of the roots (by 11.2–75.8%) and shoots (by 12.5–47.0%) under osmotic stress, when the strains Cap 03D *Arthrobacter* sp. And App 11D *Paenibacillus* sp. Showed the highest efficiency. The increase in the DW of wheat plants upon PGPB inoculation was accompanied by an increase in dry matter (%) by 0.6–2.2% in roots and by 0.6–3.4% shoots (Figure 6).

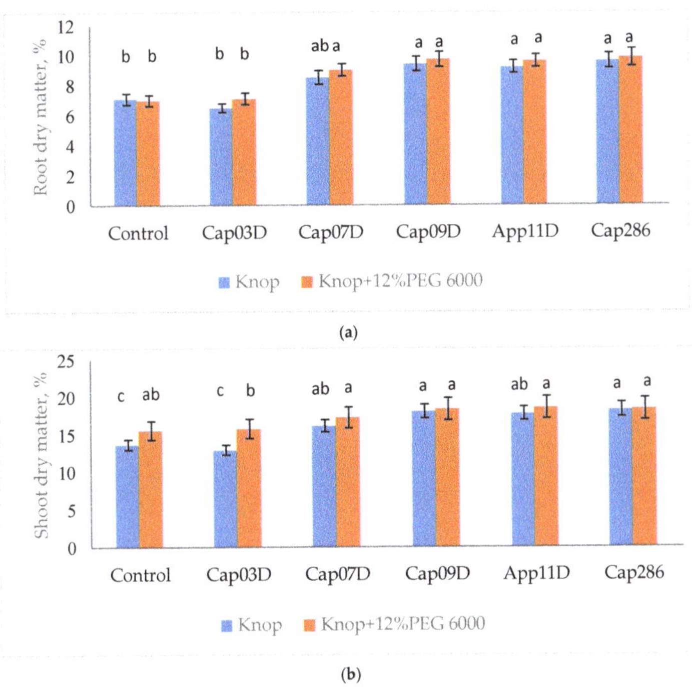

Figure 6. Effect of inoculation with PGPB on dry matter content in roots (**a**) and shoots (**b**) of wheat cv. Leningradskaya 6, grown under normal and osmotic stress conditions. Notes: Control: non-inoculated

wheat plants, grown under control conditions (Knop) and under osmotic stress (the addition of 12% PEG 6000). Cap 03D, Cap 07D, Cap 09D, App 11D, and Cap 286: wheat plants inoculated with appropriate strains. Wheat plants cv. Leningradskaya 6 were grown under hydroponic conditions for 14 days (7 days of exposure to osmotic stress). Bars show ±SEM. Values in columns followed by different letters are significantly different at $p \leq 0.05$, as determined by Duncan's multiple-range test.

2.8.2. Content of Photosynthetic Pigments in Wheat Leaves

Inoculation with Cap 07D, Cap 09D, and Cap 286 *Bacillus* spp. Led to a significant increase in chlorophyll a content in wheat leaves in the control variants, whereas the strains Cap 03D *Arthrobacter* sp. And App 11D *Paenibacillus* sp. Had no effect on the chlorophyll a content (Figure 7a). The content of chlorophyll b did not increase significantly and the strains Cap 09D *Bacillus* sp. And App 11D *Paenibacillus* sp. Saw a decrease in chlorophyll b content compared to the control variant (without the addition of 12% PEG6000) (Figure 7b).

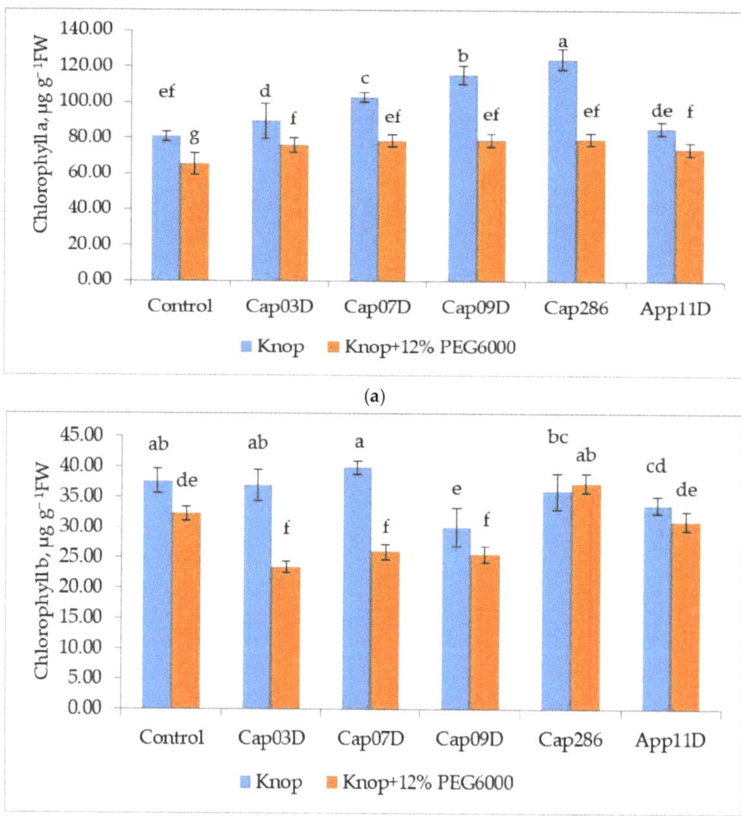

Figure 7. Effect of inoculation with PGPB on chlorophyll a (**a**) and chlorophyll b (**b**) content in wheat leaves cv. Leningradskaya 6, grown under normal and osmotic stress conditions. Notes: Control: non-inoculated wheat plants, growing under control conditions (Knop solution) and under osmotic stress (the addition of 12% PEG6000). Cap03D, Cap07D, Cap09D, App11D, and Cap286: wheat plants inoculated with appropriate strains. Wheat plants cv. Leningradskaya 6 were grown under hydroponic conditions for 14 days (7 days of exposure to osmotic stress). Bars show ±SEM. Values in columns followed by different letters (a–g) are significantly different at $p \leq 0.05$, as determined by Duncan's multiple-range test.

Osmotic stress significantly decreased the content of total chlorophyll by 17.5% (mainly due to a decrease in chlorophyll a) (Figure 8a) and the content of carotenoids by 14% in wheat leaves (Figure 8b).

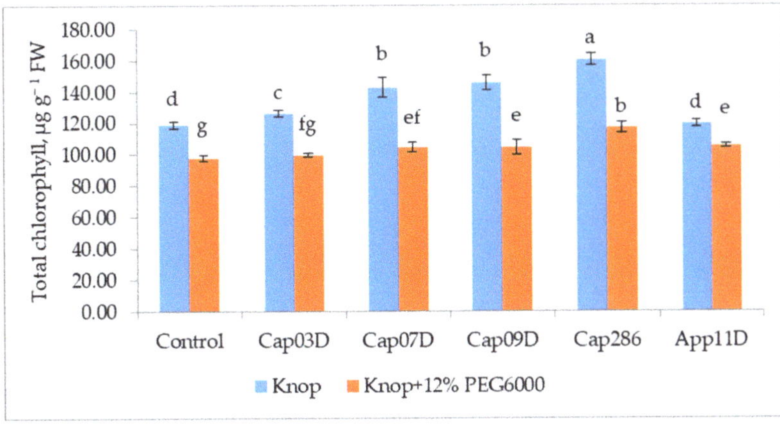

(**a**)

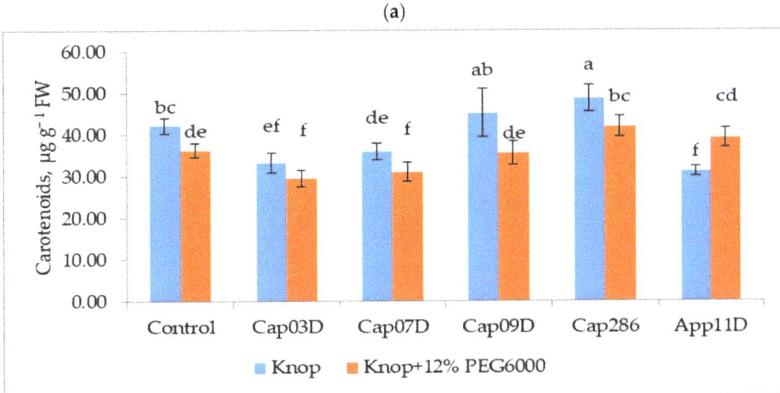

(**b**)

Figure 8. Effect of inoculation with PGPB on total chlorophyll (**a**) and carotenoids (**b**) contents in wheat leaves cv. Leningradskaya 6, grown under normal and osmotic stress conditions. Notes: Control: non-inoculated wheat plants, growing under control conditions (Knop solution) and under osmotic stress (the addition of 12% PEG6000). Cap03D, Cap07D, Cap09D, App11D, and Cap286: wheat plants inoculated with appropriate strains. Wheat plants cv. Leningradskaya 6 were grown under hydroponic conditions for 14 days (7 days of exposure to osmotic stress). Bars show ±SEM. Values in columns followed by different letters (a–g) are significantly different at $p \leq 0.05$, as determined by Duncan's multiple-range test.

The differences of the bacteria action on the photosynthetic pigments of the plants both in the control conditions and under osmotic stress were revealed. Bacteria under control conditions increased the content of total chlorophyll in wheat leaves by 5.5–13.4. The strain App 11D *Paenibacillus* sp. did not show its efficiency on the content of total chlorophyll and carotenoids in wheat leaves. The strain Cap 286 *Bacillus* sp. showed the highest efficiency in control conditions on the number of photosynthetic pigments: total chlorophyll and carotenoids. Inoculation with this strain under osmotic stress led to an increase in the content of chlorophyll and carotenoids to the level of the control variant.

2.8.3. Effect of PGPB on Enzyme Activities of Wheat cv. Leningradskaya 6

The inoculation with PGPB of wheat plants modulated the peroxidase (POX) and catalase (CAT) activities (Table 6).

Table 6. Effect of inoculation with PGPB on POX (U s^{-1} g^{-1} FW) and CAT (μM H_2O_2 mg^{-1} protein min^{-1}) activities of wheat cv. Leningradskaya 6 grown under normal and osmotic stress conditions (7 days).

Treatment	POX in Roots U s^{-1} g^{-1} FW	POX in Shoots U s^{-1} g^{-1} FW	CAT in Roots μM H_2O_2 mg^{-1} Protein min^{-1}	CAT in Shoots μM H_2O_2 mg^{-1} Protein min^{-1}
Control	58.94 ± 2.01 ^{de}	31.06 ± 1.41 ^{e}	203.82 ± 4.65 ^{a}	525.16 ± 12.31 ^{a}
12PEG	83.81 ± 3.90 ^{a}	53.85 ± 3.76 ^{b}	150.9 ± 5.24 ^{g}	456.66 ± 19.16 ^{b}
Cap 03D	74.15 ± 2.40 ^{b}	37.69 ± 3.07 ^{d}	173.73 ± 4.60 ^{ef}	509.86 ± 11.17 ^{a}
Cap 03D-12PEG	77.41 ± 3.78 ^{b}	44.28 ± 3.13 ^{c}	183.92 ± 3.06 ^{bc}	425.37 ± 6.86 ^{de}
App 11D	55.95 ± 2.96 ^{e}	35.58 ± 1.88 ^{d}	179.86 ± 4.48 ^{cde}	464.74 ± 7.08 ^{b}
App 11D-12PEG	59.51 ± 4.89 ^{de}	45.31 ± 3.24 ^{c}	188.06 ± 4.47 ^{b}	452.72 ± 6.88 ^{bc}
Cap 07D	68.8 ± 4.79 ^{c}	31.57 ± 1.36 ^{e}	175.92 ± 5.04 ^{def}	421.43 ± 9.94 ^{de}
Cap 07D-12PEG	75.01 ± 2.59 ^{b}	72.99 ± 4.63 ^{a}	184.31 ± 4.96 ^{bc}	422.98 ± 10.02 ^{de}
Cap 09D	60.13 ± 2.65 ^{de}	28.76 ± 1.59 ^{e}	169.58 ± 4.14 ^{f}	459.78 ± 16.44 ^{b}
Cap 09D-12PEG	64.24 ± 5.49 ^{cd}	48.14 ± 2.04 ^{c}	182.3 ± 5.15 ^{bcd}	436.74 ± 12.42 ^{cd}
Cap 286	59.49 ± 4.57 ^{de}	48.0 ± 1.5 ^{c}	189.89 ± 5.42 ^{b}	412.05 ± 8.43 ^{e}
Cap 286-12PEG	61.51 ± 2.50 ^{de}	56.4 ± 3.3 ^{b}	202.27 ± 4.45 ^{a}	466.49 ± 10.73 ^{b}

Notes: Control: non-inoculated wheat plants, growing under normal conditions; 12 PEG: wheat plants growing under osmotic stress conditions (the addition of 12% PEG6000). Bars show SEM and different letters show a significant difference among treatments at $p < 0.05$ level as determined by Duncan's multiple-range test. Wheat plants cv. Leningradskaya 6 were grown under hydroponic conditions for 14 days (7 days of exposure to osmotic stress).

The peroxidase (POX) activity in the shoots of wheat var. Leningradskaya 6 in the control variant was 1.9 times lower than in the roots. The inoculation with Cap 03D *Arthrobacter* sp. and Cap 07D *Bacillus* sp. under control conditions (without PEG 6000) increased the POX activity in wheat roots by 25.8 and 16.7%, respectively. The inoculation with other studied strains under control conditions had no significant effect on the POX activity in the roots of wheat plants.

Under the conditions of osmotic stress, the POX activity in plant roots increased by 42.2%. The inoculation with bacteria under stress conditions decreased the POX activity in wheat roots (by 7.6–29%). When inoculated with bacteria under osmotic stress, the POX activity in shoots also decreased, except for strain Cap 286 *Bacillus* sp., which did not affect the POX activity in the shoots, and strain Cap 07D *Bacillus* sp., which, on the contrary, increased the POX activity in the shoots by 1.4 times.

The shoots of wheat plants have shown a higher catalase (CAT) activity, which was 2.57 times higher than its activity in roots. Inoculation with PGPB under control conditions resulted in a decrease in CAT activity in the roots (by 6.8–16.7% and shoots (by 3.1–19.8%) of wheat plants, respectively. Osmotic stress significantly decreased the CAT activity in the roots and shoots by 26.3% and 13%, respectively. The inoculation with PGPB under osmotic stress led to an increase in CAT activity in the roots (by 20.8–34.0%) and, on the contrary, a decrease in CAT activity in the shoots (by 0.9–9.7%), except for strain Cap 286 *Bacillus* sp., which did not significantly change the CAT activity in shoots.

2.8.4. Effect of PGPB on MDA Content

One of the metabolites of lipid peroxidation is malondialdehyde (MDA), an increase in which shows that plants are subjected to oxidative stress. The inoculation with PGPB under control conditions did not affect the change in MDA in wheat roots, but decreased its content in the shoots. When 12% PEG 6000 was added, the MDA content increased significantly in both the roots (by 13%) and shoots (by 37%) (Figure 9a). This indicates that wheat shoots were more susceptible to osmotic stress. The strains studied had a protective effect on wheat plants. The content of MDA decreased significantly in roots (by 11–13%) and shoots (by 28–33%) when inoculated with PGPB under osmotic stress. This indicates the protective function of PGPB. The strains Cap 07D and Cap286 *Bacillus* spp. had the greatest effect on reducing the amount of MDA.

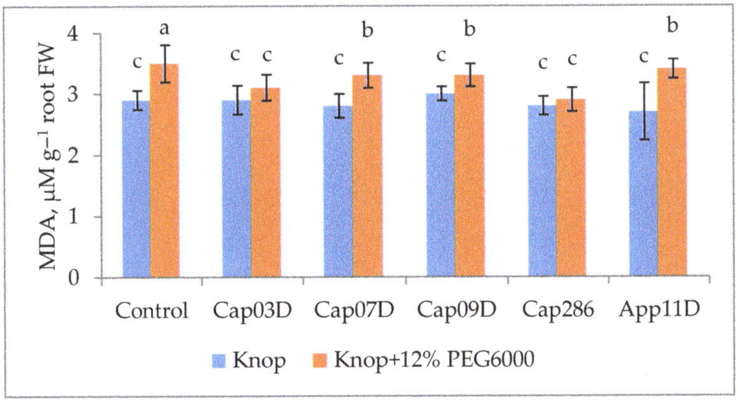

(**a**)

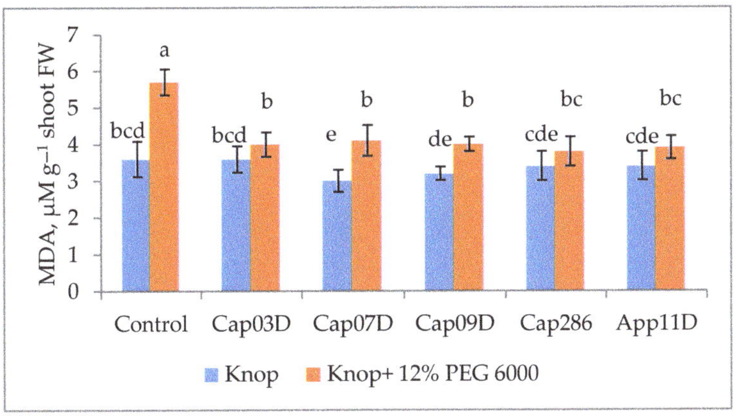

(**b**)

Figure 9. Effect of with PGPB on MDA content in roots (**a**) and shoots (**b**) of wheat cv. Leningradskaya 6, grown under normal and osmotic stress conditions. Notes: Control: non-inoculated wheat plants, grown under normal conditions (Knop) and under osmotic stress (the addition of 12% PEG6000). Cap03D, Cap07D, Cap09D, App11D, and Cap286: wheat plants inoculated with appropriate strains. Wheat plants cv. Leningradskaya 6 were grown under hydroponic conditions for 14 days (7 days of exposure to osmotic stress). Bars with different letters are significantly different at $p \leq 0.05$, as determined by Duncan's multiple-range test.

2.8.5. Effects of PGPB on Endogenous Wheat Hormone Contents

We analyzed the levels of phytohormones IAA, GAS_3, and ABA in the roots of plants inoculated with PGPB strains both under normal conditions and under osmotic stress. The influences of PGPB on the plant hormone contents in roots are shown in Figure 10.

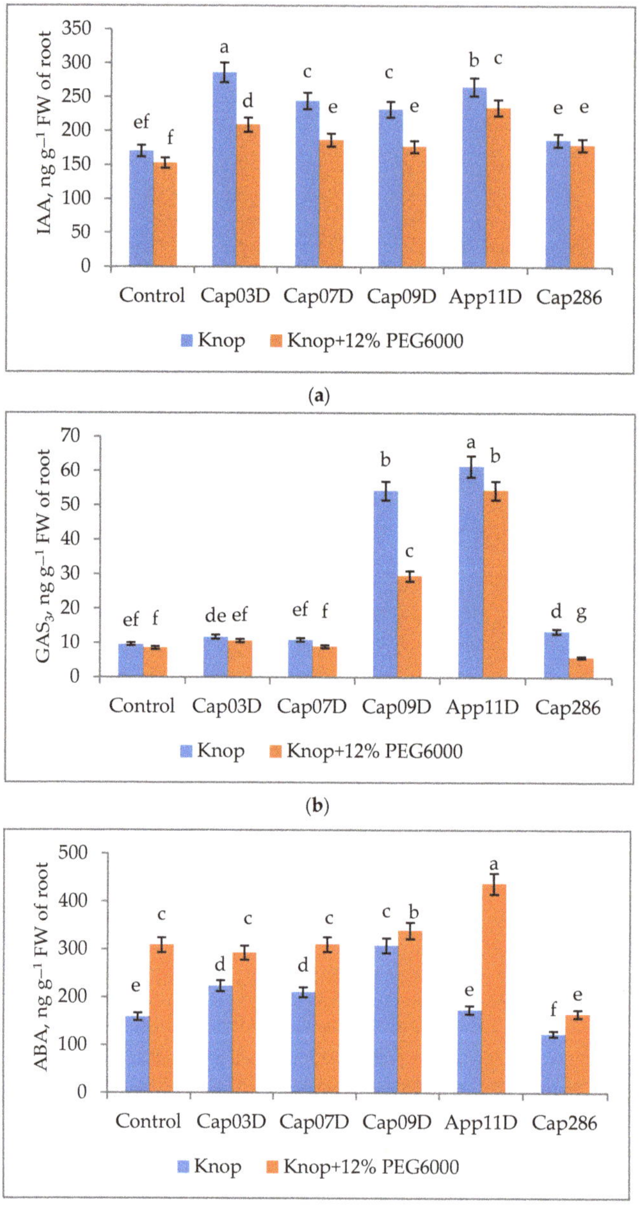

Figure 10. Effect of inoculation with PGPB on concentrations of plant hormones: (**a**) indole-3-acetic acid (IAA); (**b**) gibberellic acid (GAS_3), and (**c**) abscisic acid (ABA) in roots of wheat plants grown under normal and osmotic stress conditions. Notes: Control: non-inoculated wheat plants, growing under normal conditions (Knop solution) and under osmotic stress (the addition of 12% PEG6000). Cap03D, Cap07D, Cap09D, App11D, and Cap286: wheat plants inoculated with appropriate strains. Bars with different letters are significantly different at $p \leq 0.05$, as determined by Duncan's multiple-range test. Wheat plants cv. Leningradskaya 6 were grown under hydroponic conditions for 14 days (7 days of exposure to osmotic stress).

In the control variant (uninoculated), the concentrations of IAA and GAS_3 in wheat roots decreased under osmotic stress by 10.5% and by 11%, accordingly (Figure 10a,b). The ABA concentration in the control variant, on the contrary, increased by 48.4% under osmotic stress (compared to the uninoculated control variant) (Figure 10c).

The inoculation with PGPB mainly increased the contents of IAA, GAS_3, and ABA compared to the uninoculated variant. However, the inoculation with strain Cap 07D *Bacillus* sp. did not change the GAS_3 content in both variants and the inoculation with Cap 286 *Bacillus* sp. decreased the GAS_3 in wheat roots under osmotic stress. The highest level of GAS_3 was found in the roots of wheat plants inoculated with strains Cap 09D *Bacillus* sp. and App 11D *Paenibacillus* sp.

Under osmotic stress, the concentrations of IAA and GAS3 decreased when bacteria were inoculated, while the concentration of ABA, on the contrary, increased by 25–60.2%. At the same time, under the conditions of osmotic stress, strains Cap 03D *Arthrobacter* sp. and Cap 07D *Bacillus* sp. did not change the ABA content compared to the control without inoculation, and the strain Cap 286 *Bacillus* sp. significantly reduced the ABA content in wheat roots.

3. Discussion

Drought is a serious problem, limiting the production of crops, including cereals. PGPGs are known to help wheat plants overcome drought stress [50–53]. The wheat root system is severely depressed in times of drought stress, including the root biomass, surface area, and length. Moreover, the level of photosynthetic pigments and wheat yield decreased under drought stress [52,54]. The inoculation with PGPB of wheat plants under drought conditions helps to overcome plant stress, because bacterial phytohormones improve the plant growth, plasma membrane integrity, and photosynthesis of plants [34,55].

The bacteria selected for our experiments preliminarily showed high growth-stimulating activity on wheat plants and were identified as *Bacillus* spp. (strains Cap 07D, Cap 09 D and App 11D), *Paenibacillus* sp. (strain Cap 286), and *Arthrobacter* sp. (strain Cap 03D). All the PGPB studied (with exception Cap 03D) produced exopolysaccharides, which protect plants' roots from desiccation [56]. The strains *Bacillus* spp. Cap 09D and App 11D were able to grow at 50 °C and 10% NaCl, which allows them to be used not only under drought conditions but also under high soil salinity.

All the strains were found capable of producing IAA, ABA, and GAS_3. It is known that PGPB from these genera produce various phytohormones, such as IAA, ABA, GA, and CK [57–60]. Indeed, all our strains studied produced total indoles (including IAA), according to the Salkowskiy test, but the most effective producers were Cap 07D and Cap 286. Indeed, most PGPB produced auxin, which is a major component of the bacterial promotion of plant growth and development [61]. There are several IAA biosynthetic pathways in PGPB and these pathways can intersect with each other [62]. IAA promotes plant growth, the differentiation of vascular tissues, the stimulation of cell division [63], and bacterial colonization of plant roots [64,65]. IAA is involved in the regulation of the adaptive response of plants to drought stress. Auxin acts on gene expression through a family of functionally distinct transcription factors, the DNA-binding auxin response factors (ARFs) [66]. Different ARFs regulate soluble sugars, promote root development, and maintain the chlorophyll content under drought and salt stress conditions, helping plants to adapt to these stresses [66]; IAA is one of the most studied hormones in various plant crops, including grain cultures [61].

It was shown that IAA-producing bacteria reduced osmotic stress in different plants [67–69]. In our experiments, drought stress had a negative effect on the wheat length and weight. Inoculation with PGPB increased the root growth and lateral root formation. It is known that IAA-producing bacteria control endogenous IAA levels in plant roots by regulating auxin-responsive genes, which changes the rice root architecture [70]. Moreover, auxin-producing bacteria enhance both the root and shoot growth [71]. IAA biosynthesis by PGPB helps in successful root colonization, which leads to a decrease in the drought stress in different plants [72]. Our results are in agreement with the previously obtained data [34,73]. It was found that the bacteria *Bacillus amyloliquefaciens* 5113 and *Azospirillum brasilense* NO40 also influenced the wheat root system and improved the drought stress tolerance in wheat [74].

Along with auxin synthesis, all the studied bacteria synthesize gibberelic (GAS_3) and abscisic acids (ABA), which are hormone antagonists in various physiological processes [75]. ABA is the plant hormone which is synthesized in response to abiotic stresses and activates the genes responsible for stress resistance [76]. Under drought stress, ABA affects different aspects of plant growth, such as reducing photosynthesis, inhibiting leaf surface area expansion, decreasing lateral root initiation [77], promoting stomatal closure [76], increasing hydraulic conductivity, stimulating root cell elongation [78], and promoting plant resistance to drought [79]. It is known that inoculation with ABA-producing bacteria against drought conditions can lead to an increase in ABA content in plants, which become more resistant to stress [35,80–82].

Gibberellic acids (GASs) modulate various developmental processes in plants, including seed germination, stem elongation, flowering, ripening [83], the chlorophyll content, and the photosynthetic activity [84]. It is known that the exogenous application of gibberellic acid mitigates drought stress in wheat plants [85]. It was shown that GAS-producing PGPB *Leifsonia soli* SE134 and *Enterococcus faecium* LKE12 stimulated shoot growth in mutant rice plants (deficient in gibberellin synthesis), compensating for the absence of plant gibberellins [86,87]. Inoculation with GAS-producing PGPB decreased drought stress [88–90].

Under drought conditions, the CO_2 influx is restricted by limiting the stomatal and mesophyll conductance [91], resulting in reduced carbon assimilation by the photosynthetic apparatus [12,81]. It was shown that, under drought stress in wheat plants, ATP synthesis was impaired due to decreased electron transport and membrane damage, and the content and efficiency of the key photosynthetic enzyme Rubisco (Ribulose bisphosphate carboxylase/oxygenase) were reduced [92]. Bacterial inoculation under drought stress led to an increase in photosynthetic pigments [93] and maximized the efficiency of photosystem II by improving the stomatal conductance, CO_2 assimilation, and transpiration rate [81]. In our experiments, osmotic stress significantly decreased the content of total chlorophyll by 17.5% and the content of carotenoids by 14% in wheat leaves in our experiment. PGPB under osmotic stress increased the total chlorophyll content by increasing chlorophyll *a*. It is likely that this favorable effect on the chlorophyll content of the PGPB was due to the production of GAS, which regulates the chlorophyll content. The strain Cap 286 *Paenibacillus* sp., in addition to synthesizing GAS_3, has nitrogen-fixing activity and a maximally increased chlorophyll content in wheat plants. The strong linear relationships between nitrogen and both ribulose 1,5-bisphosphate (RuBP) carboxylase and chlorophyll were established early [94]. Therefore, the N-fixing bacterium *Paenibacillus* sp. Cap 286 increased the chlorophyll content in wheat leaves due to the nitrogen accumulation in plants.

Catalase (CAT) and peroxidase (POX) are known antioxidant enzymes, whose activity leads to a decrease in the production of highly reactive hydroxide ions. Peroxidases promote the conversion of hydrogen peroxide (H_2O_2) to water. H_2O_2 is a main component of stress response regulation in wheat [95] and other cereals [96,97]. Both antioxidant enzymes CAT and POX take part in counteracting the oxidative damage in wheat during drought stress [73,98–100]. In some studies, the POX activity increased under drought conditions indicating the adverse effects on the morphological and physiological characters in several plants [101–104]. The increase in the POX activity is indicative of an excessive formation of ROS during photorespiration and photosynthesis in peroxisomes and chloroplasts [105]. The POX activity in the wheat roots and shoots in our experiments increased under osmotic stress, which is consistent with earlier results obtained in wheat cultivation [74,104]. However, there are contrary results indicating a decrease in wheat the POX activity under drought stress [106,107]. Perhaps the contradictions in the results are explained by the different wheat genotypes used in the experiments and the different durations of stress.

In our experiments, PGPB modulate the POX and CAT activities after 7 days of osmotic stress influence. The POX activity in the roots and shoots of wheat plants mainly decreased after the inoculation with PGPB under drought conditions (with the exception of Cap 07D *Bacillus* sp.). Similar results were received early [74]. The authors attributed the decrease in peroxidase activity to the protective effect of the bacteria. In general, PGPB under osmotic stress conditions resulted in an increase in CAT activity in the roots and, conversely, a decrease in CAT in the shoots. Bacterial inoculation under drought has different effects on the CAT activity in different plants, both increasing and decreasing the activity of the enzyme [41,108–110]. Our studies are in accordance with the results obtained early [44], where it was reported that the inoculation of wheat plants with PGPB (*Bacillus* sp. D13) under drought stress decreased the CAT activity. At the same time, there is an opposite result regarding the increase in CAT activity in wheat plants when inoculated with the bacteria *Bacillus subtilis* under drought conditions [41]. These differences in the CAT response to the inoculation of wheat with *Bacillus* sp. under drought stress may be explained by the genetic differences in wheat plants and PGPB.

In our experiments, osmotic stress leads to an increase in MDA content in wheat roots and shoots. It was previously shown that the MDA content in various plants such as *Brassica napus* [111], *Medicago sativa* [112], *Zea mays* [113], and *Triticum aestivum* [34,114] increased under drought stress. In general, the MDA content (a product of lipid peroxidation) is a reliable indicator of oxidative membrane damage due to oxidative stress [115]. However, there are studies that did not reveal a correlation between the MDA content in wheat and wheat drought tolerance [116–118]. This appears to indicate the complex mechanisms of drought tolerance in wheat that may vary among genotypes. Bacterial inoculation significantly reduced the MDA content in the roots and shoots under drought stress conditions in our experiments. Of all the bacteria studied, the most effective were the strains Cap 286 and Cap 07D *Bacillus* spp. Other studies have found that the PGPB significantly decreased the MDA content under drought conditions (compared to non-inoculated plants) and protected plants against stress [34,119]. It follows that PGPB prevent the damage of plant cell membranes and it is one of the key points for a plant to resist drought stress in the plant–bacterial association.

The PGPB studied also modulated the wheat hormonal system, which plays a pivotal role in regulating the plant stress resistance [120]. PGPB mitigate the adverse effects of osmotic stress by increasing the ABA levels in wheat roots by 25–60.2%. Our results on the increase in ABA in wheat plant roots when plants are inoculated with ABA-producing bacteria under osmotic stress conditions are consistent with the previous results, obtained for other plants [80,81]. However, information on the effect of PGPB on the ABA levels in plants is limited and contradictory [121], as both increases and decreases in ABA levels have been reported [80,81,122]. These results suggest a complex hormonal regulation of plant drought stress processes by bacteria.

In general, PGPB isolated from drought-tolerant plants and producing ABA well protected wheat plants under osmotic stress. Bacterial ABA, which is a key phytohormone in the plant response to drought stress, probably influenced the change in the plant hormone background and contributed to the protection of wheat plants. However, the bacteria also produced other phytohormones (IAA and GAS_3) and had other beneficial properties such as the production of exopolysaccharides and siderophores, which also provided the resistance of wheat to osmotic stress. It should be noted that strains Cap 09D *Bacillus* sp. and App 11D *Paenibacillus* sp. were capable of growing at temperatures of 50 °C and under saline conditions, which is very important for the selection of bioproducts for use in arid climate conditions, since high temperatures and salinity are often found together in natural arid conditions. Furthermore, we are planning to test the efficiency of PGPB strains in the pot and field experiments on wheat cultivation in arid regions.

Recently, a number of successful experiments have been conducted on the use of PGP bacteria of different species in wheat cultivation, both in model experiments under osmotic stress [34,44,48] and in vegetation and field experiments [28,49,52,123]. Effective inoculation with *Bacillus rugosus* MMH101under drought stress resulted in an increase in the shoot length (108.5%), and root length (134.9%) of wheat cultivar Gimmeza-9 [49]. Rashid et al. [28] studied the effect of *Bacillus megaterium* and *B. licheniformis* on wheat plants under drought conditions. The bacteria possessed ACC deaminase activities and IAA production. In response to drought stress, *B. megaterium* induced three new polypeptides, indicating proteomic changes. Bacterial inoculation increased photosynthetic pigments and osmolytes in wheat plants. Moreover, bacteria modulated the expression of genes involved in hormone signaling and the stress response. By analyzing the literature, we can conclude that the PGPB-mediated drought tolerance in wheat involves complex interactions between microbes, hormones, and proteomic dynamics.

Currently, there is a huge need for robust and environmentally friendly farming strategies under biotic and abiotic stresses. Products aimed to protect agricultural crops from stresses such as drought, waterlogging, salinization, heavy metals, and pathogenicity will become increasingly important in the future due to the scenario of global climate change. In this regard, the use of PGPB could be a means of protecting plants from stress and lead to promising solutions for sustainable and environmentally friendly agriculture. There is a huge growing market worldwide for microbiological preparations to improve plant growth and increase yields under biotic and abiotic stress conditions with an annual growth rate of about 10% [124]. However, researchers must provide to the market the strains of PGPB which should be resistant to the extreme conditions of temperature and humidity, be susceptible to applied agrochemicals and pesticides, and demonstrate a high colonization and competition ability with the resident microorganisms.

4. Materials and Methods

4.1. Plant Samples for Bacterial Isolation

Plants and seeds of *C. album* were collected in Izobilny district of Stavropol Territory (near the town of Izobilny, N_45°23′36″, E_41°45′52″). Plants and seeds of *A. pseudoalhagi* were collected in Neftekumskiy district of Stavropol Territory (near the village Achikulak, N_44°33′43″, E_44°49′59″). Samples were taken at five points in each place. Plants were taken together with roots and adjacent soil. Each plant was taken with root and placed into craft bag, then transferred to the laboratory.

4.2. Isolation and Screening of PGPB for Hydroponic Experiments

PGPB were initially isolated from plant seeds, leaves, and stems of *A. pseudoalhagi* and *C. album* according to the protocol described by Pishchik et al. [58].

Plant material (10 g of seeds or aboveground parts) was placed to flasks (with 100 mL sterile water). The flasks were set on the ultrasonic bath (Bandelin; 50 Hz) for 10 min. Then, 0.1 mL of various dilutions was inoculated on Petri dishes containing LB (Luria Bertani, Sigma-Aldrich, St. Louis, USA) agar medium and Ashby medium (g L^{-1}: sucrose—20.0, K_2HP0_4—0.2, $MgSO_4{\cdot}7H_2O$—0.2, NaCl—0.2, K_2SO_4—0.2, $CaCO_3$—5.0, and agar—18.0). Bacteria were grown for 3–5 days. Then, the strains were isolated and screened for PGP activity via the method described earlier [58]. The seeds of cv. Leningradskaya 6 (30 seeds per variant) were dipped for 30 min into bacterial cultures (1–3 $\times$ 10^5 cells per mL) or sterile water, and were used as a control. Then, the seeds were replaced in sterile Petri dishes (in triplicate) with filter paper and 10 mL of sterile water. Grain germination was taken into account on the 4th day of the incubation at 22 °C. Wheat seedlings were measured and weighted.

The strains which showed the greatest plant growth promotion activity were selected for the hydroponic experiments. To study the effect of 12% PEG 6000 on bacterial growth, we grew the bacteria on LB medium and on LB medium supplemented with 12% PEG 6000 for 72 h at 28 °C at 180 rpm. Then, we measured the absorbances of the bacteria on a Biomate 160 spectrophotometer (Thermo Scientific™, Madison, WI, USA) at 600 nm.

4.3. Bacterial Identification

Total DNA from bacterial strains was isolated with use of Monarch® Genomic DNA Purification Kit (New England BioLabs® Inc., Ipswich, MA, USA) according to the producer's instructions. The universal primers 27f (50-AGA GTT TGATCM TGG CTC AG-30) and 1525r (50-AAG GAG GTG WTC CAR CC-30) [125] were used for amplification of DNA fragments in C1000 TouchTM Thermal Cycler (Bio-Rad, Hercules, CA, USA). PCR was performed in 30 μL reaction mixtures containing 10× Taq Turbo buffer (Evrogen, Moscow, Russia), 150 μM dNTPs (Evrogen, Moscow, Russia), 5 pM of each primer, and 2 U of Taq polymerase (Evrogen, Moscow, Russia). The initial DNA denaturation was performed at 95 °C for 3 min, then followed by 30 cycles of denaturation at 95 °C for 30 s; the primer annealing at 54 °C for 30 s; the elongation at 72 °C for 30 s; and final elongation at 72 °C for 5 min. The obtained PCR fragments were isolated from 1% agarose gel using Monarch® DNA Gel Extraction Kit (New England BioLabs® Inc., Ipswich, MA, USA) and sequenced using an ABI PRISM 3500xl capillary electrophoresis sequencing station (Applied Biosystems, Waltham, MA, USA) according to the manufacturer's protocol.

The strains were identified by comparing the obtained nucleotide sequences of the 16S rRNA gene with the RDP [126] and GenBank [127] databases. The 16S rRNA gene sequences of all the isolates were deposited in the GenBank database with the accession numbers PQ516836–PQ516840. Multiple sequence alignment was conducted by using the Clustal X 1.8 software package. Phylogenetic trees were generated by the Neighbor-Joining method using the MEGA 11 software package, and confidence was tested by bootstrap analysis with 500 repeats.

4.4. Nitrogen-Fixing Ability

NFA was assayed on the bacterial growth on nitrogen-free Ashby medium (g L^{-1}: sucrose—20.0, K_2HPO_4—0.2, $MgSO_4 \cdot 7H_2O$—0.2, NaCl—0.2, K_2SO_4—0.2, $CaCO_3$—5.0, and agar—18.0). The bacteria were incubated at a temperature of 28 °C for 48 h.

4.5. Indole Production Activity

Indole production activity was studied using Salkowski reagent [128]. The studied bacteria were grown in 100 mL R2A medium (yeast extract 0.5; proteose peptone 0.5; casein hydrolysate 0.5; glucose 0.5; starch soluble 0.5; sodium pyruvate 0.3; dipotassium hydrogen phosphate 0.3; and magnesium sulfate 0.024) with 500 mg L^{-1} tryptophan for 72 h at 28 °C at 180 rpm. Then, bacterial cells were centrifuged twice during 90 s at 13,500 on Eppendorf AG 5804 R (Hamburg, Germany) for 15 min at 5000 rpm, and the supernatant fluid was subsequently filtered through 0.22 μm. Then, 1 mL Salkowski reagent (1 mL 0.5 M $FeCl_3$ in 50 mL of 35% $HClO_4$) was added to 0.5 mL of supernatant. After 30 min of storage in the dark, the absorbances of colored supernatants were measured on a spectrophotometer Biomate 160 (Thermo Scientific™, Madison, USA) at 540 nm. The indole contents were calculated used IAA standard curve based on dilution series of authentic IAA (0.5; 1.0; 5.0; and 50.0 μg/mL).

4.6. Phosphorus Solubilization

The ability of isolated endophytic bacteria to dissolve calcium orthophosphates was assayed on Muromtsev (g L^{-1}: glucose—10; asparagine—1; K_2SO_4—0.2; $MgSO_4x7H_2O$—0.2; agar—20; yeast extract—0.02; and tap water, pH 6.8) as described by Chebotar et al. [129]. Calcium phosphates $Ca_3(PO_4)_2$ were introduced into the nutrient medium by precipitation. To do this, 3.4 g of $CaCl_2$ and 3.8 g Na_3PO_4 were added to the sterile molten agar medium (based on one liter) as they dissolved. Then, 1.5 g of freshly formed $Ca_3(PO_4)_2$ was formed per one liter of medium. The nutrient medium was poured into Petri dishes and tested strains were spread on Petri dishes. The Petri dishes were incubated at a temperature of 28 °C for 120–216 h. The phosphorus-solubilizing activity was judged as the appearance of clear zones around the growth area of a bacterial sample spotted on the plate.

4.7. Production of Exopolysaccharides

Bacterial exopolysaccharides were assayed as described by Shaffique et al. [27] with the modification. The Congo red agar assay was used to identify the presence of exopolysaccharides if produced by the bacterial isolate. Assay plates were prepared by mixing in LB broth (25 g L^{-1}), agar (20 g L^{-1}), sucrose (10 g L^{-1}), and Congo red (0.8 g L^{-1}). All chemicals were well-mixed and then autoclaved. Petri dishes with Congo red agar were inoculated with the strains studied. PGPR grow for 5 days at 30 °C. The black colonies are visible against the red background, indicating the presence of polysaccharides.

4.8. Bacterial Phytohormone Assay

The strains studied were grown on liquid R2A medium for 72 h at 28 °C at 180 rpm. The control was uninoculated R2A medium. The concentrations of bacterial phytohormones—indole acetic acid (IAA), abscisic acid (ABA), and gibberellic acid (GAS_3) were determined using the high-performance liquid chromatograph VARIAN 212 LC with a mass-selective detector (Varian 500 MS system) as described early [58]. Detection of IAA, ABA, and GAS_3 was carried out using ESI- (electrospray) ion at 174, 265, and 345 m/z, respectively. For phytohormones determination, 50 mL of liquid culture medium (and 50 mL of sterile liquid medium used as control) was taken and centrifuged at 5000 rpm for 5 min. The supernatant was poured into a separating funnel. The combined supernatant in the separating funnel was acidified with 10% acetic acid solution to pH 2, after which phytohormones were extracted three times with 10 mL of ethyl acetate. The upper ethyl acetate layer was drained through anhydrous sodium sulfate and evaporated to dry residue on a rotary vacuum evaporator at a temperature not exceeding 40°C. The residue in the distillation flask was washed with 2 mL of mobile phase A. Chromatography was carried out in gradient mode (phase A, methanol + 0.1% formic acid; and phase B, deionized water +0.1% formic acid). A Cosmosil C18 4.6 ID × 150 mm column was used in the chromatographic system. The chromatograph was calibrated using SIGMA-ALDRICH internal standards for pure phytohormones. The identification of hormones was carried out in mass–mass mode.

4.9. Hydroponic Experiments

Seeds of wheat plant cv. Leningradskaya 6 were obtained from Leningrad Scientific Research Institute of Agricultural Science "Belogorka" (St. Petersburg, Leningrad Region, Russia). The wheat seeds were sterilized in 70% ethyl alcohol for 60 s, washed with tap water, and then placed into 2% sodium hypochlorite solution for 15 min and rinsed in sterile water 7 times. After that, the seeds were placed in Petri dishes in the thermostat at 25 °C for 48 h for the germination. The germinated seeds (three-day-old seedlings) were placed between two layers of hydrophilic tissue at a distance of 2 cm from each other. The tissue was rolled up and placed in 300 mL vessels with Knop nutrient solution (containing $CaNO_3$—1 g, KH_2PO_4—0.25 g, $MgSO_4$—0.25 g, KCl—0.125 g, and $FeCl_3$—0.0125 g per 1 L). Bacterial inoculation was carried out immediately after placing the seedlings in Knop's solution. PGPB strains Cap 03D, Cap 07D, Cap 09D, App 11D, and Cap 286 were used as inoculants. Bacteria were added to the Knop solution in concentrations of 3×10^5 CFU in mL. Bacteria were precultured on liquid medium; their titer was determined and diluted to the required concentration.

Osmotic stress was created by addition of polyethylene glycol (PEG 6000) at a concentration of 12% on the 7th day of experiment. PEG concentration was chosen in our preliminary experiments (see Figure S3), taking into account the recommendations of other researchers [34]. The vessels with wheat seedlings were placed on an installation for vegetation experiments. The vegetation was maintained in a 16 h photoperiod (16 h of light and 8 h of darkness) under a constant temperature of 22 °C during the day and 18 °C at night, with a light intensity of 23–25 klx per m^2. The vegetation period lasted for 14 days (7 days with osmotic stress exposure and inoculation with PGPB (10 days of exposure)). After the end of experiments, the 14-day-old seedlings were taken to determine the contents of photosynthetic pigments and plant ferments and the levels of hormones.

Seedlings were measured and weighted. To determine dry weight (DW), plants were dried for 5 h at 70 °C, and then for 30 min at 105 °C. Dry matter (DM) was calculated as % of plant DW in plant fresh weight FW.

4.10. Analysis of Plant Phytohormones

Plants phytohormones (IAA, ABA, and GAS_3) were determined as described early [59]. First, 10 g of leaves were homogenized with 80% methanol at 4 °C and evaporated. One half of remaining aqueous phase was acidified with 10% solution of muriatic acid to pH 2.5–3 and extracted three times in a separating funnel with 30 mL diethyl ether (phase A). The final extracts were evaporated to a dry residue and dissolved in 2 mL of mobile phase A. Detection of IAA, ABA, and GAS_3 was carried out using ESI- (electrospray) ion at 174, 265, and 345 m/z, respectively. Chromatography was performed in gradient mode.

Cosmosil C18 4.6 ID × 150 mm column was used in the chromatographic system. The liquid chromatograph VARIAN 212 LC with mass selective detector (Varian 500 MS system) was calibrated using Sigma-Aldrich internal standards for pure hormones. The hormones were identified in MS/MS mode.

4.11. Chlorophyll and Carotenoids Analysis

For the determination of chlorophyll a (Chl a), chlorophyll b (Chl b), and carotenoids (Car), concentrations of 0.2 g of fresh leaves were ground in a porcelain mortar with a small amount of acetone and sand. The ground mass was placed in tubes containing 10 mL of 80% acetone for 24 h at 20 °C. The tubes were centrifuged at 20,000× g for 20 min. Then, supernatant was transferred into a 50 mL volumetric flask and made up to the mark with acetone. The absorbance of the resulting supernatant was measured at 663, 645, and 470 nm using a spectrophotometer (ModelSpekol 1500 (Jena, Germany). The contents of chlorophyll a, chlorophyll b, and total carotenoid were calculated after [130]. The pigments concentrations in mg/dm^3 were calculated using the following formulae:

$$\text{C chl a } (mg/dm^3) = (9.784 \times OD662) - (0.99 \times OD644)$$

$$\text{C chl b } (mg/dm^3) = (21.426 \times OD644) - (4.65 \times OD662)$$

$$\text{C car } (mg/dm^3) = (4.695 \times OD440.5) - (0.268\ (\text{C chl} + \text{C chl b}))$$

where: OD—absorbtion value at the corresponding wawelenghth

To recalculate the pigments concentrations X in mg/g, the following formula was used:

$$X = C \times V/n \times 1000$$

where C—pigment content (mg/dm^3), V—extract volume in cm^3, and n—sample weight in grams.

4.12. Malondialdehyde (MDA) Content

Lipid peroxidation (LPO) occurs in response to oxidative stress, giving rise to unsaturated aldehydes as malondialdehyde (MDA). The malondialdehyde (MDA) content was measured using the method described by Uchiyama and Mihara [131]. First, 0.3 g of fresh leaves was homogenized and filtered. The reaction medium consists of 0.3 mL plant homogenate, 3 mL 1% H_3PO_4, and 1 mL 0.6% thiobarbituric acid (TBA) aqueous solution. The homogenate was put in a water bath at 95–100 °C for 60 min, after which the samples were cooled down. Then, 4 mL of n-butanol were added and centrifuged for 10 min at 10,000× g. The absorbance of butanol extract was measured at 532 nm and 600 nm using the spectrophotometer Model Spekol 1500, Jena, Germany. The concentration of TBA-reactive products was expressed in MDA $\mu M\ g^{-1}$ FW.

4.13. Catalase (CAT)

The CAT (EC1.11.1.6) was measured using the method described by Aebi [132]. This method is based on the effect of CAT on hydrogen peroxide (H_2O_2) and the measurement of the ultraviolet absorption of H_2O_2 at 240 nm. The reaction mixture at a volume of 3 mL contained 0.1 M of sodium phosphate buffer (pH 7.0), 2 mM of H_2O_2, and 0.2 mL of enzyme extract. For the calculation of the activity, the extinction coefficient of 0.036 mM^{-1} cm^{-1} was employed. CAT activity was expressed in μM of H_2O_2 per mg^{-1} protein min^{-1}.

4.14. POX Activity

Activity of POX in plants was determined as described by Ermakov et al. [133]. A 200–500 mg weight of plant material is finely ground in a porcelain mortar with acetate buffer with pH 5.4 and transferred into a 50 cm^3 measuring flask. After 10 min of infusion, the extract is centrifuged at 4000 rpm. Two quartz cuvettes with a working length of 20 mm are filled with 2 cm^3 of fermentation fume or centrifugate and 2 cm^3 of benzidine. To prepare a solution of benzidine, 100 cm^3 of distilled water is poured into a 200 cm^3 measuring flask, and 2.3 cm^3 of glacial acetic acid and 184 mg of benzidine are added. The flask is heated in a water bath at 60 °C, shaking constantly; after complete dissolution of the benzidine, 5.45 g sodium acetic acid is added to the flask, cooled, and brought to the mark with water. The cuvettes in the photoelectrocolorimeter are placed against the red filters (590 nm). The measurement is carried out using a red-light filter (at 590 nm) in the following sequence. At the beginning, with the left drum, set the galvanometer pointer to zero; then, with the right drum, move the galvanometer pointer to the leftmost position (D = 0.125 or 0.250). After that, 2 cm^3 of water is poured into the control right cuvette, and 2 cm^3 of 0.3% hydrogen peroxide from a wide-bore pipette is poured into the left experimental (measuring) cuvette. When the first drop of hydrogen peroxide is added, the stopwatch is started. The galvanometer hand starts to deviate from the edge of the scale to the zero division. The stopwatch is stopped when the galvanometer hand reaches zero. From the reaction rate found, the activity A of the enzyme is calculated as follows:

$$A = D \times a \times b \times c/f \times t$$

where:

D—optical density, equal to 0.125 or 0.250;
a—ratio of the amount of liquid taken for preparation of the extract to the mass of raw tissue, cm^3/g;
b—degree of additional dilution of the extract after centrifugation;
c—degree of constant dilution of the extract in the reaction mixture in the cuvette;
f—layer thickness (2 cm);
t—time, s.

POX activity was expressed in units, each representing the absorbance value per second per g^{-1} FW.

4.15. Statistical Analysis

The independent experiments were performed to assess the wheat growth parameters. We carried these out three times with 15 seedlings of Leningradskaya 6 ($n = 45$). We also used three biological replicates, with 30 seedlings per replicate, to assess the biochemical parameters of plants ($n = 9$). The data were statistically evaluated using STATISTICA-11 and subjected to a one-way analysis of variance (ANOVA). The mean values are shown as error bars representing standard errors of the means in all the figures. The data are presented as average mean standard error (SEM). Duncan's multiple-range test was performed to determine significant differences between individual means. The differences between the means were determined at the level of significance $p < 0.95$.

5. Conclusions

PGPB from the genera *Bacillus* spp., *Paenibacillus* sp. and *Arthrobacter* sp. isolated from arid plants *A. pseudoalhagi* and *C. album* stimulated growth and induced wheat (cv. Leningradskaya 6) drought tolerance. All bacteria produced IAA, GAS3, ABA, and siderophores. Most of them produced exopolysaccharides (with the exception of *Arthrobacter* sp. (Cap 03D). The strain Cap 07 *Bacillus* sp. possessed phosphate-mobilizing activity. The strain Cap 286 *Paenibacillus* sp. was able to fix nitrogen. Two strains from the genus *Bacillus* (Cap 09D and App 11D) grew at 50 °C and 10% NaCl. All PGPB significantly protected wheat plants under drought conditions in vegetative experiments. PGPB significantly increased the length and weight of wheat roots and shoots and decreased the content of MDA, indicating the protective effect of bacteria. PGPB modulated enzyme activities, mainly reducing the POX activity in wheat roots and shoots and differently affecting the CAT activity, increasing it in roots and decreasing it in shoots. PGPB increased the total chlorophyll content, predominantly increasing the chlorophyll *a* content in the leaves of wheat plants. PGPB modulated endogenous plant hormones, increasing the ABA content in wheat roots by 25–60.2% and, thus, helping plants overcome osmotic stress. All the bacteria studied in our experiments can be offered for further study in vegetative and field experiments on wheat cultivation under drought stress, and strains Cap 09D *Bacillus* sp. and App 11D *Paenibacillus* sp. also under soil salinization.

Supplementary Materials: The following supporting information can be downloaded at: https://www.mdpi.com/article/10.3390/plants13233381/s1, Table S1: Molecular identification of PGPB; Figure S1: Metabolic activities of PGPB isolated from drought-tolerant plants; Figure S2 Effect of 12% PEG6000 on the bacterial growth; Figure S3: Effect of different concentrations of PEG 6000 (14 days of vegetation and 7 days of osmotic stress exposure) and inoculation with PGPB on growth of wheat plants cv. Leningradskaya 6.

Author Contributions: Conceptualization, I.A.T. and V.N.P.; methodology, Y.V.K., V.N.P., G.V.M. and E.P.C. software, G.V.M. and V.N.P.; validation, E.P.C., G.V.M. and V.K.C.; formal analysis, V.N.P. and V.K.C.; investigation, V.N.P., E.P.C., G.V.M., Y.V.K., V.E.V., D.V.K., O.A.B. and P.Y.K.; resources, N.G.L. and V.N.P.; data curation, G.V.M., Y.V.K., V.N.P. and D.V.K.; writing—original draft preparation, V.N.P.; writing—review and editing, V.N.P. and V.K.C.; visualization, D.V.K. and O.A.B.; supervision, I.A.T.; project administration, I.A.T.; funding acquisition, I.A.T., V.K.C., E.P.C. and V.N.P. All authors have read and agreed to the published version of the manuscript.

Funding: This work was supported by the Russian Science Foundation Project No. 23-66-10013.

Data Availability Statement: The data presented in this study are available upon request from the corresponding authors.

Acknowledgments: This microbiological research was performed using the equipment of the Core Centrum 'Genomic Technologies, Proteomics and Cell Biology' in ARRIAM.

Conflicts of Interest: The authors declare no conflicts of interest.

References

1. Rojas, O. Agricultural extreme drought assessment at global level using the FAO-Agricultural Stress Index System (ASIS). *Weather Clim. Extrem.* **2020**, *27*, 100184. [CrossRef]
2. Mansoor, S.; Khan, T.; Farooq, I.; Shah, L.R.; Sharma, V.; Sonne, C.; Rinklebe, J.; Ahmad, P. Drought and global hunger: Biotechnological interventions in sustainability and management. *Planta* **2022**, *256*, 97. [CrossRef] [PubMed]
3. Nyaupane, S.; Poudel, M.R.; Panthi, B.; Dhakal, A.; Paudel, H.; Bhandari, R. Drought stress effect, tolerance, and management in wheat—A review. *Cogent Food Agric.* **2024**, *10*, 2296094. [CrossRef]
4. Wang, F.; He, Z.; Sayre, K.; Li, S.; Si, J.; Feng, B.; Kong, L. Wheat cropping systems and technologies in China. *Field Crops Res.* **2009**, *111*, 181–188. [CrossRef]
5. Singh, S.K.; Kumar, S.; Kashyap, P.L.; Sendhil, R.; Gupta, O.P. Wheat. In *Trajectory of 75 Years of Indian Agriculture After Independence*; Ghosh, P.K., Das, A., Saxena, R., Banerjee, K., Kar, G., Vijay, D., Eds.; Springer Nature: Singapore, 2023; pp. 137–162. [CrossRef]
6. Le Gouis, J.; Oury, F.X.; Charmet, G. How changes in climate and agricultural practices influenced wheat production in Western Europe. *J. Cereal Sci.* **2020**, *93*, 102960. [CrossRef]

7. Senapati, N.; Stratonovitch, P.; Paul, M.J.; Semenov, M. Drought tolerance during reproductive development is important for increasing wheat yield potential under climate change in Europe. *J. Exp. Bot.* **2019**, *70*, 2549–2560. [CrossRef]
8. Zhao, W.; Liu, L.; Shen, Q.; Yang, J.; Han, X.; Tian, F.; Wu, J. Effects of water stress on photosynthesis, yield, and water use efficiency in winter wheat. *Water* **2020**, *12*, 2127. [CrossRef]
9. Farooq, M.; Wahid, A.; Kobayashi, N.; Fujita, D.; Basra, S.M.A. Plant drought stress: Effects, mechanisms and management. *Agron. Sustain. Dev.* **2009**, *29*, 185–212. [CrossRef]
10. Gupta, A.; Rico-Medina, A.; Caño-Delgado, A.I. The physiology of plant responses to drought. *Science* **2020**, *368*, 266–269. [CrossRef]
11. Yang, X.; Lu, M.; Wang, Y.; Wang, Y.; Liu, Z.; Chen, S. Response mechanism of plants to drought stress. *Horticulturae* **2021**, *7*, 50. [CrossRef]
12. Camaille, M.; Fabre, N.; Clément, C.; Ait Barka, E. Advances in wheat physiology in response to drought and the role of plant growth promoting rhizobacteria to trigger drought tolerance. *Microorganisms* **2021**, *9*, 687. [CrossRef] [PubMed]
13. Sunkar, R.; Li, Y.F.; Jagadeeswaran, G. Functions of microRNAs in plant stress responses. *Trends Plant Sci.* **2012**, *17*, 196–203. [CrossRef]
14. Patel, P.; Yadav, K.; Ganapathi, T.R.; Penna, S. Plant miRNAome: Cross talk in abiotic stressful times. In *Genetic Enhancement of Crops for Tolerance to Abiotic Stress: Mechanisms and Approaches. Sustainable Development and Biodiversity*; Rajpal, V., Sehgal, D., Kumar, A., Raina, S., Eds.; Springer: Cham, Switzerland, 2019; Volume 20, pp. 25–52. [CrossRef]
15. Quarrie, S.A.; Gulli, M.; Calestani, C.; Steed, A.; Marmiroli, N. Location of a gene regulating drought-induced abscisic acid production on the long arm of chromosome 5A of wheat. *Theor. Appl. Genet.* **1994**, *89*, 794–800. [CrossRef] [PubMed]
16. Mwadzingeni, L.; Shimelis, H.; Dube, E.; Laing, M.D.; Tsilo, T.J. Breeding wheat for drought tolerance: Progress and technologies. *J. Integr. Agric.* **2016**, *15*, 935–943. [CrossRef]
17. Iquebal, M.A.; Sharma, P.; Jasrotia, R.S.; Jaiswal, S.; Kaur, A.; Saroha, M.; Angadi, U.B.; Sheoran, S.; Singh, R.; Singh, G.P.; et al. RNAseq analysis reveals drought-responsive molecular pathways with candidate genes and putative molecular markers in root tissue of wheat. *Sci Rep.* **2019**, *9*, 13917. [CrossRef] [PubMed]
18. Pshenichnikova, T.A.; Osipova, S.V.; Smirnova, O.G.; Leonova, I.N.; Permyakova, M.D.; Permyakov, A.V.; Rudikovskaya, E.G.; Konstantinov, D.K.; Verkhoturov, V.V.; Lohwasser, U.; et al. Regions of chromosome 2A of bread wheat (*Triticum aestivum* L.) associated with variation in physiological and agronomical traits under contrasting water regimes. *Plants* **2021**, *10*, 1023. [CrossRef]
19. Wang, X.; Li, Q.; Xie, J.; Huang, M.; Cai, J.; Zhou, Q.; Dai, T.; Jiang, D. Abscisic acid and jasmonic acid are involved in drought priming-induced tolerance to drought in wheat. *Crop J.* **2020**, *9*, 120–132. [CrossRef]
20. Hasanuzzaman, M.; Bhuyan, M.H.M.B.; Zulfiqar, F.; Raza, A.; Mohsin, S.M.; Mahmud, J.A.; Fujita, M.; Fotopoulos, V. Reactive Oxygen species and antioxidant defense in plants under abiotic stress: Revisiting the crucial role of a universal defense regulator. *Antioxidants* **2020**, *9*, 681. [CrossRef]
21. Noctor, G.; Foyer, C.H. Intracellular redox compartmentation and ROS-related communication in regulation and signaling. *Plant Physiol.* **2016**, *171*, 1581–1592. [CrossRef]
22. Choudhury, F.K.; Rivero, R.M.; Blumwald, E.; Mittler, R. Reactive oxygen species, abiotic stress and stress combination. *Plant J.* **2017**, *90*, 856–867. [CrossRef]
23. Chakraborty, U.; Pradhan, B. Drought stress-induced oxidative stress and antioxidative responses in four wheat (*Triticum aestivum* L.) varieties. *Arch. Agron. Soil Sci.* **2011**, *58*, 617–630. [CrossRef]
24. Nasirzadeh, L.; Sorkhilaleloo, B.; Majidi Hervan, E.; Fatehi, F. Changes in antioxidant enzyme activities and gene expression profiles under drought stress in tolerant, intermediate, and susceptible wheat genotypes. *Cereal Res. Commun.* **2021**, *49*, 83–89. [CrossRef]
25. Ukkola, A.M.; Martin, G.; De Kauwe, M.G.; Roderick, M.L.; Abramowitz, G.; Pitman, A.J. Robust future changes in meteorological drought in CMIP6 projections despite uncertainty precipitation. *Geophys. Res. Lett.* **2020**, *47*, e2020GL087820. [CrossRef]
26. Muhu-Din, A.H.G.; Zeng Yawen, Z.; Noor, S.A.; Majid, Y.M.; Ullah Aziz, U.; Muhammad, A. Conferring of drought tolerance in wheat (*Triticum aestivum* L.) genotypes using seedling indices. *Front. Plant Sci.* **2022**, *13*, 961049. [CrossRef]
27. Shaffique, S.; Imran, M.; Kang, S.-M.; Khan, M.A.; Asaf, S.; Kim, W.-C.; Lee, I.-J. Seed Bio-priming of wheat with a novel bacterial strain to modulate drought stress in Daegu, South Korea. *Front. Plant Sci.* **2023**, *14*, 1118941. [CrossRef] [PubMed]
28. Rashid, U.; Yasmin, H.; Hassan, M.N.; Naz, R.; Nosheen, A.; Sajjad, M.; Ilyas, N.; Keyani, R.; Jabeen, Z.; Mumtaz, S.; et al. Drought-tolerant *Bacillus megaterium* isolated from semi-arid conditions induces systemic tolerance of wheat under drought conditions. *Plant Cell* **2022**, *41*, 549–569. [CrossRef] [PubMed]
29. Blake, C.; Christensen, M.N.; Kovács, Á. Molecular aspects of plant growth promotion and protection by *Bacillus subtilis*. *Mol. Plant-Microbe Interact.* **2021**, *34*, 15–25. [CrossRef]
30. Kumar, A.; Ram Maurya, B.; Raghuwanshi, R. The microbial consortium of indigenous rhizobacteria improving plant health, yield and nutrient content in wheat (*Triticum aestivum*). *J. Plant Nutr.* **2021**, *44*, 1942–1956. [CrossRef]
31. Miljakovic, D.; Marinkovic, J.; Balesevic-Tubic, S. The significance of *Bacillus* spp. in disease suppression and growth promotion of field and vegetable crops. *Microorganisms* **2020**, *8*, 1037. [CrossRef]

32. Mesa, J.; Mateos-Naranjo, E.; Caviedes, M.A.; Redondo-Gómez, S.; Pajuelo, E.; Rodríguez-Llorente, I.D. Endophytic cultivable bacteria of the metal bioaccumulator *Spartina maritima* improve plant growth but not metal uptake in polluted marshes soils. *Front. Microbiol.* **2015**, *6*, 1450. [CrossRef]
33. Nadeem, S.M.; Ahmad, M.; Zahir, Z.A.; Javaid, A.; Ashraf, M. The role of mycorrhizae and plant growth promoting rhizobacteria (PGPR) in improving crop productivity under stressful environments. *Biotechnol. Adv.* **2014**, *32*, 429–448. [CrossRef] [PubMed]
34. Lastochkina, O.; Yuldashev, R.; Avalbaev, A.; Allagulova, C.; Veselova, S. The contribution of hormonal changes to the protective effect of endophytic bacterium *Bacillus subtilis* on two wheat genotypes with contrasting drought sensitivities under osmotic stress. *Microorganisms* **2023**, *11*, 2955. [CrossRef] [PubMed]
35. Chieb, M.; Gachomo, E.W. The role of plant growth promoting rhizobacteria in plant drought stress responses. *BMC Plant Biol.* **2023**, *23*, 407. [CrossRef]
36. Wang, S.; Na, X.; Yang, L.; Liang, C.; He, L.; Jin, J.; Ziyu, L.; Qin, J.; Li, J.; Wang, X.; et al. *Bacillus megaterium* strain WW1211 promotes plant growth and lateral root initiation via regulation of auxin biosynthesis and redistribution. *Plant Soil* **2021**, *466*, 491–504. [CrossRef]
37. Salem, G.; Stromberger, M.E.; Byrne, P.F.; Manter, D.K.; El-Feki, W.; Tiffany, L.; Weir, T.L. Genotype-specific response of winter wheat (*Triticum aestivum* L.) to irrigation and inoculation with ACC deaminase bacteria. *Rhizosphere* **2018**, *8*, 1–7. [CrossRef]
38. Maslennikova, D.; Lastochkina, O. Contribution of ascorbate and glutathione in endobacteria *Bacillus subtilis*—Mediated drought tolerance in two *Triticum aestivum* L. genotypes contrasting in drought sensitivity. *Plants* **2021**, *10*, 2557. [CrossRef]
39. Chen, C.; Xin, K.; Liu, H.; Cheng, J.; Shen, X.; Wang, Y.; Zhang, L. *Pantoea alhagi*, a novel endophytic bacterium with ability to improve growth and drought tolerance in wheat. *Sci. Rep.* **2021**, *11*, 8160. [CrossRef]
40. Ilyas, N.; Mumtaz, K.; Akhtar, N.; Yasmin, H.; Sayyed, R.Z.; Khan, W.; Enshasy, H.A.E.; Dailin, D.J.; Elsayed, E.A.; Ali, Z. Exopolysaccharides producing bacteria for the amelioration of drought stress in wheat. *Sustainability* **2020**, *12*, 8876. [CrossRef]
41. Sood, G.; Kaushal, R.; Sharma, M. Significance of inoculation with *Bacillus subtilis* to alleviate drought stress in wheat (*Triticum aestivum* L.). *Vegetos* **2020**, *33*, 782–792. [CrossRef]
42. Bakaeva, M.; Chetverikov, S.; Timergalin, M.; Feoktistova, A.; Rameev, T.; Chetverikova, D.; Kenjieva, A.; Starikov, S.; Sharipov, D.; Hkudaygulov, G. PGP-bacterium *Pseudomonas protegens* improves bread wheat growth and mitigates herbicide and drought stress. *Plants* **2022**, *11*, 3289. [CrossRef]
43. Lastochkina, O.V.; Allagulova, C.R. Mechanisms of growth promotional and protective effects of endophytic PGP-bacteria in wheat plants under the impact of drought (Review). *Appl. Biochem. Microbiol.* **2023**, *59*, 14–32. [CrossRef]
44. Bouremani, N.; Cherif-Silini, H.; Silini, A.; El Houda Rabhi, N.; Bouket, A.C.; Belbahri, L. Osmotolerant plant growth promoting bacteria mitigate adverse effects of drought stress on wheat growth. *AIMS Microbiol.* **2024**, *10*, 507–541. [CrossRef]
45. Ji, C.; Tian, H.; Wang, X.; Song, X.; Ju, R.; Li, H.; Liu, X. *Bacillus subtilis* HG-15, a halotolerant rhizoplane bacterium, promotes growth and salinity tolerance in wheat (*Triticum aestivum*). *Front. Plant Sci.* **2022**, *13*, 9506227. [CrossRef]
46. Liu, H.; Brettell, L.E.; Qiu, Z.; Singh, B.K. Microbiome-mediated stress resistance in plant. *Science* **2020**, *25*, 733–743. [CrossRef] [PubMed]
47. Sandrini, M.; Nerva, L.; Sillo, F.; Balestrini, R.; Chitarra, W.; Zampieri, E. Abiotics stress and belowground microbiome: The Potential of Omics Approaches. *Int. J. Mol. Sci.* **2022**, *23*, 1091. [CrossRef]
48. Coutinho, R.; Gokarn, K.; Chavan, S.; Chakraborty, P. Use of IAA-producing plant growth-promoting bacteria for the amelioration of salt and drought stress in wheat. *IJANS* **2024**, *17*, 274–291. [CrossRef]
49. Salem, M.A.; Ismail, M.A.; Radwan, K.H.; Abd-Elhalim, H.M. Unlocking the potential of plantgrowth-promoting rhizobacteria to enhance drought tolerance in egyptian wheat (*Triticum aestivum*). *Sustainability* **2024**, *16*, 4605. [CrossRef]
50. Chandra, D.; Srivastava, R.; Vadakattu, V.S.R.G.; Franco, C.M.M.; Sharma, A.K. Evaluation of ACC-deaminase-producing rhizobacteria to alleviate water-stress impacts in wheat (*Triticum aestivum* L.) plants. *Can. J. Microbiol.* **2019**, *65*, 387–403. [CrossRef]
51. Kour, D.; Yadav, A.N. Bacterial mitigation of drought stress in plants: Current perspectives and future challenges. *Curr. Microbiol.* **2022**, *79*, 248. [CrossRef]
52. Rostamian, A.; Moaveni, P.; Sadeghi-Shoae, M.; Mozafari, H.; Rajabzadeh, F. Effective drought mitigation by rhizobacteria consortium in wheat field trials. *Rhizosphere* **2023**, *25*, 100653. [CrossRef]
53. Noureen, S.; Iqbal, A.; Muqeet, H.A. Potential of drought tolerant rhizobacteria amended with biochar on growth promotion in wheat. *Plants* **2024**, *13*, 1183. [CrossRef] [PubMed]
54. Robin, A.H.K.; Ghosh, S.; Shahed, M.A. PEG-induced osmotic stress alters root morphology and root hair traits in wheat genotypes. *Plants* **2021**, *10*, 1042. [CrossRef] [PubMed]
55. Orozco-Mosqueda, M.D.C.; Santoyo, G.; Glick, B.R. Recent advances in the bacterial phytohormone modulation of plant growth. *Plants* **2023**, *12*, 606. [CrossRef]
56. Naseem, H.; Ahsan, M.; Shahid, M.A.; Khan, N. Exopolysaccharides producing rhizobacteria and their role in plant growth and drought tolerance. *J. Basic Microbiol.* **2018**, *58*, 1009–1022. [CrossRef] [PubMed]
57. Vanissa, T.T.G.; Berger, B.; Patz, S.; Becker, M.; Turečková, V.; Novák, O.; Tarkowská, D.; Henri, F.; Ruppel, S. The response of maize to inoculation with *Arthrobacter* sp. and *Bacillus* sp. in phosphorus-deficient, salinity-affected soil. *Microorganisms* **2020**, *8*, 1005. [CrossRef] [PubMed]

58. Pishchik, V.N.; Filippova, P.S.; Mirskaya, G.V.; Khomyakov, Y.V.; Vertebny, V.E.; Dubovitskaya, V.I.; Ostankova, Y.V.; Semenov, A.V.; Chakrabarty, D.; Zuev, E.V. Epiphytic PGPB *Bacillus megaterium* AFI1 and *Paenibacillus nicotianae* AFI2 improve wheat growth and antioxidant status under Ni stress. *Plants* **2021**, *10*, 2334. [CrossRef]
59. Mirskaya, G.V.; Khomyakov, Y.V.; Rushina, N.A.; Vertebny, V.E.; Chizhevskaya, E.P.; Chebotar, V.K.; Chesnokov, Y.V.; Pishchik, V.N. Plant development of early-maturing spring wheat (*Triticum aestivum* L.) under inoculation with *Bacillus* sp. V2026. *Plants* **2022**, *14*, 1817. [CrossRef]
60. Chebotar, V.K.; Zaplatkin, A.N.; Chizhevskaya, E.P.; Gancheva, M.S.; Voshol, G.P.; Malfanova, N.V.; Baganova, M.E.; Khomyakov, Y.V.; Pishchik, V.N. Phytohormone production by the endophyte *Bacillus safensis* TS3 increases plant yield and alleviates salt stress. *Plants* **2024**, *13*, 75. [CrossRef]
61. Keswani, C.; Singh, S.P.; Cueto, L.; García-Estrada, C.; Mezaache-Aichour, S.; Glare, T.R.; Borriss, R.; Singh, S.P.; Blázquez, M.A.; Sansinenea, E. Auxins of microbial origin and their use in agriculture. *Appl. Microbiol. Biotechnol.* **2020**, *104*, 8549–8565. [CrossRef]
62. Duan, J.; Jiang, W.; Cheng, Z.; Heikkila, J.J.; Glick, B.R. The complete genome sequence of the plant growth-promoting bacterium *Pseudomonas* sp. UW4. *PLoS ONE* **2013**, *8*, e58640. [CrossRef]
63. Alarcon, M.V.; Salguero, J.; Lloret, P. Auxin modulated initiation of lateral roots is linked to pericycle cell length in maiz. *Front. Plant Sci.* **2019**, *10*, 11. [CrossRef] [PubMed]
64. Duca, D.; Lorv, J.; Patten, C.L.; Rose, D.; Glick, B.R. Indole-3-acetic acid in plant-microbe interactions. *Antonie Van Leeuwenhoek* **2014**, *106*, 85–125. [CrossRef] [PubMed]
65. Qian, J.-M.; Yang, B. Bacterial-auxin-mediated bacterial colonization of plant roots. *Cell Host Microbe* **2021**, *29*, 1471–1473. [CrossRef]
66. Verma, S.; Negi, N.P.; Pareek, S.; Mudgal, G.; Kumar, D. Auxin response factors in plant adaptation to drought and salinity stress. *Physiol. Plant.* **2022**, *174*, e13714. [CrossRef]
67. Raheem, A.; Shaposhnikov, A.; Belimov, A.A.; Dodd, I.C.; Ali, B. Auxin production by rhizobacteria was associated with improved yield of wheat (*Triticum aestivum* L.) under drought stress. *Arch. Agron. Soil Sci.* **2018**, *64*, 574–587. [CrossRef]
68. Pereira, L.B.; Andrade, G.S.; Meneghin, S.P.; Vicentini, R.; Ottoboni, L.M. Prospecting plant growth-promoting bacteria isolated from the rhizosphere of sugarcane under drought stress. *Curr. Microbiol.* **2019**, *76*, 1345–1354. [CrossRef] [PubMed]
69. Zhang, M.; Yang, L.; Hao, R.; Bai, X.; Wang, Y.; Yu, X. Drought-tolerant plant growth-promoting rhizobacteria isolated from jujube (*Ziziphus jujuba*) and their potential to enhance drought tolerance. *Plant Soil* **2020**, *452*, 423–440. [CrossRef]
70. Ambreetha, S.; Chinnadurai, C.; Marimuthu, P.; Balachandar, D. Plant-associated *Bacillus modulates* the expression of auxin-responsive genes of rice and modifies the root architecture. *Rhizosphere* **2018**, *5*, 57–66. [CrossRef]
71. Ali, B.; Sabri, A.N.; Ljung, K.; Hasnain, S. Auxin production by plant associated bacteria: Impact on endogenous IAA content and growth of *Triticum aestivum* L. *Lett. Appl. Microbiol.* **2009**, *48*, 542–547. [CrossRef]
72. Ahmad, E.; Sharma, P.K.; Khan, M.S. IAA Biosynthesis in bacteria and its role in plant-microbe interaction for drought stress management. In *Plant Stress Mitigators*; Vaishnav, A., Arya, S., Choudhary, D.K., Eds.; Springer: Singapore, 2022; pp. 235–258. [CrossRef]
73. Qayyum, A.; Ayoubi, S.A.; Sher, A.; Bibi, Y.; Ahmad, S.; Shen, Z.; Jenks, M.A. Improvement in drought tolerance in bread wheat is related to an improvement in osmolyte production, antioxidant enzyme activities, and gaseous exchange. *Saudi J. Biol. Sci.* **2021**, *28*, 5238–5249. [CrossRef]
74. Kasim, W.A.; Osman, M.E.H.; Omar, M.N.; Salama, S. Enhancement of drought tolerancein *Triticum aestivum* L. seedlingsusing *Azospirillum brasilense* NO40 and *Stenotrophomonas maltophilia* B11. *Bull. Natl. Res. Cent.* **2021**, *45*, 95. [CrossRef]
75. Tsukanova, K.A.; Chebotar, V.K.; Meyer, J.J.M.; Bibikova, T.N. Effect of plant growth-promoting rhizobacteria on plant hormone homeostasis. *S. Afr. J. Bot.* **2017**, *113*, 91–102. [CrossRef]
76. Sah, S.K.; Reddy, K.R.; Li, J. Abscisic acid and abiotic stress tolerance in crop plants. *Front. Plant Sci.* **2016**, *7*, 571. [CrossRef] [PubMed]
77. Vysotskaya, L.B.; Korobova, A.V.; Veselov, S.Y.; Dodd, I.C.; Kudoyarova, G.R. ABA mediation of shoot cytokinin oxidase activity: Assessing its impacts on cytokinin status and biomass allocation of nutrient deprived durum wheat. *Funct. Plant Biol.* **2009**, *36*, 66–72. [CrossRef]
78. Daszkowska-Golec, A. The role of abscisic acid in drought stress: How aba helps plants to cope with drought stress. In *Drought Stress Tolerance in Plants*; Hossain, M., Wani, S., Bhattacharjee, S., Burritt, D., Tran, L.S., Eds.; Springer: Cham, Switzerland; Berlin/Heidelberg, Germany, 2016; Volume 2, pp. 123–151. [CrossRef]
79. Aslam, M.M.; Waseem, M.; Jakada, B.H.; Okal, E.J.; Lei, Z.; Saqib, H.S.A.; Yuan, W.; Xu, W.; Zhang, Q. Mechanisms of abscisic acid-mediated drought stress responses in plants. *Int. J. Mol. Sci.* **2022**, *23*, 1084. [CrossRef]
80. Cohen, A.C.; Bottini, R.; Pontin, M.; Berli, F.J.; Moreno, D.; Boccanlandro, H.; Travaglia, C.N.; Piccoli, P.N. *Azospirillum brasilense* ameliorates the response of *A. thaliana* to drought mainly via enhancement of ABA levels. *Physiol. Plant.* **2015**, *153*, 79–90. [CrossRef] [PubMed]
81. Barnawal, D.; Bharti, N.; Pandey, S.S.; Pandey, A.; Chanotiya, C.S.; Kalra, A. Plant growth-promoting rhizobacteria enhance wheat salt and drought stress tolerance by altering endogenous phytohormone levels and TaCTR1/TaDREB2 expression. *Physiol. Plant.* **2017**, *161*, 502–514. [CrossRef]
82. Ali, S.; Khan, N. Delineation of mechanistic approaches employed by plant growth promoting microorganisms for improving drought stress tolerance in plants. *Microbiol. Res.* **2021**, *249*, 126771. [CrossRef]

83. Wu, W.; Sun, N.; Xu, Y.; Chen, Y.; Liu, X.; Shi, L.; Chen, W.; Zhu, Q.; Gong, B.; Yin, X.; et al. Exogenous gibberellin delays maturation in persimmon fruit through transcriptional activators and repressors. *Plant Physiol.* **2023**, *193*, 840–854. [CrossRef]
84. Alharby, H.F.; Rizwan, M.; Iftikhar, A.; Hussaini, K.M.; Zia ur Rehman, M.; Bamagoos, A.A.; Alharbi, B.M.; Asrar, M.; Yasmeen, T.; Ali, S. Effect of gibberellic acid and titanium dioxide nanoparticles on growth, antioxidant defense system and mineral nutrient uptake in wheat, *Ecotoxicol. Environ. Saf.* **2021**, *221*, 112436. [CrossRef]
85. Al Mahmud, J.; Biswas, P.K.; Nahar, K.; Fujita, M.; Hasanuzzaman, M. Exogenous application of gibberellic acid mitigates drought-induced damage in spring wheat. *Acta Agrobot.* **2019**, *72*, 2. [CrossRef]
86. Kang, S.M.; Khan, A.L.; You, Y.H.; Kim, J.G.; Kamran, M.; Lee, I.J. Gibberellin production by newly isolated strain Leifsonia soli SE134 and its potential to promote plant growth. *J. Microbiol. Biotechnol.* **2014**, *24*, 106–112. [CrossRef] [PubMed]
87. Lee, K.E.; Radhakrishnan, R.; Kang, S.M.; You, Y.H.; Joo, G.J.; Lee, I.J.; Ko, J.H.; Kim, J.H. *Enterococcus faecium* LKE12 cell-free extract accelerates host plant growth via gibberellin and indole-3-acetic acid secretion. *J. Microbiol. Biotechnol.* **2015**, *25*, 1467–1475. [CrossRef] [PubMed]
88. Yasmin, H.; Nosheen, A.; Naz, R.; Bano, A.; Keyani, R. L-tryptophan-assisted PGPR-mediated induction of drought tolerance in maize (*Zea mays* L.). *J. Plant Interact.* **2017**, *12*, 567–578. [CrossRef]
89. Lotfi, N.; Soleimani, A.; Çakmakçı, R.; Vahdati, K.; Mohammadi, P. Characterization of plant growth-promoting rhizobacteria (PGPR) in *Persian walnut* associated with drought stress tolerance. *Sci. Rep.* **2022**, *12*, 12725. [CrossRef]
90. Ashry, N.M.; Alaidaroos, B.A.; Mohammed, S.A.; Badr, O.A.; El-Saadony, M.T.; Esmael, A. Utilization of drought-tolerant bacterial strains isolated from harsh soils as a plant growth-promoting rhizobacteria (PGPR). *Saudi J. Biol. Sci.* **2022**, *29*, 1760–1769. [CrossRef]
91. Chaves, M.M.; Flexas, J.; Pinheiro, C. Photosynthesis under drought and salt stress: Regulation mechanisms from whole plant to cell. *Ann. Bot.* **2009**, *103*, 551–560. [CrossRef]
92. Farooq, M.; Hussain, M.; Siddique, K.H.M. Drought stress in wheat during flowering and grain-filling periods. *Crit. Rev. Plant Sci.* **2014**, *33*, 331–349. [CrossRef]
93. Naveed, M.; Hussain, M.B.; Zahir, Z.A.; Mitter, B.; Sessitsch, A. Drought stress amelioration in wheat through inoculation with *Burkholderia phytofirmans* strain PsJN. *Plant Growth Regul.* **2014**, *73*, 121–131. [CrossRef]
94. Evans, J.R. Photosynthesis and nitrogen relationships in leaves of C3 plants. *Oecologia* **1989**, *78*, 9–19. [CrossRef]
95. Habib, N.; Ali, Q.; Ali, S.; Javed, M.T.; Zulqurnain Haider, M.; Perveen, R.; Shahid, M.R.; Rizwan, M.; Abdel-Daim, M.M.; Elkelish, A. Use of Nitric oxide and hydrogen peroxide for better yield of wheat (*Triticum aestivum* L.) under water deficit conditions: Growth, osmoregulation, and antioxidative defense mechanism. *Plants* **2020**, *9*, 285. [CrossRef] [PubMed]
96. Terzi, R.; Kadioglu, A.; Kalaycioglu, E.; Saglam, A. Hydrogen peroxide pretreatment induces osmotic stress tolerance by influencing osmolyte and abscisic acid levels in maize leaves. *J. Plant Interact.* **2014**, *9*, 559–565. [CrossRef]
97. Sohag, A.A.M.; Tahjib-Ul-Arif, M.; Brestic, M.; Afrin, S.; Sakil, M.A.; Hossain, M.T.; Hossain, M.A.; Hossain, M.A. Exogenous salicylic acid and hydrogen peroxide attenuate drought stress in rice. *Plant Soil Environ.* **2020**, *66*, 7–13. [CrossRef]
98. Luna, C.M.; Pastori, G.M.; Driscoll, S.; Groten, K.; Bernard, S.; Foyer, C.H. Drought controls on H2O2 accumulation, catalase (CAT) activity and CAT gene expression in wheat. *J. Exp. Bot.* **2005**, *56*, 417–423. [CrossRef] [PubMed]
99. Singh, S.; Gupta, A.; Kaur, N. Diferential Responses of antioxidative defense system to long-term field drought in wheat (*Triticum aestivum* L.) genotypes differing in drought tolerance. *J. Agron. Crop Sci.* **2012**, *198*, 185–195. [CrossRef]
100. Su, P.; Sui, C.; Niu, Y.; Li, J.; Wang, S.; Sun, F.; Yan, J.; Guo, S. Comparative transcriptomic analysis and functional characterization reveals that the class III peroxidase gene TaPRX-2A regulates drought stress tolerance in transgenic wheat. *Front. Plant Sci.* **2023**, *14*, 1119162. [CrossRef]
101. Patel, P.K.; Hemantaranjan, A.; Sarma, B.K.; Radha, S. Growth and antioxidant system under drought stress in chickpea (*Cicer arietinum* L.) as sustained by salicylic acid. *J. Stress Physiol. Biochem.* **2011**, *7*, 130–144.
102. Outoukarte, I.; El Keroumi, A.; Dihazi, A.; Naamani, K. Use of morpho-physiological parameters and biochemical markers to select drought tolerant genotypes of durum wheat. *J. Plant Stress Physiol.* **2019**, 5. [CrossRef]
103. Abdelaal, K.A.A.; Attia, K.A.; Alamery, S.F.; El-Afry, M.M.; Ghazy, A.I.; Tantawy, D.S.; Al-Doss, A.A.; El-Shawy, E.-S.E.; Abu-Elsaoud, A.M.; Hafez, Y.M. Exogenous application of proline and salicylic acid can mitigate the injurious impacts of drought stress on barley plants associated with physiological and histological characters. *Sustainability* **2020**, *12*, 1736. [CrossRef]
104. Kirova, E.; Pecheva, D.; Simova-Stoilova, L. Drought response in winter wheat: Protection from oxidative stress and mutagenesis effect. *Acta Physiol. Plant.* **2021**, *43*, 8. [CrossRef]
105. Abdelaal, K.; AlKahtani, M.; Attia, K.; Hafez, Y.; Király, L.; Künstler, A. The role of plant growth-promoting bacteria in alleviating the adverse effects of drought on plants. *Biology* **2021**, *10*, 520. [CrossRef] [PubMed]
106. Khamssi, N.N. Physiological and biochemical responses of durum wheat under mild terminal drought stress. *Cell. Mol. Biol.* **2018**, *64*, 59–63. [CrossRef]
107. Gholamin, R.; Khayatnezhad, M. Study of bread wheat genotype physiological and biochemical responses to drought stress. *Helix Sci. Explor. Peer Rev. Bimon. Int. J.* **2020**, *10*, 93–97. [CrossRef]
108. Mansour, E.; Mahgoub, H.A.M.; Mahgoub, S.A.; El-Sobky, E.-S.E.A.; Abdul-Hamid, M.I.; Kamara, M.M.; Qamar, S.F.A.; El-Tarabily, K.A.; Desoky, E.-S.M. Enhancement of drought tolerance in diverse *Vicia faba* cultivars by inoculation with plant growth-promoting rhizobacteria under newly reclaimed soil conditions. *Sci. Rep.* **2021**, *11*, 24142. [CrossRef]

109. Azeem, M.; Haider, M.Z.; Javed, S.; Saleem, M.H.; Alatawi, A. Drought stress amelioration in maize (*Zea mays* L.) by inoculation of *Bacillus spp.* strains under sterile soil conditions. *Agriculture* **2022**, *12*, 50. [CrossRef]
110. Zhao, X.; Yuan, X.; Xing, Y.; Dao, J.; Zhao, D.; Li, Y.; Li, W.; Wang, Z. A meta-analysis on morphological, physiological and biochemical responses of plants with PGPR inoculation under drought stress. *Plants Cells Environ.* **2022**, *46*, 199–214. [CrossRef]
111. Hasanuzzaman, M.; Nahar, K.; Hossain, M.S.; Anee, T.I.; Parvin, K.; Fujita, M. Nitric oxide pretreatment enhances antioxidant defense and glyoxalase systems to confer PEG-induced oxidative stress in rapeseed. *J. Plant Interact.* **2017**, *12*, 323–331. [CrossRef]
112. Antoniou, C.; Chatzimichail, G.; Xenofontos, R.; Pavlou, J.J.; Panagiotou, E.; Christou, A.; Fotopoulos, V. Melatonin systemically ameliorates drought stress-induced damage in *Medicago sativa* plants by modulating nitro-oxidative homeostasis and proline metabolism. *J. Pineal Res.* **2017**, *62*, e12401. [CrossRef]
113. Anjum, S.A.; Ashraf, U.; Tanveer, M.; Khan, I.; Hussain, S.; Shahzad, B.; Zohaib, A.; Abbas, F.; Saleem, M.F.; Ali, I. Drought induced changes in growth, osmolyte accumulation and antioxidant metabolism of three maize hybrids. *Front. Plant Sci.* **2017**, *8*, 69. [CrossRef]
114. Abbas, T.; Rizwan, M.; Ali, S.; Adrees, M.; Mahmood, A.; Zia-ur-Rehman, M.; Ibrahim, M.; Arshad, M.; Qayyum, M.F. Biochar application increased the growth and yield and reduced cadmium in drought stressed wheat grown in an aged contaminated soil. *Ecotoxicol. Environ. Saf.* **2018**, *148*, 825–833. [CrossRef]
115. Choudhary, A.; Kumar, A.; Kaur, N. ROS and oxidative burst: Roots in plant development. *Plant Divers.* **2020**, *42*, 33–43. [CrossRef] [PubMed]
116. Marcek, T.; Hamow, K.A.; Végh, B.; Janda, T.; Darko, E. Metabolic response to drought in six winter wheat genotypes. *PLoS ONE* **2019**, *14*, e0212411. [CrossRef] [PubMed]
117. Abid, M.; Ali, S.; Qi, L.K.; Zahoor, R.; Tian, Z.; Jiang, D.; Snider, J.L.; Dai, T. Physiological and biochemical changes during drought and recovery periods at tillering and jointing stages in wheat (*Triticum aestivum* L.). *Sci. Rep.* **2018**, *8*, 4615. [CrossRef] [PubMed]
118. Vassileva, V.; Georgieva, M.; Zehirov, G.; Dimitrova, A. Exploring the genotype-dependent toolbox of wheat under drought stress. *Agriculture* **2023**, *13*, 1823. [CrossRef]
119. Gontia-Mishra, I.; Sapre, S.; Sharma, A.; Tiwari, S. Amelioration of drought tolerance in wheat by the interaction of plant growth-promoting rhizobacteria. *Plant Biol.* **2016**, *18*, 992–1000. [CrossRef]
120. Shakirova, F.M.; Avalbaev, A.M.; Bezrukova, M.V.; Kudoyarova, G.R. Role of endogenous hormonal system in the realization of the antistress action of plant growth regulators on plants. *Plant Stress* **2010**, *4*, 32–38.
121. Akhtyamova, Z.; Arkhipova, T.; Sharipova, G.; Ivanov, R.; Nuzhnaya, T.; Kudoyarova, G.; Veselov, D. The Effect of plant growth-promoting bacteria *Bacillus subtilis* IB-22 on the hydraulic conductivity and abundance of PIP2 aquaporins in the roots of an abscisic acid-deficient barley mutant. *Int. J. Mol. Sci.* **2024**, *25*, 10706. [CrossRef]
122. Curá, J.A.; Franz, D.R.; Filosofía, J.E.; Balestrasse, K.B.; Burgueño, L.E. Inoculation with *Azospirillum* sp. and *Herbaspirillum* sp. bacteria increases the tolerance of maize to drought stress. *Microorganisms* **2017**, *5*, 41. [CrossRef]
123. Chandra, D.; Srivastava, R.; Gupta, V.V.S.R.; Franco, C.M.M.; Paasricha, N.; Saifi, S.K.; Tuteja, N.; Sharma, A.K. Field performance of bacterial inoculants to alleviate water stress effects in wheat (*Triticum aestivum* L.). *Plant Soil* **2019**, *441*, 261–281. [CrossRef]
124. Timmusk, S.; Behers, L.; Muthoni, J.; Muraya, A.; Aronsson, A.-C. Perspectives and challenges of microbial application for crop improvement. *Front. Plant Sci.* **2017**, *8*, 49. [CrossRef]
125. Lane, D.J. 16S/23S rRNA sequencing. In *Nucleic Acid Techniques in Bacterial Systematics*; Stackebrandt, E., Goodfellow, M., Eds.; John Wiley and Sons: Chichester, UK, 1991; pp. 115–175.
126. RDP. Available online: https://rdp.cme.msu.edu (accessed on 10 October 2024).
127. GenBank. Available online: https://blast.ncbi.nlm.nih.gov (accessed on 10 October 2024).
128. Chrastil, J. Colorimetric estimation of indole-3-acetic acid. *Anal. Biochem.* **1976**, *72*, 134–138. [CrossRef] [PubMed]
129. Chebotar, V.K.; Chizhevskaya, E.P.; Baganova, M.E.; Keleinikova, O.V.; Yuzikhin, O.S.; Zaplatkin, A.N.; Khonina, O.V.; Kostitsin, R.D.; Lapenko, N.G. Endophytes from halotolerant plants aimed to overcome salinity and drought. *Plants* **2022**, *11*, 2992. [CrossRef] [PubMed]
130. Wellburn, A.R. The spectral determination of chlorophylls a and b, as well as total carotenoids, using various solvents with spectrophotometers of different resolution. *J. Plant Physiol.* **1994**, *144*, 307–313. [CrossRef]
131. Uchiyama, M.; Mihara, M. Determination of malonaldehyde precursor in tissues by thiobarbituric acid test. *Anal. Biochem.* **1978**, *86*, 287–297. [CrossRef]
132. Aebi, H. Catalases. *Methods Enzym. Anal.* **1971**, *3*, 273–286.
133. Ermakov, A.I.; Arasimovich, V.V.; Yarosh, N.P.; Peruansky, Y.V.; Lukovnikova, G.A.; Ikonnikova, M.I. *Methods of Biochemicalanalysis of Plants*, 3rd ed.; Agropromizdat: Leningrad, Russia, 1987; pp. 42–43, (430). (In Russian)

plants

Brief Report

Effect of Biopesticide Novochizol on Development of Stem Rust *Puccinia graminis* f. sp. *tritici* in Wheat, *T. aestivum* L.

Andrey B. Shcherban [1,2,*], Ekaterina S. Skolotneva [2], Anna V. Fedyaeva [2], Natalya I. Boyko [2] and Vladislav V. Fomenko [3]

1 Kurchatov Genomics Center, Institute of Cytology and Genetics SB RAS, Lavrentiev ave. 10, 630090 Novosibirsk, Russia
2 Institute of Cytology and Genetics SB RAS, Lavrentiev ave. 10, 630090 Novosibirsk, Russia; sk-ska@yandex.ru (E.S.S.); fedyaevaav@bionet.nsc.ru (A.V.F.); bojkoni@bionet.nsc.ru (N.I.B.)
3 N.N. Vorozhtsov Institute of Organic Chemistry SB RAS, Lavrentiev ave. 9, 630090 Novosibirsk, Russia; chemistnsk@yandex.ru
* Correspondence: atos@bionet.nsc.ru

Abstract: The use of biological plant protection products is promising for agriculture. In particular, chitosan-based biopesticides have become widespread for stimulating growth and protecting plants from a wide range of pathogens. Novochizol is a product obtained by intramolecular cross-linking of linear chitosan molecules and has a globular shape, which provides it with a number of advantages over chitosan. Novochizol has previously been shown to have a stimulating effect on the growth and development of common wheat (*Triticum aestivum* L.). However, the effect of this preparation on the protective mechanisms against rust diseases has not been studied before. Our studies have revealed the dose effect of the preparation on the development of stem rust of wheat. When treating plants with novochizol at a concentration of 0.125% four days before infection, the best results were obtained, namely: a stable reaction was observed and the number of pustules decreased. To identify critical points of the drug's effect on the protective mechanism against stem rust, we used an adrenaline test, which allows for a quick assessment of the pro/antioxidant status of plant extracts. We also assessed the activity of the major antioxidant enzymes, peroxidase and catalase, using commercial kits and the Folin–Ciocalteu reaction to assess the concentration of phenolic compounds. As a result, two stages were identified in infected plants pretreated with novochizol: early (up to 10 h after inoculation), characterized by antioxidant activity, and late (10–244 h), with prooxidant activity. These stages correspond to two peaks of accumulation of reactive oxygen species (ROS) in response to pathogen infection. The first peak is associated with the accumulation of superoxide anion O_2-, which is converted into oxygen and hydrogen peroxide under the action of the enzyme SOD (superoxide dismutase). The second peak is associated with the accumulation of H_2O_2. Hydrogen peroxide performs a protective function leading to the death of pathogen mycelial cells. In comparison with infected plants without novochizol treatment, we found a decrease in the activity of catalase (an enzyme that breaks down H_2O_2) at both stages, as well as peroxidase in the interval from 10 to 144 h after inoculation. Also, an increase in the concentration of phenolic compounds was found in the treated infected plants. We suggest that these changes under the influence of pretreatment with novochizol contribute to enhancements in plant defense functions against stem rust. Taking into account the physicochemical advantages of novochizol over chitosan, which provide a very low effective dose of the drug, the obtained results indicate its promise and safety as a biological plant protection product. This work is a preliminary stage for an extended analysis of the effect of novochizol on plant immunity using biochemical and molecular genetic approaches.

Keywords: wheat; *Triticum aestivum*; stem rust; resistance; biopesticide; chitosan; novochizol; antioxidant activity; reactive oxygen species (ROS); total phenolic content (TPC); peroxidase; catalase

1. Introduction

Genetic breeding is a successful for of modern technology for increasing productivity and stress resistance in agricultural plants; however, the development and use of highly effective plant protection products (PPPs) remains relevant. Traditional, chemical pesticides have significant drawbacks: toxicity and allergenicity for humans and animals, reduced nutritional value of food products, development of resistance in harmful organisms, and pollution of the environment. Biological PPPs based on natural compounds are increasingly attracting attention, since, without these drawbacks, they are often not inferior to chemical preparations in effectiveness.

The range of biopesticides and their application methods are quite diverse. But they all target the interaction between the pathogen and the plant, acting either by increasing the plant's immunity or by interrupting the pathogen's life cycle. The outcome of this interaction depends on the pathogen, including lifestyle and nutrition strategy: biotrophy, hemibiotrophy, and necrotrophy. In the process of evolution, the host plant has developed hormone-dependent defense mechanisms corresponding to each type of pathogen. Currently, two classical mechanisms are generally recognized: (1) involving salicylic acid (SA) for the host defense response against biotrophic and hemibiotrophic pathogens; (2) involving jasmonic acid (JA) and ethylene, which promote the defense response against necrotrophs and hemibiotrophs [1].

Chitin and chitosan derivatives are widely used as biological PPPs [2,3]. The polymeric carbohydrate chitin is widespread in nature and is a component of the integuments of arthropods (including crustaceans and insects) and fungi. Chitosan is obtained by hydrolysis and deacetylation of chitin. Chitosan-based preparations have a stimulating effect on plant growth and development, and also enhance resistance to abiotic stress [4,5]. Chitosan derivatives are also of particular interest as inducers of resistance to fungal, bacterial, and viral diseases [6,7]. As for biotrophic fungal pathogens, it was shown that chitosan treatment of plants activates the SA-dependent cascade, leading to the accumulation of reactive oxygen species (ROS), hypersensitivity reaction (HSR), synthesis of phenolic substances, and specific protective PR (pathogenesis-related) proteins [8,9]. Increasingly, evidence suggests that other hormones and signaling molecules, especially ROS and nitric oxide (NO), are involved in the regulation of plant defense responses under the action of chitosan [1,8]. The efficiency of chitosan derivatives can be significantly enhanced by their modification: introduction of functional groups of Schiff bases, halogen atoms (Cl or F), metal nanoparticles, urea groups, etc. [7]. A positive protective effect of combining chitosan preparations with other biologically active substances (BASs) and beneficial microorganisms (plant growth-promoting bacteria—PGPB) has been established [10–12].

A new chitosan derivative, novochizol, obtained by intramolecular cross-linking of linear chitosan molecules, has great potential. The novochizol molecule has a globular shape, which gives it a number of advantages over chitosan: increased solubility in aqueous solutions, chemical stability and resistance to biodegradation, high penetrating ability and adhesion, as well as the ability to absorb various BASs and slowly release them [13]. The latter property is especially important in terms of creating complex formulations with any BAS, regardless of their structural and physical parameters. Another important difference between novochizol and chitosan is the consistent and identical quality of the product. This allows for both novochizol itself and formulations to be used in a predictable and reproducible manner [14]. Novochizol has a growth-stimulating effect when treated with seeds and leaves. Using common wheat, *T. aestivum*, as an example, it was shown that this preparation enhanced seed germination and contributed to an increase in root mass and the total mass of plants [15]. However, there are still no studies devoted to the influence of novochizol on plant sensitivity to biotic stress factors.

Since chitosan preparations have been previously shown to have a protective effect against various pathogens, it was of interest to evaluate the similar effect of novochizol, as a chitosan derivative. This article is devoted to the laboratory evaluation of the effect of novochizol on the susceptibility of common wheat plants to stem rust caused by *P.*

graminis f. sp. *tritici* (Pgt). The latter is a very harmful fungal pathogen that in some years affects up to 40% of wheat crops in Russia [16]. Since *P. graminis* is a biotrophic pathogen, we hypothesize that novochizol enhances wheat resistance to stem rust through an SA-dependent cascade that affects the antioxidant system (see above). This system is known to maintain the level of ROS required for protection against the pathogen and serves as a convenient marker to assess the action of biopesticides [17]. At the initial step, it was necessary to find the critical points at which the reaction to the pathogen occurs under the influence of novochizol. For this purpose, biochemical tests were used in this work: an adrenaline test for antioxidant activity and an assessment of peroxidase and catalase activity. Additionally, an express test for phenolic compounds was used. In the future, a more detailed study of the effect of novochizol on wheat immunity will be carried out with an analysis of the expression of various known PR genes and cytophysiological studies.

2. Results

2.1. Visual Assessment of Novochizol's Effect on Stem Rust Development

During the first stage, the effect of different novochizol concentrations on the disease development was studied on the susceptible variety Novosibirskaya 29. A wide range of drug concentrations was used in the experiments—from 0.125 to 2.5%. Pustules with IT "4" (pustules of type 4) were formed on the plants treated with water (k). The results showed that novochizol in any concentration affected the development of the disease, which was manifested in a decrease in the density of urediniopustules, a reduction in their size, and the appearance of chlorosis zones of various sizes around sporulation (Figure 1). The minimal IT decrease was noted when using 2.5% solution of novochizol (IT "3","3+"). The preparation in concentrations of 0.125, 0.75, and 1.5% induced a stable reaction of plants. The maximum number of pustules was observed on the leaves of control plants, but the minimum was found when treating plants with novochizol at a dose of 0.125% (20 and 8 pustules per leaf, respectively) (Figure 1). This concentration was used for further study of plant defense reactions.

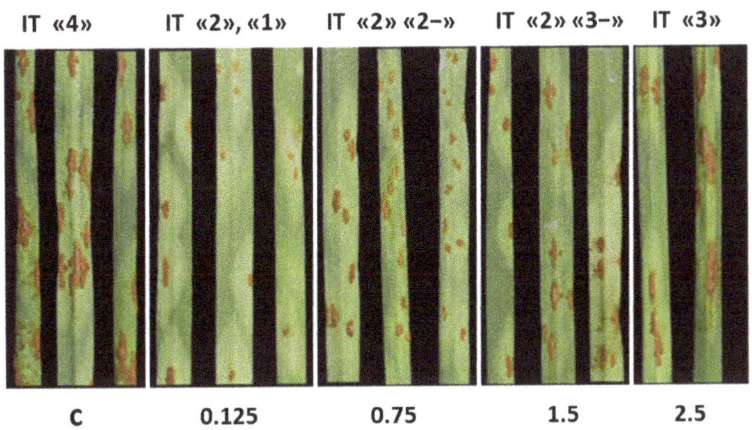

Figure 1. Visual assessment of the novochizol effect on the stem rust development. c—control plants treated with water; 0.125; 0.75; 1.5; 2.5—the different concentrations of novochizol (in %) used to treat plants before infection with stem rust pathogen. The sensitivity indices of IT are indicated above the figure.

2.2. AA Determination

As described in Materials and Methods, AA was assessed by the degree of inhibition of superoxide radical by plant extracts in the adrenaline autooxidation reaction. The AA level of control plants changed from prooxidant activity at 2–10 hp/in to antioxidant activity (72–240 hp/in) (Figure 2A). We assume that this change was caused by stress due to the limited size of the pot and the root ball. Uninfected novochizol(+) plants showed increased antioxidant activity up to 24 hp/in compared to control, which may be due to the effect of novochizol (Figure 2A; Table S1). In infected plants, compared to the control, we observed an increased level of AA at 2–10 hp/in, most pronounced in plants not treated with novochizol at 7 hp/in. After 10 hp/in, the curve shifts towards prooxidant activity, especially in plants pretreated with novochizol (Figure 2A).

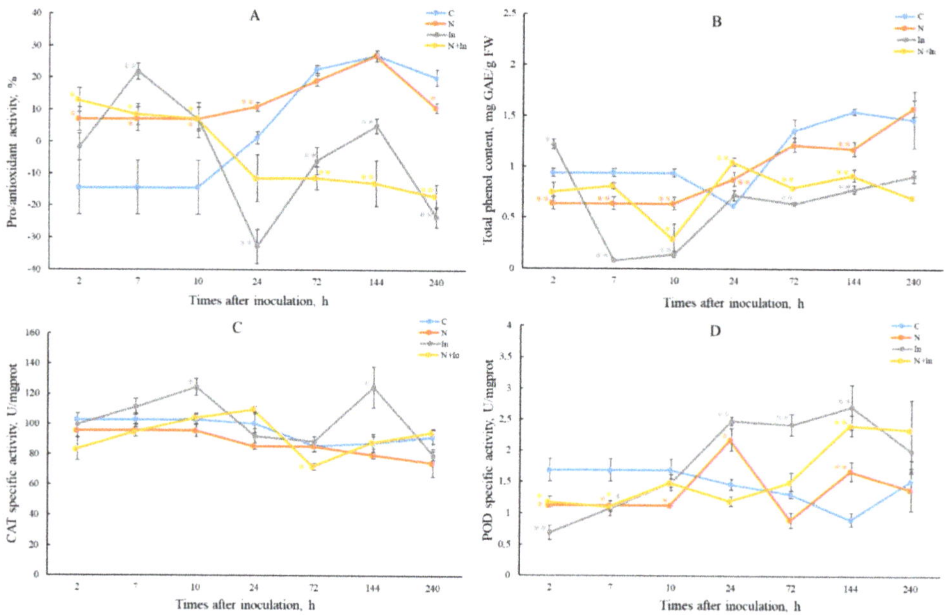

Figure 2. Changes in pro/antioxidant activity (**A**), total phenol content (**B**), catalase activity (**C**) and peroxidase activity (**D**) in leaves of common wheat after treatment with 0.125% novochizol and inoculation with urediniospores of the West Siberian population *Pgt*. Designations: C—control; N—novochizol treatment; In—inoculation; N+In—novochizol treatment and inoculation. *p* values * $p \leq 0.05$, ** $p \leq 0.01$, when compared with control plants at the same time point.

2.3. *TPC Determination*

TPC was assessed using Folin–Ciocalteu reagent (see Section 4). On average, control plants have a higher TPC value compared to other groups, except for the 24 hp/in point (Figure 2B) (Figure 2B; Table S1). Infected novochizol(−) plants demonstrated a lower TPC value at the interval about 5–144 hp/in, compared to infected novochizol(+) plants.

2.4. *Estimation of Peroxidase and Catalase Activity*

The level of catalase (CAT) activity in uninfected control and novochizol(+) plants remained more or less stable during the experiment (Figure 2C; Table S1). In infected novochizol(−) plants, catalase activity increased at 10 hp/in and then decreased to control level at 72 hp/in. After the last point, the activity increased again, reaching its highest value at 144 hp/in. Unlike the previous group, no significant changes in the level of catalase activity were observed in the infected novochizol(+) plants compared to control plants, with the exception of a slight decrease in this level below the control at 72 hp/in.

In uninfected novochizol(+) plants at 24 hp/in, a significant increase in peroxidase (POD) activity was observed relative to the control (Figure 2D; Table S1). But after 2 days, the activity dropped below the control level. In infected plants, a reduced activity of this enzyme was observed within about 2–10 hp/in. After that, the infected novochizol(−) plants demonstrated an increase in peroxidase activity up to 144 hp/in, whereas novochizol(+) plants showed a later increase in this activity after 72 hp/in reaching the level of the previous group at 144 hp/in.

3. Discussion

This paper presents preliminary data on the effect of novochizol on the development of stem rust in common wheat. A concentration-dependent effect of novochizol on the development of stem rust was established (Figure 1). This is consistent with the results of previous studies on the effect of chitosan doses on defense responses [4]. In the variant with 0.125% novochizol treatment, the best ratio of pustules by infectious type and condition of leaves was revealed. We used this preparation in the following stages of work.

In the 2000s, a hypothesis of a two-level organization of plant immunity was formulated, which was called "PTI-ETI" [18]. It was assumed that plants have PRRs (pattern recognition receptors) that recognize molecules of pathogenic microorganisms (PAMPs—pathogen-associated molecular patterns). As a result of recognition of these molecules, the first level of PTI (PAMP-triggered immunity) protection is launched. After overcoming PTI, the second level of protection associated with the recognition of specific effectors is activated—ETI (effector-triggered immunity).

According to earlier studies, two peaks of ROS formation in response to pathogen infection were revealed in plants. The first peak occurs within minutes or few hours after recognition of PAMPs and was associated with activation of the enzyme NADPH oxidase, which constitutively exists in the membrane, and formation of superoxide anion O_2-, which is quickly converted into H_2O_2 by the enzyme superoxide dismutase (SOD) [19]. To study the pro/antioxidant activity associated with SOD, we used the adrenaline test (www.csun.edu/~hcchm001/sodassay.pdf accessed on 5 December 2024) [20,21]. As a result, we observed an increase in AA at 2–10 hp/in, which corresponds to the first peak of ROS accumulation (Figure 2A). A higher level of AA is characteristic of infected plants without treatment with novochizol, while the level of infected novochizol(+) plants is almost the same as that of uninfected novochizol(+) plants. This may indicate a lower level of stress sensitivity provided by the novochizol treatment. After 10 hp/in, we observed a decrease in AA in plants, both treated and untreated with novochizol (prooxidant activity).

The second peak appears after 3–5 days and is associated with de novo synthesis of enzymes of the pro/antioxidant system (peroxidases, catalase, oxalate oxidases, etc.). The pro/antioxidant system maintains an optimal level of ROS in tissues. Catalase breaks down H_2O_2 into water and molecular oxygen; various peroxidases, polyphenol oxidase, and ascorbate oxidase utilize ROS in oxidative reactions [22]. Next, we analyzed the activity of antioxidant enzymes peroxidase and catalase.

The level of catalase activity in infected plants without novochizol treatment demonstrated two peaks at 10 and 144 hp/in (Figure 2C). These peaks probably correspond to the main peaks of ROS accumulation. In infected plants with treatment, this level was generally lower, except for the point 24 hp/in. Along with the data on the AA estimation (see above), this may indicate a higher stress resistance of the latter group. As for peroxidase, its activity level was lower in the same group in the period from 10 to 144 hp/in (Figure 2D). During the assessment of the effect of complex chitosan preparations on wheat resistance to fungal diseases, the ability of these preparations to inhibit the activity of catalase and peroxidase was shown [23]. Suppression of the activity of these enzymes leads to an increased level of H_2O_2 and stimulation of defense reactions against pathogens. It is known that ROS, including hydrogen peroxide, are directly involved in the formation and implementation of plant defense responses when infected with pathogens. The accumulation of H_2O_2 was manifested in the suppression of the development of the biotroph *Puccinia recondita* and the suppression of disease symptoms in the form of necrosis for the hemibiotroph *Cochliobolus sativus* [23]. We hypothesize that a similar mechanism operates against stem rust in wheat tissues treated with novochizol. Recently, a histochemical analysis of this plant material was carried out at 96 hp/in and an intensive accumulation of H_2O_2 in the leaves was established [24]. Apparently, this is associated with the manifestation of the second peak of the oxidative burst. Moreover, in areas of intensive accumulation of H_2O_2, the death of pathogen mycelial cells was observed. A protective role of H_2O_2 against fungal pathogens has also been confirmed by other authors [25].

Phenolic substances play an important role in protecting plants from pathogens [26]. It was previously shown that the formation of phenols and strengthening of cell walls with lignin after treatment with chitosan were the most typical defense reactions against pathogenic fungi [4,6]. Our data show that treatment with novochizol stimulated more intensive accumulation of phenols than in untreated plants throughout most of the experiment (Figure 2B). This difference is most noticeable at an early stage up to 10 hp/in.

It was previously shown that after the use of chitosan in plants infected with fungi, the expression of genes encoding PR proteins (chitinase, glucanase, peroxidase, polyphenoloxidase, PR-1, PR-5, etc.) increased [8,9,27,28]. Some of these PR proteins, like peroxidase and polyphenoloxidase, are involved in the metabolism of phenolic compounds. We further plan to study the expression of various PR genes in infected wheat treated with novochizol.

It seems interesting to compare the efficiency of novochizol with other chitosan- based preparations in field conditions. In the Russian Federation, a number of these preparations containing PGPB (Vitaplan, Gamair), as well as complex preparations with organic acids chitosan I and II, have been developed [29]. Comparison of their effects with the effect of novochizol on the development of root rot in wheat showed similar efficiency (unpublished data). Unfortunately, we were unable to properly evaluate the effect of novochizol on stem rust in the field due to unfavorable conditions for this disease in the last 2 years. Nevertheless, novochizol demonstrated a more pronounced effect on grain productivity, compared with the above chitosan preparations [29,30].

Thus, we have identified for, the first time, the critical points associated with the effect of novochizol in response to infection of wheat with the fungal pathogen of stem rust. The identified changes in pro/antioxidant activity, activity of the main antioxidant enzymes, and concentration of phenolic compounds indicate the similarity of the protective effects of novochizol and some other forms of chitosan. However, given the cost-effectiveness provided by the low active concentration and ease of application of novochizol, one can assume its high potential as a safe and effective biological plant protection product.

4. Material and Methods

4.1. Plant Material

The research objects were 10-day-old seedlings of the common wheat variety Novosibirskaya 29, susceptible to stem rust. The plants were grown in vessels with soil, which is recommended for experiments with rust fungi by international protocols [31].

4.2. Stem Rust Pathogen

To infect the seedlings, we used urediniospores of the West Siberian population *Pgt* from the collection of the Institute of Cytology and Genetics of SB RAS. Urediniospores were stored at −70 °C before the experiment and re-cultivated on the susceptible variety of common wheat "Khakasskaya" [32].

4.3. Novochizol Treatment and Plant Infection

Novochizol™ base, having a degree of deacetylation of 90% or above and a molecular weight in the 500 kD range, was provided by NOVOCHIZOL SA, Monthey, Switzerland (www.novochizol.ch accessed on 5 December 2024). Novochizol aqueous suspensions were obtained as described in [15]. In laboratory conditions, greater biomass accumulation and more uniform seedling development were observed with 0.1% novochizol treatment [15]. However, the effect of any concentration of novochizol on the development of stem rust has not been previously assessed, so we selected the range 0.1–2.5%, since concentrations above 2.5% complicate both the processes of dissolution and application of the preparation by spraying. The seedlings were treated with novochizol solutions at concentrations of 0.125, 0.75, 1.5, and 2.5%. The solutions were applied to the plants using a sprayer (15 mL/100 plants) four days before infection with *Pgt*.

A suspension of urediniospores with a concentration of 0.8 mg/mL Novec 7100 (3MNovec^{TM}, St. Paul, MN, USA) [33] was applied to the seedlings using a sprayer. Control plants were sprayed with bidistilled water. Inoculated plants were incubated for 24 h in a humidified chamber in the dark at a temperature of 15–20 °C to ensure maximum spore germination. Then, the plants were transferred to growth chambers and incubated under 16 h illumination with phytolamps with an intensity of 10,000 lux at a temperature of 26–28 °C. This temperature is necessary for the formation of appressoria, penetration of the pathogen into the stomata, and the development of infectious hyphae in the intercellular spaces of the plant [34].

4.4. Evaluation of the Susceptibility Level of Plants to Infection

The disease scoring was assessed in accordance with the recommendations of Roelfs et al. (1992) for seedlings [34]. Fourteen days after inoculation, we evaluated both the infection type and the number of pustules on the leaf blade. The infection types of plant response (IT) was determined 12–14 days after inoculation using the modified Stackman scale, where IT "0", ";", "1", and "2" were interpreted as resistant (low—L), and "3", "3+", and "4" were interpreted as susceptible (high—H) [35].

4.5. Evaluation of Antioxidant Activity of Plant Extracts

Taking into account the previous assessment of the effect of novochizol on disease development, biochemical analyses were performed on the plants treated with 0.125% novochizol. The material was collected 2, 7, 10, 24, 72, 144, and 240 h after inoculation (hp/in). Leaf pieces were frozen in liquid nitrogen and stored at −80 °C until analysis. Extracts were obtained using a modified method [36]. Frozen leaves were ground in cooled sterile mortars. A quantity of 0.1 g of tissue was placed in cooled Eppendorf tubes and 0.6 mL of cooled (4 °C) 0.2 M K-phosphate buffer pH 7.8 was added. The tubes were thoroughly shaken and centrifuged for 20 min at 10,000 rpm at 4 °C. The supernatant was collected and stored on ice until analysis. Antioxidant activity (AA) of plant extracts was assessed by their ability to inhibit the reaction of adrenaline autooxidation with its conversion to adrenochrome, according to a modified method [24]. The reaction mixture was obtained by sequentially adding 0.045 mL of 0.2 M K-phosphate buffer (pH 7.8) and 0.15 mL of 0.1% (5.46 mM) adrenaline hydrochloride solution to 3 mL of 0.2 M Na-carbonate buffer (pH 10.65). The control sample (D1) was mixed, placed in a spectrophotometer SPEKS SSP 705 ("Spectroscopic Systems", Moscow, Russia), and the optical density was measured for 30 s. at a wavelength of 347 nm (absorption spectrum of adrenochrome). When analyzing plant extract samples, the K-phosphate buffer was replaced with 0.045 mL of the extract (D2). AA (%) was assessed as a percentage of adrenaline autooxidation inhibition using the following formula: AA = ((D1 − D2) × 100)/D1.

4.6. Determination of Total Phenolic Content

Total phenolic content (TPC) was estimated using Folin–Ciocalteu reagent [37]. Frozen leaves were ground in cooled sterile mortars. Samples of 0.02 g of ground tissue were placed in cooled Eppendorf tubes; 2 mL of ice-cold 95% methanol was added in each sample, vortexed, and left in the dark for 48 h at room temperature for extraction. The tubes were centrifuged at 14,500 rpm for 5 min and the supernatant was transferred to new tubes. A quantity of 0.1 mL of each sample was used for the analysis. Standard solutions of gallic acid and 95% methanol for the blank probe were used to construct the calibration curve. Gallic acid solutions were prepared in 95% methanol; the final concentrations were 10, 50, 100, 150, 200, and 240 μg/mL; 0.2 mL of 10% (*v*/*v*) Folin–Ciocalteu reagent was added to the analyzed samples and mixed in a vortex, and after 3–4 min, 0.8 mL of 700 mM Na_2CO_3 was added. Immediately after that, the samples were vortexed and left in the dark at room temperature for 2 h. After 2 h, the samples were centrifuged as described above. The optical density of the solutions was determined at $\lambda = 765$ nm using a spectrophotometer. TPC was determined from a calibration graph using the regression equation between gallic acid standards and optical density of solutions at $\lambda = 765$ nm and expressed as gallic acid equivalents.

4.7. Evaluation of Enzymatic Activity

The collected plant material was stored until analysis at −80 °C as described above. Samples were ground in cooled mortars, 0.05 g was collected, and 0.45 mL of cold PBS buffer (pH 7.4) with 1 mM EDTA was added. The samples were then centrifuged at 7000 rpm for 10 min at 4 °C. The supernatant was collected and stored on ice until analysis. The analyses were performed on the same day. The protein level in the samples was estimated using the Bradford protein determination kit (Catalog No: E-BC-K168-S, Elabscience Inc., Houston, TX, USA). No sample dilution was required to determine catalase activity. To determine peroxidase activity, the samples were additionally diluted with 0.01 M PBS buffer (pH 7.4) containing 1 mM EDTA with a dilution factor of 1:10. The enzymatic activity of each sample was determined according to the manufacturer's protocols using the Plant Peroxidase (POD) Activity Assay Kit (Catalog No.: E-BC-K227-S) and Catalase (CAT) Activity Assay Kit (Catalog No.: E-BC-K031-M) from Elabscience Inc.

4.8. Statistical Analysis

All experiments were conducted in 3 biological and 2 analytical replicates. Data obtained from each analysis were first evaluated for normality. Data were analyzed using Statistica 12 software from StatSoft (Tulsa, OK, USA). To test for the significant difference between the two samples, we used the Mann–Whitney U test (Table S1). Figure 2 presents the mean values and their standard errors.

Supplementary Materials: The following supporting information can be downloaded at: https://www.mdpi.com/article/10.3390/plants13233455/s1, Table S1: Summary of the significant difference (p-values) obtained with Mann-Whitney U Test. At p-values <0.05—significant difference between the compared groups, p-values >0.05—not significant difference. Designations as in Figure 2.

Author Contributions: Conceptualization, A.B.S. and E.S.S.; methodology, A.B.S., E.S.S. and A.V.F.; investigation, E.S.S., A.V.F. and N.I.B.; chemical resources, V.V.F.; supervision, A.B.S.; writing, A.B.S., E.S.S. and A.V.F.; project administration, A.B.S. All authors have read and agreed to the published version of the manuscript.

Funding: This research was funded by the Russian Science Foundation Project No. 23-16-00119.

Data Availability Statement: The data presented in this study are available upon request from the corresponding authors.

Conflicts of Interest: The authors declare no conflicts of interest.

References

1. Jones, J.; Dangl, J. The plant immune system. *Nature* **2006**, *444*, 323–329. [CrossRef] [PubMed]
2. Rinaudo, M. Chitin and chitosan: Properties and applications. *Prog. Polym. Sci.* **2006**, *31*, 603–632. [CrossRef]
3. Malerba, M.; Cerana, R. Chitosan effects on plant systems. *Int. J. Mol. Sci.* **2016**, *17*, 996. [CrossRef] [PubMed]
4. Orzali, L.; Corsi, B.; Forni, C.; Riccioni, L. Chitosan in Agriculture: A new challenge for managing plant disease. In *Biological Activities and Application of Marine Polysaccharides*; Shalaby, E.A., Ed.; InTech: London, UK, 2017; pp. 87–96. [CrossRef]
5. Chakraborty, M.; Hasanuzzaman, M.; Rahman, M.; Khan, M.A.R.; Bhowmik, P.; Mahmud, N.U.; Tanveer, M.; Islam, T. Mechanism of plant growth promotion and disease suppression by chitosan biopolymer. *Agriculture* **2020**, *10*, 624. [CrossRef]
6. Shcherban, A.B. Chitosan and its derivatives as promising plant protection tools. *Vavilov J. Genet. Breed.* **2023**, *27*, 1010–1021. [CrossRef] [PubMed]
7. Maluin, F.N.; Hussein, M.Z. Chitosan-Based Agronanochemicals as a Sustainable Alternative in Crop Protection. *Molecules* **2020**, *25*, 1611. [CrossRef]
8. Manjunatha, G.; Niranjan-Raj, S.; Prashanth, G.N.; Deepak, S.; Amruthesh, K.N.; Shetty, H.S. Nitric oxide is involved in chitosan-induced systemic resistance in pearl millet against downy mildew disease. *Pest Manag. Sci.* **2009**, *65*, 737–743. [CrossRef]

9. Manjunatha, G.; Roopa, K.S.; Prashanth, G.N.; Shetty, H.S. Chitosan enhances disease resistance in pearl millet against downy mildew caused by *Sclerospora graminicola* and defence-related enzyme activation. *Pest Manag. Sci.* **2008**, *64*, 1250–1257. [CrossRef]
10. Rkhaila, A.; Chtouki, T.; Erguig, H.; El Haloui, N.; Ounine, K. Chemical proprieties of biopolymers (Chitin/Chitosan) and their synergic effects with endophytic bacillus species: Unlimited applications in Agriculture. *Molecules* **2021**, *26*, 1117. [CrossRef]
11. Yarullina, L.G.; Burkhanova, G.F.; Tsvetkov, V.O.; Cherepanova, E.A.; Sorokan, A.V.; Zaikina, E.A.; Mardanshin, I.S.; Fatkullin, I.Y.; Maksimov, I.V.; Kalatskaja, J.N.; et al. The effect of chitosan conjugates with hydroxycinnamic acids and *Bacillus subtilis* bacteria on the activity of protective proteins and resistance of potato plants to *Phytophthora infestans*. *Appl. Biochem. Microbiol.* **2024**, *60*, 231–240. [CrossRef]
12. Yarullina, L.; Kalatskaja, J.; Tsvetkov, V.; Burkhanova, G.; Yalouskaya, N.; Rybinskaya, K.; Zaikina, E.; Cherepanova, E.; Hileuskaya, K.; Nikalaichuk, V. The influence of chitosan derivatives in combination with *Bacillus subtilis* bacteria on the development of systemic resistance in potato plants with viral infection and drought. *Plants* **2024**, *13*, 2210. [CrossRef]
13. Fomenko, V.; Loroch, V. Novochizol: A new type of cross- linked chitosan particles for formulation and parsimonious delivery of copper compounds. In Proceedings of the 6th European Copper Conference in Plant Protection, Berlin, Germany, 17–18 November 2021; IFOAM Organics Europe: Berlin, Germany. Available online: https://www.boelw.de/fileadmin/user_upload/Dokumente/Veranstaltungen/Kupfertagung_2021/copper_novochizol.pdf (accessed on 5 December 2024).
14. Lipovka, A.; Kharchenko, A.; Dubovoy, A.; Filipenko, M.; Stupak, V.; Mayorov, A.; Fomenko, V.; Geydt, P.; Parshin, D. The Effect of Adding Modified Chitosan on the Strength Properties of Bacterial Cellulose for Clinical Applications. *Polymers* **2021**, *13*, 1995. [CrossRef]
15. Teplyakova, O.I.; Fomenko, V.V.; Salakhutdinov, N.F.; Vlasenko, N.G. Novochizol™ seed treatment: Effects on germination, growth and development in soft spring wheat. *Nat. Prod. Chem. Res.* **2022**, *10*, 1–4.
16. Skolotneva, E.S.; Lekomtseva, S.N.; Kosman, E. The wheat stem rust pathogen in the central region of the Russian Federation. *Plant Pathol.* **2013**, *5*, 1003–1010. [CrossRef]
17. Grant, J.J.; Loake, G.J. Role of reactive oxygen intermediates and cognate redox signaling in disease resistance. *Plant Physiol.* **2000**, *124*, 21–29. [CrossRef] [PubMed]
18. Gill, U.S.; Lee, S.; Mysore, K.S. Host versus nonhost resistance: Distinct wars with similar arsenals. *Phytopathology* **2015**, *105*, 580–587. [CrossRef] [PubMed]
19. Boller, T.; Keen, N.T. Perception and transduction of elisitor signals in host-pathogen interactions. In *Mechanisms of Resistance to Plant Diseases*; Slusarenko, A.J., Frazer, R.S.S., van Loon, L.S., Eds.; Kluwer: Dordrecht, The Netherlands, 2000; pp. 189–230.
20. Ryabinina, E.E.; Zotova, E.N.; Vetrova, N.I.; Ponomareva, T.N.; Ilyushina, A. New approach to assessing the antioxidant activity of plant raw materials in studying the process of adrenaline autooxidation. *Chem. Plant Raw Mater.* **2011**, *3*, 117–121.
21. Sirota, T.V. Method for Determining the Antioxidant Activity of Superoxide Dismutase and Chemical Compounds. Russia Patent No. 2,144,674, 20 January 2000.
22. Maksimov, I.V.; Cherepanova, E.A. Pro-/antioxidant system and plant resistance to pathogens. *Adv. Mod. Biol.* **2006**, *126*, 250–261.
23. Popova, E.V.; Domnina, N.S.; Sokornova, S.V.; Kovalenko, N.M.; Tyuterev, S.L. Innovative hybrid plant immunomodulators based on chitosan and bioactive antioxidants and prooxidants. *Agric. Biol.* **2021**, *56*, 158–170. [CrossRef]
24. Shcherban, A.V.; Plotnikova, L.Y.; Skolotneva, E.S.; Fedyaeva, A.V.; Fomenko, V.V. Cyto-physiological manifestations of protective reactions of wheat against stem rust, induced by the biofungicide Novochizol. *Vavilov J. Genet. Breed.* **2024**, *in press*.
25. Shetty, N.P.; Mehrabi, R.; Lütken, H.; Haldrup, A.; Kema, G.H.J.; Collinge, D.B.; Jørgensen, H.J.L. Role of hydrogen peroxide during the interaction between the hemibiotrophic fungal pathogen *Septoria tritici* and wheat. *New Phytol.* **2007**, *174*, 637–647. [CrossRef] [PubMed]
26. Caverzan, A.; Casassola, A.; Brammer, S.P. Antioxidant responses of wheat plants under stress. *Genet. Mol. Biol.* **2016**, *39*, 1–6. [CrossRef] [PubMed]
27. Orzali, L.; Forni, C.; Riccioni, L. Effect of chitosan seed treatment as elicitor of resistance to *Fusarium graminearum* in wheat. *Seed Sci. Technol.* **2014**, *42*, 132–149. [CrossRef]
28. Nandeeshkumar, P.; Sudisha, J.; Ramachandra, K.K.; Prakash, H.; Niranjana, S.; Shekar, S.H. Chitosan induced resistance to downy mildew in sunflower caused by *Plasmopara halstedii*. *Physiol. Mol. Plant Pathol.* **2008**, *72*, 188–194. [CrossRef]
29. Kolesnikov, L.E.; Popova, E.V.; Novikova, I.I.; Priyatkin, N.S.; Arkhipov, M.V.; Kolesnikova, Y.R.; Potrakhov, N.N.; van Duijn, B.; Gusarenko, A.S. Joint use of microorganisms strains and chitosan complexes to increase wheat (*Triticum aestivum* L.) yield. *Agric. Biol.* **2019**, *5*, 1024–1040. [CrossRef]
30. Burlakova, S.V.; Egorycheva, M.T.; Fomenko, V.V.; Salakhutdinov, N.F.; Shcherban, A.B. Biological justification of the use of Novochizol with natural fungicides in the cultivation of bread wheat. *Chem. Sustain. Dev.* **2024**. *in press* (In Russian)
31. Woldeab, G.; Hailu, E.; Bacha, N. Protocols for Race Analysis of Wheat Stem Rust (*Puccinia graminis* f. sp. *tritici*). Ethiopian Institute of Agricultural Research, Ambo Plant Protection Research Center, Ethiopia. 2017. Available online: https://bgri.cornell.edu/wp-content/uploads/2020/09/Ambo-Manual-revised-2019FINAL.pdf (accessed on 5 December 2024).
32. Rsaliyev, A.S.; Rsaliyev, S.S. Principal approaches and achievements in studying race composition of wheat stem rust. *Vavilov J. Genet. Breed.* **2018**, *22*, 967–977. [CrossRef]
33. Sørensen, C.K.; Thach, T.; Hovmøller, M.S. Evaluation of spray and point inoculation methods for the phenotyping of *Puccinia striiformis* on wheat. *Plant Dis.* **2016**, *100*, 1064–1070. [CrossRef]

34. Roelfs, A.P.; Singh, R.P.; Saari, E.E. *Rust Diseases of Wheat: Concepts and Methods of Disease Management*; Cimmyt: Mexico City, Mexico, 1992; 81p.
35. Roelfs, A.P.; Martens, J.W.M. An international system of nomenclature for *Puccinia graminis* f. sp. tritici. *Phytopathology* **1988**, *78*, 526–533. [CrossRef]
36. Sunkar, R. Plant stress tolerance. *Methods Mol. Biol.* **2010**, *639*, 401. [CrossRef]
37. Ainsworth, E.A.; Gillespie, K.M. Estimation of total phenolic content and other oxidation substrates in plant tissues using Folin–Ciocalteu reagent. *Nat. Protoc.* **2007**, *2*, 875–877. [CrossRef] [PubMed]

plants

MDPI

Article

Mechanisms of Increase of Winter Wheat Frost Resistance Under Tebuconazole Treatment at Early Stage of Growth: Role of Hormone- and Reactive Oxygen Species-Mediated Signaling Pathways

Anna V. Korsukova, Irina V. Lyubushkina, Natalya S. Zabanova, Ekaterina V. Berezhnaya, Elizaveta A. Polyakova, Tamara P. Pobezhimova, Kuzma A. Kirichenko, Nikolay V. Dorofeev, Lyubov V. Dudareva and Olga I. Grabelnych *

Siberian Institute of Plant Physiology and Biochemistry, Siberian Branch of the Russian Academy of Sciences, Irkutsk 664033, Russia; avkorsukova@gmail.com (A.V.K.); ostrov1873@yandex.ru (I.V.L.); pavnatser@mail.ru (N.S.Z.); ekaterina809@mail.ru (E.V.B.); polyackova.elizaveta727@yandex.ru (E.A.P.); pobezhimova@sifibr.irk.ru (T.P.P.); gargamelle@yandex.ru (K.A.K.); nikolay.v.dorofeev@gmail.com (N.V.D.); laser@sifibr.irk.ru (L.V.D.)

* Correspondence: grolga@sifibr.irk.ru

Abstract: 1, 2, 4-triazole derivatives, including tebuconazole, have been reported to show positive physiological effects in cereals apart from fungicidal activity and to increase plants' tolerance against temperature stress. This study investigates the mechanisms of increasing frost resistance of etiolated winter wheat (*Triticum aestivum* L., "Irkutskaya" variety) seedlings by tebuconazole-based seed dresser "Bunker" (1.5 μL g^{-1} of seeds) and tebuconazole (30 μg g^{-1} of seeds). To identify ABA-dependent and ABA-independent pathways of frost resistance, we used fluridone (FLD, 5 mg L^{-1}), an inhibitor of endogenous abscisic acid (ABA) synthesis. FLD effectively inhibited the accumulation of carotenoids in the shoots and prevented the formation of carotenoids caused by the "Bunker" and tebuconazole. In non-hardened seedlings, FLD stimulated coleoptile and first leaf growth, but did not suppress the growth inhibitory effects of "Bunker" and tebuconazole. In shoots of hardened seedlings, FLD reduced the retarding effect of tebuconazole. Regardless of seedling age, temperature, and the protectant treatment, FLD had no effect on the sugar content in the shoots. FLD did not essentially influence frost resistance induced by "Bunker" and tebuconazole in cold-hardened seedlings. Fluridone increased H_2O_2 content and guaiacol peroxidase activity under control conditions (both with tebuconazole and without tebuconazole) and during cold hardening (in seedlings from seeds treated with tebuconazole). ABA levels in cold-hardened seedlings treated with FLD alone, tebuconazole alone, or a combination of the two were two to three times lower than in untreated hardened seedlings. Changes in indole-3-acetic and salicylic acids in response to FLD and tebuconazole treatment indicate complex interactions with signaling cellular systems. Our results suggest that tebuconazole activates ABA-independent pathways more strongly than ABA-dependent pathways in enhancing frost resistance. The potential mechanisms of tebuconazole action in plant cells are discussed.

Keywords: growth; cold hardening; frost resistance; fluridone; tebuconazole; water-soluble carbohydrates; reactive oxygen species; hormones; *Triticum aestivum* L.

1. Introduction

Low temperature is one of the important environmental factors that limits the growth, productivity, and distribution of plants around the world. An increase in the cold and frost resistance of plants to unfavorable low temperatures occurs as a result of cold hardening (cold acclimation)—a complex of structural and functional rearrangements, including changes in lipid and fatty acid compositions, synthesis of stress proteins, and accumulation of compounds with a cryoprotective function [1–4]. There is evidence that plant resistance to low temperatures can be increased by treatment with a variety of chemicals, including plant growth and development regulators and fungicides. Currently, systemic fungicides, which are inhibitors of C14 demethylation, are extensively used in agriculture. Triazole-based plant growth regulators are widely used in plants for growth, stress, and disease control [5]. Triazoles are characterized by low phytotoxicity, efficacy at small doses, environmental friendliness, and a wide spectrum of antifungal activity [6,7].

1, 2, 4-triazole derivatives with fungicidal properties, which are inhibitors of sterol and terpenoid synthesis, are the most widely used ones [7]. In addition, 1, 2, 4-triazole derivatives are retardants, since they inhibit internode elongation in cereal crops and the growth of plant axial organs by disrupting the biosynthesis of gibberellins (GB), namely gibberellic acid (GA3) [8,9]. Their other effects on the balance of phytohormones were also found. In particular, their ability to increase the content of abscisic acid (ABA) was shown [8,10–14]. GB and ABA biosynthesis is carried out from a common precursor—mevalonic acid (isoprenoid pathway). GBs are formed from acetate and mevalonate via ent-kaurene. GA12 synthesis (a common precursor of GB) from ent-kaurene is catalyzed by enzymes of the cytochrome P450 monooxygenase group. The enzyme activity is inhibited by 1, 2, 4-triazole derivatives, leading to the suppression of GB synthesis; mevalonic acid begins to be consumed for the synthesis of other phytohormones, in particular, ABA [9]. The mode of action of paclobutrazole, which is one of 1, 2, 4-triazole derivatives, has been well studied. Paclobutrazole was shown to inhibit the oxidation of ent-kaurene to ent-kaurenoic acid, thereby reducing GA production and increasing ABA content [10,12]. Paclobutrazole has a stress-protective effect and protects plants from abiotic stressors by maintaining the water ratio, cell membrane stability, and pro/antioxidant balance [12]. Triazole fungicides have been observed to play a role in the protection of the photosynthetic apparatus, an increase in photosynthetic pigments, soluble protein, and sugars, the effect on respiration, changes in the composition of fatty acids, as well as the enhancement of plant resistance to temperature stress, water deficiency, chloride salinization, and oxidative stress [9,12,15–22].

One of the representatives of 1, 2, 4-triazoles is tebuconazole (1-(4-chlorophenyl)-4,4-dimethyl-3-(1H-1,2,4-triazol-1-ylmethyl)-3-pentanol), which is contained in preparations widely used in agriculture. Tebuconazole is mainly used as a preventive and curative systemic fungicide, being effective against all types of cereal rust, septoria, smut, root rot, and seed mold. Like the other triazole derivatives, tebuconazole has, in addition to its fungicidal properties, a retardant effect by inhibiting GB biosynthesis [7,23]. In rapeseed cultivation, tebuconazole is used as a retardant to enhance branching and reduce overall plant height [24,25]. An increase in grain yield was one of the positive effects of treating wheat plants with tebuconazole [26].

According to Korsukova et al. [27,28] and Pobezhimova et al. [21,29], the tebuconazole-based seed protectant, "Bunker" (water suspension concentrate, tebuconazole content 60 g/L, JSC "Avgust" Inc., Russia) increases the resistance of etiolated seedlings of the cereals (spring wheat, winter wheat, and rye) to low temperatures. "Bunker" is systemic fungicide with both preventive and curative properties. It increases cold and frost resistance through its retardant effect and its impacts on carbohydrate, protein, fatty acid, and respiratory metabolism [27–29]. The "Bunker" effectively decontaminates the seeds [30]. A different effect of "Bunker" and pure tebuconazole was shown in winter wheat mitochondria [29]. Therefore, it is crucial to confirm that the increased frost resistance observed in wheat treated with tebuconazole-based seed dresser is specifically associated with the action of tebuconazole. It is known that triazole derivatives, including tebuconazole, increase the contents of ABA [8,10,12,14]. ABA is a polyfunctional plant hormone that plays an important role in mechanisms of increasing resistance to cold stress [31]. Increasing ABA synthesis might be the cause of high cold and frost resistance of cereals observed after treatment by a tebuconazole-based protectant.

To test the hypothesis regarding the involvement of endogenous ABA in the tebuconazole effect, we examined the influence of the endogenous ABA synthesis inhibitor (fluridone herbicide, FLD) on the physiological action of the tebuconazole-based protectant "Bunker" and pure tebuconazole in the early stages of growth of winter wheat (3-day-old etiolated seedlings). Fluridone inhibits synthesis and the accumulation of carotenoids and ABA biosynthesis [32]. The use of fluridone may reveal ABA-dependent and ABA-independent pathways of increasing frost resistance. Morpho-physiological (length and biomass of shoots and frost resistance of seedlings) and biochemical (content of carotenoids, sugars, hormones, hydrogen peroxide, lipid peroxidation, catalase, and peroxidase activity) analyses were carried out.

We showed that the fluridone action blocking the ABA synthesis weakened the signal transduction through the ABA-dependent pathway, reducing freezing tolerance. The protective effect of tebuconazole under freezing conditions was associated mainly with the activation of ABA-independent signal transduction.

2. Results

2.1. Carotenoids Content

Fluridone is known to inhibit the synthesis of endogenous ABA by inhibiting the formation of carotenoids [32]. Figure 1a presents clearly seen differences in the synthesis of carotenoids in shoots of control (Control) and fluridone-treated seedlings (CFLD) of winter wheat, as well as differences in the shoots of seedlings grown from seeds treated with the tebuconazole-containing protectant (Bunker) and tebuconazole (Tebuconazole). Shoots grown from seeds treated with the "Bunker" and tebuconazole are bright yellow, while control fluridone-treated (CFLD), Bunker-fluridone-treated (BFLD), and tebuconazole-fluridone-treated (TFLD) shoots are colorless.

Fluridone was found to effectively reduce (by 87%) carotenoid content in the shoots of 3-day-old seedlings (CFLD) grown from untreated seeds (Figure 1b). Seed treatment with "Bunker" and tebuconazole led to the increase in the content of carotenoids in the shoots. The content was 1.4 and 1.8 times higher in BFLD and TFLD shoots, respectively, compared to control seedlings (Figure 1b). Fluridone almost completely blocked the "Bunker"- and tebuconazole-induced increase in carotenoids, reducing levels by 83% and 87% in BFLD and TFLD treatments, respectively (Figure 1b).

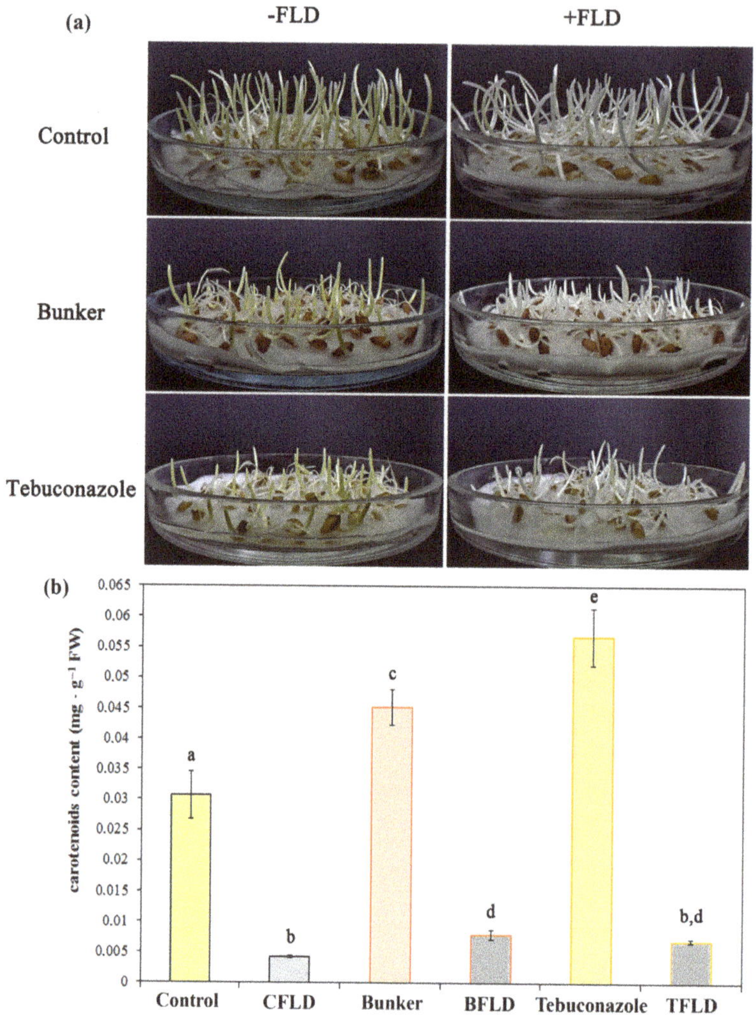

Figure 1. Influence of the fluridone, tebuconazole-based protectant "Bunker", and tebuconazole on the content of carotenoids in shoots of 3-day-old etiolated winter wheat seedlings. (**a**) Photo of 3-day-old seedlings. (**b**) Quantitative content of carotenoid: control—plants grown from untreated seeds; CFLD—plants grown from untreated seeds in fluridone solution (5 mg·L^{-1}); Bunker—plants grown from "Bunker"-treated seeds (1.5·µL g^{-1} of seeds); BFLD—plants grown from "Bunker"-treated seeds in fluridone solution; tebuconazole—plants grown from tebuconazole-treated seeds (30 µg·g^{-1} of seeds); and TFLD—plants grown from tebuconazole-treated seeds in fluridone solution. Data are presented as mean ± SD (n = 6). Different letters in the graph indicate the presence of statistically significant differences between the mean values of variants according to Fisher's LSD method ($p \leq 0.05$).

2.2. Growth Parameters Under Control and Cold Hardening

To assess the role of endogenous ABA in tebuconazole's retardant effect, we studied how fluridone affected coleoptiles and first leaf length (Table 1) in winter wheat seedlings under different conditions. The study included both unhardened etiolated seedlings of various ages (3, 5, and 10 days) and hardened seedlings (+2 °C, 7 days), with seeds treated with either "Bunker" or tebuconazole alone.

"Bunker" showed a significant growth-inhibiting effect on seedlings (Table 1). The protectant had a more pronounced effect on the growth of coleoptiles, with inhibition ranging from 31 to 46% (Table 1). The strongest coleoptile growth inhibition (46%) was observed in 3-day-old seedlings and remained at this level in cold-hardened seedlings (Table 1). Unlike coleoptiles, first leaf growth (and consequently, overall shoot growth) was less sensitive to the tebuconazole-containing protectant. Shoot growth inhibition decreased with seedling age: 43% in 3-day-old seedlings, 25% in 5-day-old seedlings, and 12% in 10-day-old seedlings compared to the control (Table 1).

Table 1. Influence of fluridone, tebuconazole-based protectant "Bunker", and tebuconazole on the length of coleoptiles and shoots of 3-, 5-, and 10-day-old etiolated winter wheat seedlings and after cold hardening (2 °C, 7 days). Data are presented as median (Me) and interquartile range [25%; 75%] (n = 3–4). The values following the same letters do not statistically differ significantly in plants of the same group (3 days, 5 days, 10 days, and cold hardening). Kruskal–Wallis one-way analysis of variance on ranks and all pairwise multiple comparison procedures (Student–Newman–Keuls method), $p \leq 0.05$. Designations of variants coincide with designations in Figure 1.

Variants	Growth Period, Days			Cold Hardening
	3	5	10	
		Coleoptile, mm		
Control	39 [35; 43] c	49 [46; 53] d	46 [43; 48] d	42 [38; 45] c
CFLD	45 [42; 47] d	58 [54; 62] e	50 [46; 53] e	46 [41; 49] d
Bunker	21 [18; 24] a	34 [31; 37] a, b	27 [25; 29] a, b	22 [20; 25] a
BFLD	24 [22; 27] b	38 [36; 43] a	30 [28; 33] c	25 [24; 29] b
Tebuconazole	19 [13; 22] a	28 [26; 30] c	25 [23; 27] a	22 [17; 24] a
TFLD	27 [23; 29] b	31 [29; 33] b, c	30 [25; 32] b, c	29 [26; 31] b
		Shoot, mm		
Control	37 [31; 40] c	83 [75; 95] b	156 [131; 178] a	40 [34; 45] c
CFLD	39 [36; 43] d	93 [85; 104] c	182 [145; 211] b	36 [30; 41] a
Bunker	21 [18; 24] a	62 [58; 67] a	138 [121; 149] c	22 [20; 26] b
BFLD	25 [22; 27] b	71 [65; 78] d	165 [148; 181] a, b	27 [24; 30] d
Tebuconazole	19 [13; 22] a	63 [52; 71] a	125 [111; 137] d	23 [17; 27] b
TFLD	27 [23; 29] b	79 [69; 85] e	151 [136; 165] a	35 [32; 37] a

During cold hardening of 3-day-old seedlings, the growth-inhibiting effect of "Bunker" remained at approximately the same level as it was before hardening (about 45%). The growth-inhibiting effect of "Bunker" was associated with the tebuconazole effect (Table 1). Inhibition of the length of coleoptiles of non-hardened seedlings by pure tebuconazole was also most pronounced in 3-day-old seedlings and amounted to 51%, which was not significantly different from the inhibition caused by "Bunker". Inhibition of the length of coleoptiles of 5- and 10-day-old seedlings was 43% and 46%, respectively. In cold-hardened seedlings, tebuconazole inhibited the growth of coleoptiles by 48%, which also did not differ significantly from inhibition with "Bunker" (Table 1).

In 3- and 5-day old seedlings, the growth inhibitory effect of tebuconazole on the length of shoots did not significantly differ from the effect of "Bunker" and amounted to 49% and 24%, respectively, while in 10-day-old seedlings it was 20%. In cold-hardened seedlings, the growth-inhibiting effect of tebuconazole was 43%, and did not significantly differ from the effect of "Bunker" (Table 1).

Treatment with fluridone led to a statistically significant increase in the length of coleoptiles and shoots both in unhardened seedlings of winter wheat of different ages (3, 5, and 10 days) and hardened (+2 °C, 7 days) ones grown from untreated (Control) and treated with "Bunker" and tebuconazole seeds (Table 1). The only exception is the lack of a stimulating effect of fluridone on the shoot length of hardened seedlings grown from seeds untreated with the protectant. Thus, treatment with fluridone led to the increase in the length of coleoptiles of winter wheat grown from untreated seeds by 9–18%, and the increase in the length of shoots by 5–20% (Table 1). The increase in coleoptile length under the action of fluridone in seedlings grown from seeds treated with "Bunker" was 11–14%, and the increase in shoot length was 14–23%. Fluridone led to an increase in the length of seedlings from seeds treated with tebuconazole, so the increase in the length of coleoptiles was 11–32%, and that of shoots was 21–52% (Table 1).

Shoot growth turned out to be more sensitive to fluridone. Under the action of fluridone, the length of shoots inhibited by the protectant reached the level of control seedlings (165 mm in BFLD, 151 mm in TFLD, and 156 mm in the Control) on the 10th day of growth (Table 1).

Significant inhibition of shoot length was observed in 3-day-old seedlings grown from seeds treated with "Bunker" and tebuconazole. This was accompanied by a decrease in the fresh and dry biomass of shoots (Table 2). Treatment with fluridone did not lead to changes in the fresh and dry biomass of shoots (coleoptile and first leaf) in control plants or in seedlings from seeds treated with "Bunker" (Table 2). Fluridone treatment resulted in a statistically significant increase in the fresh and dry weight of 3-day-old and cold-hardened plants from tebuconazole-treated seeds (Table 2).

Table 2. Influence of fluridone, tebuconazole-based protectant "Bunker", and tebuconazole on the fresh and dry biomasses of shoots of 3- and 10-day-old etiolated winter wheat seedlings and after cold hardening (2 °C, 7 days). Data are presented as mean ± SD (n = 3). Fresh and dry biomasses contain the weight of 30 seedlings. Values followed by different letters are statistically significantly different ($p \leq 0.05$). One-way ANOVA followed by Fisher's LSD multiple comparisons within plants of the same group (3 days, 10 days, and cold hardening). Designations of variants coincide with designations in Figure 1.

Variant	Growth Period, Days		Cold Hardening
	3	10	
	Fresh Weight, g		
Control	1.047 ± 0.034 a	1.624 ± 0.667 a	1.042 ± 0.025 a
CFLD	1.016 ± 0.046 a	1.591 ± 0.624 a	0.965 ± 0.138 a
Bunker	0.573 ± 0.114 b	1.721 ± 0.359 a	0.638 ± 0.229 a
BFLD	0.633 ± 0.098 b	1.734 ± 0.507 a	0.788 ± 0.280 a
Tebuconazole	0.480 ± 0.142 c	2.760 ± 0.020 b	0.616 ± 0.149 b
TFLD	0.823 ± 0.033 d	2.462 ± 0.393 b	0.998 ± 0.211 c
	Dry Weight, g		
Control	0.0871 ± 0.0036 a	0.1490 ± 0.0588 a	0.1200 ± 0.0115 a
CFLD	0.0944 ± 0.0078 a	0.1510 ± 0.0610 a	0.1250 ± 0.0281 a
Bunker	0.0591 ± 0.0094 b	0.1710 ± 0.0354 a	0.0889 ± 0.0361 a
BFLD	0.0667 ± 0.0069 b	0.1680 ± 0.0517 a	0.1080 ± 0.0437 a
Tebuconazole	0.0555 ± 0.0111 c	0.2750 ± 0.0059 b	0.0879 ± 0.0139 b
TFLD	0.0842 ± 0.0036 d	0.2500 ± 0.0412 b	0.1260 ± 0.0268 c

2.3. Water-Soluble Carbohydrate Content in Shoots Under Control and Cold Hardening

The further studies examined changes in water-soluble carbohydrate content in shoots of etiolated winter wheat seedlings under fluridone treatment. This included seedlings grown from untreated (Control) and treated ("Bunker" and tebuconazole) seeds at three stages: 3-day-old, 10-day-old, and cold-hardened (7 days at 2 °C) seedlings.

As shown in Figure 2, treatment of winter wheat seeds with the tebuconazole-containing protectant "Bunker" and tebuconazole led to a statistically significant increase in water-soluble carbohydrate content in shoots of unhardened and hardened seedlings. By day 3 of growth, shoot water-soluble carbohydrate content increased by 28% and 35% under "Bunker" and tebuconazole treatment, respectively, compared to control; during cold hardening, these increases were 21% and 35%. Fluridone treatment did not significantly affect water-soluble carbohydrate content in any experimental variants, except in seedlings from tebuconazole-treated seeds (TFLD) (Figure 2).

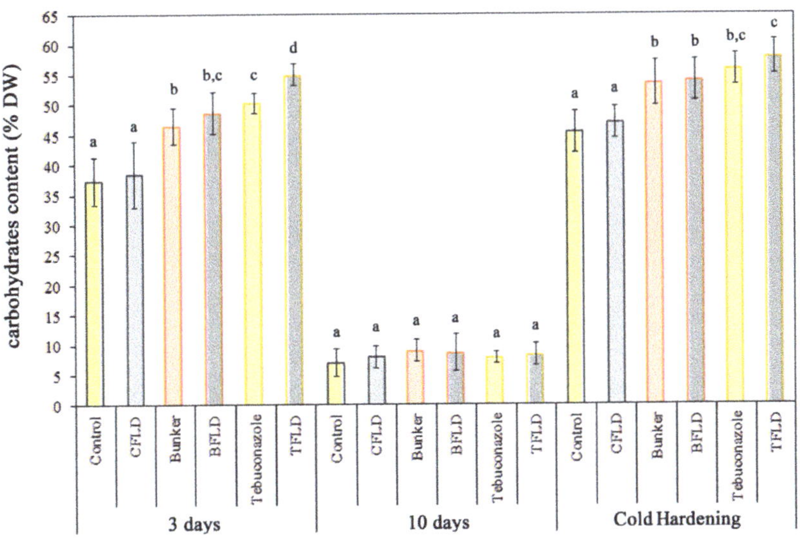

Figure 2. Influence of fluridone, tebuconazole-based protectant "Bunker", and tebuconazole on the content of water-soluble carbohydrates in 3- and 10-day-old shoots of etiolated winter wheat seedlings and after cold hardening (2 °C, 7 days). Data are presented as mean ± SD (n = 3). Various lowercase letters in the graph indicate the presence of statistically significant differences between the mean values of variants within the plants of the same group (3 days, 10 days, and cold hardening) according to Fisher's LSD method ($p \leq 0.05$). Designations of variants coincide with designations in Figure 1.

2.4. Frost Resistance of Seedlings

Korsukova et al. [27,28] showed that the tebuconazole-containing seed dresser "Bunker" increased the frost resistance of cereal seedlings. It was necessary to find out whether ABA was involved in increasing frost resistance. In order to do that, we studied the effect of fluridone on the permeability of cell membranes in shoot tissues and the survival rate of cold-hardened winter wheat seedlings (2 °C, 7 days) grown from seeds treated with "Bunker" and tebuconazole after exposure to a damaging temperature (−8 °C, 24 h). Cell membrane permeability was assessed by measuring the release of electrolytes immediately after subzero temperature exposure (Figure 3a), while seedling survival rates were evaluated 5 days after freezing during their growth under control conditions (Figure 3b).

In shoots of seedlings grown from seeds treated with "Bunker" and tebuconazole, the release of electrolytes after freezing was observed to be noticeably lower (by 47% and 37%, respectively) than in the control (Figure 3a). In seedlings grown from untreated seeds (CFLD), fluridone treatment increased electrolyte leakage from shoot tissues by 53% after freezing. Electrolyte leakage from shoot tissues after freezing increased by 48% in seedlings grown from seeds treated with both the "Bunker" and fluridone (BFLD), and by 65% in seedlings treated with both tebuconazole and fluridone (TFLD). In seeds treated with "Bunker", the release of electrolytes from shoot tissues was significantly lower than in control seedlings, whereas in seeds treated with tebuconazole, the yield of electrolytes did not significantly differ compared to the control group (Figure 3a).

Changes in the release of electrolytes from shoot tissues immediately after exposure to negative temperatures correlated with the ability of fluridone, tebuconazole-based protectant "Bunker", and tebuconazole to modulate the survival rate of cold-hardened winter wheat seedlings after their growth under control conditions (Figure 3b). An increase in the survival rate of seedlings from seeds treated with "Bunker" and tebuconazole and a decrease in the survival rate of fluridone-treated seedlings from untreated (CFLD), "Bunker"-treated (BFLD), and tebuconazole-treated (TFLD) seeds were observed (Figure 3b).

To identify the roles of hormone and reactive oxygen species (ROS) signaling pathways, we analyzed the biochemical parameters following fluridone and tebuconazole treatments.

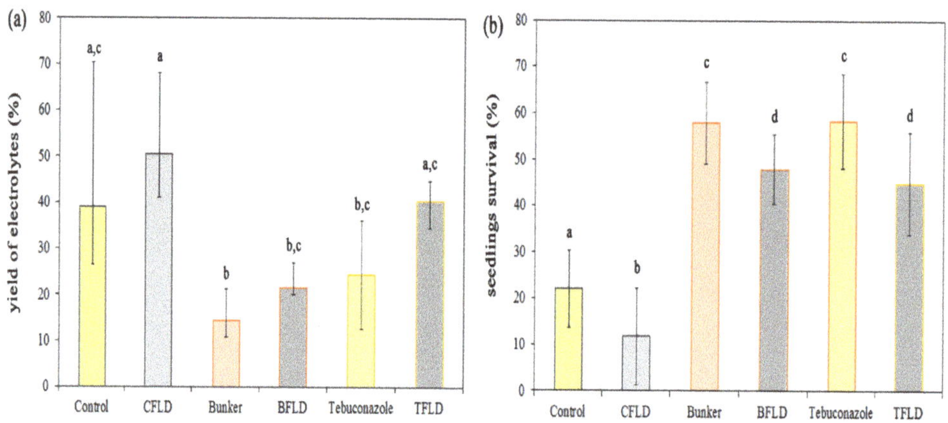

Figure 3. Influence of fluridone, tebuconazole-based protectant "Bunker", and tebuconazole on release of electrolytes and survival of cold-hardened (2 °C, 7 days) etiolated winter wheat seedlings after exposure to negative temperature (–8 °C, 24 h). (**a**) Release of electrolytes from shoot tissues, Me [25%; 75%] (n = 10). (**b**) Survival of seedlings after 5 days of recovery, M ± S.D (n = 8). Different letters in the graph indicate the presence of statistically significant differences between the values of variants according to the Kruskal–Wallis one-way analysis of variance on ranks and all pairwise multiple comparison procedures, Student–Newman–Keuls method (for (**a**)) and Fisher's LSD (for (**b**)) ($p \leq 0.05$). Designations of variants coincide with designations in Figure 1.

2.5. Hydrogen Peroxide Content, Lipid Peroxidation, and Antioxidant Enzyme Activity Under Control and Cold Hardening

Hydrogen peroxide is one of the main ROS involved in the development of oxidative stress. Previous studies have shown that triazoles regulate reactive oxygen species (ROS) content and lipid peroxidation (LPO) in plant cells under stress conditions, including low temperature stress [22]. There is also information stating that fluridone acts as a pro-oxidant [33]. In this regard, we assessed the content of hydrogen peroxide (Figure 4a) and lipid peroxidation activity, measured as malondialdehyde (MDA) content resulting from the reaction with thiobarbituric acid (TBA) (Figure 4b) in shoots of 3-day-old and cold-hardened winter wheat seedlings grown from untreated and tebuconazole-treated seeds in water, as well as in fluridone solution. We also studied the activity of catalase (CAT) (Figure 4c) and guaiacol peroxidase (GPOD) (Figure 4d) involved in the detoxification of hydrogen peroxide.

The cold hardening of seedlings resulted in a statistically significant increase in H_2O_2 content in all variants compared to 3-day-old (non-hardened) seedlings (Figure 4a). At the same time, the hydrogen peroxide content in plants grown from untreated (Control) and tebuconazole-treated seeds did not differ statistically significantly in either 3-day-old or hardened seedlings. Thus, seed treatment with tebuconazole does not result in the generation of H_2O_2 either in 3-day-old seedlings or during cold hardening, while the hydrogen peroxide content remains at the level of that in control plants (Control). At the same time, growing seedlings in fluridone solution (CFLD and TFLD) led to a statistically significant increase in H_2O_2 levels in plants grown from untreated (Control) and tebuconazole-treated seeds, both 3-day-old and cold hardened. Thus, the hydrogen peroxide content in 3-day-old seedlings from untreated (CFLD) and tebuconazole-treated (TFLD) seeds grown in fluridone solution was 39% and 27% higher, respectively, than in 3-day-old seedlings from untreated (Control) and tebuconazole-treated seeds grown in water (Figure 4a).

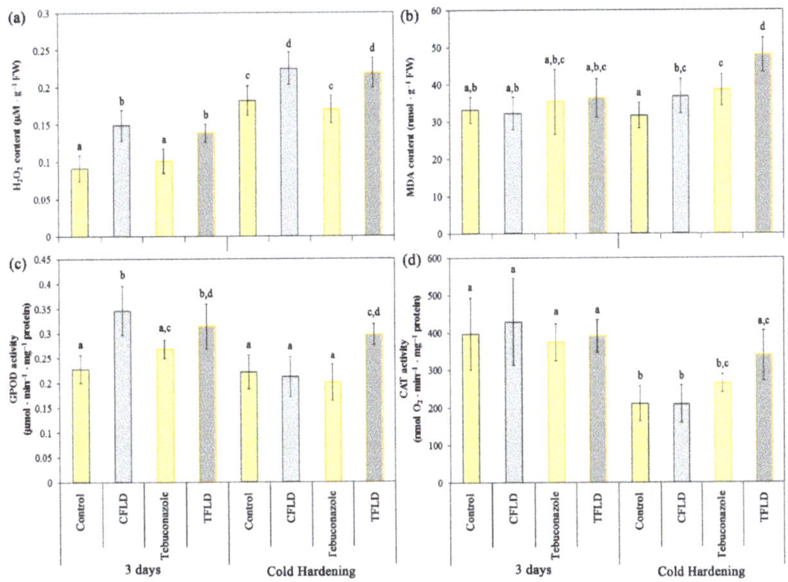

Figure 4. Influence of fluridone and tebuconazole on the hydrogen peroxide (**a**) and malondialdehyde (MDA) content (**b**), guaiacol peroxidase (GPOD) (**c**) and catalase (CAT) (**d**) activities in shoots of 3-day-old and cold-hardened (2 °C, 7 days) etiolated winter wheat seedlings. Data are presented as mean ± SD ($n = 3$). Different letters in the graph indicate the presence of statistically significant differences between the mean values of variants according to Fisher's LSD method ($p \leq 0.05$). Designations of variants coincide with designations in Figure 1.

The data we obtained on the content of the LPO product—MDA (Figure 4b)—indicate that an increase in hydrogen peroxide does not cause oxidative stress either in 3-day-old seedlings (in all experimental variants) or in cold-hardened seedlings, with the exception of the TFLD variant. Thus, in cold-hardened seedlings from tebuconazole-treated seeds grown in fluridone solution (TFLD), we observe a statistically significant increase in MDA of 34% compared to the cold-hardened control variant (Control) (Figure 4b).

It was found that in 3-day-old plants, an increase in GPOD activity occurs only in seedlings grown in fluridone solution, both in plants from untreated (CFLD) and tebuconazole-treated (TFLD) seeds, by 34% and 15%, respectively (Figure 4c). In cold-hardened plants, a statistically significant increase in GPOD activity was observed in seedlings from tebuconazole-treated seeds grown in fluridone solution (TFLD). With other treatments, GPOD activity remained at the level of that in 3-day-old unhardened plants (Control, 3 days) (Figure 4c).

The study of CAT activity (Figure 4d) showed that in 3-day-old seedlings, no treatment increased the activity of this enzyme. In cold-hardened plants, a statistically significant decrease in CAT activity was observed compared to 3-day-old plants, dropping by 40–45%. Only in the TFLD (cold-hardened seedlings from seeds treated with tebuconazole and grown in fluridone solution), did CAT activity not significantly differ from the control group (3-day-old seedlings).

2.6. Endogenous Hormone Content Under Control and Cold Hardening

When studying the effect of fluridone, tebuconazole, and their combined treatment on the hormonal status of etiolated control and hardened winter wheat seedlings, we initially focused on the change in the ABA content. On the one hand, this was because there is sufficient information in the literature on the direct and indirect effect of these substances on the ABA level in plants. Therefore, we needed to find out whether such an effect occurred in our work. On the other hand, ABA is probably the key hormone regulating the process of low-temperature hardening in plants; therefore, it is essential to thoroughly study any alterations in its concentration. It has been demonstrated that exposure of etiolated seedlings to low positive temperatures (2 °C) for a period of 7 days resulted in an almost 6-fold increase of in the content of ABA in shoot tissues (Figure 5a).

Fluridone, tebuconazole, and their combination did not affect ABA levels in the tissues of control seedlings (Figure 5a). However, in hardened seedlings, treatment with fluridone, tebuconazole, or both reduced ABA content by two to three times (Figure 5a). Thus, our study revealed significant differences in the effects of the tested substances on ABA content in etiolated winter wheat seedlings, depending on the temperature. In this regard, the following question arises: could these differences be due to the indirect effect of fluridone or tebuconazole on other components of the hormonal regulation? To answer this question, we studied changes in the concentration of active IAA in winter wheat seedlings.

The study of IAA content in the shoots of etiolated seedlings revealed a high basic level of this hormone, which remained unchanged during low-temperature hardening (Figure 5b). It was found that fluridone treatment reduced IAA content in the tissues of both control and hardened seedlings by almost three times (Figure 5b). Tebuconazole, in turn, did not significantly change IAA levels either under control conditions or during low-temperature hardening. Of particular interest was the combined effect of tebuconazole and fluridone. In control seedlings, tebuconazole-fluridone treatment increased IAA content by about 1.5 times, while in hardened seedlings, there was a decrease in IAA content of more than 3 times (Figure 5b). This finding suggests that each of the studied substances most likely has an indirect rather than direct effect on IAA synthesis, possibly through other components of the hormonal network.

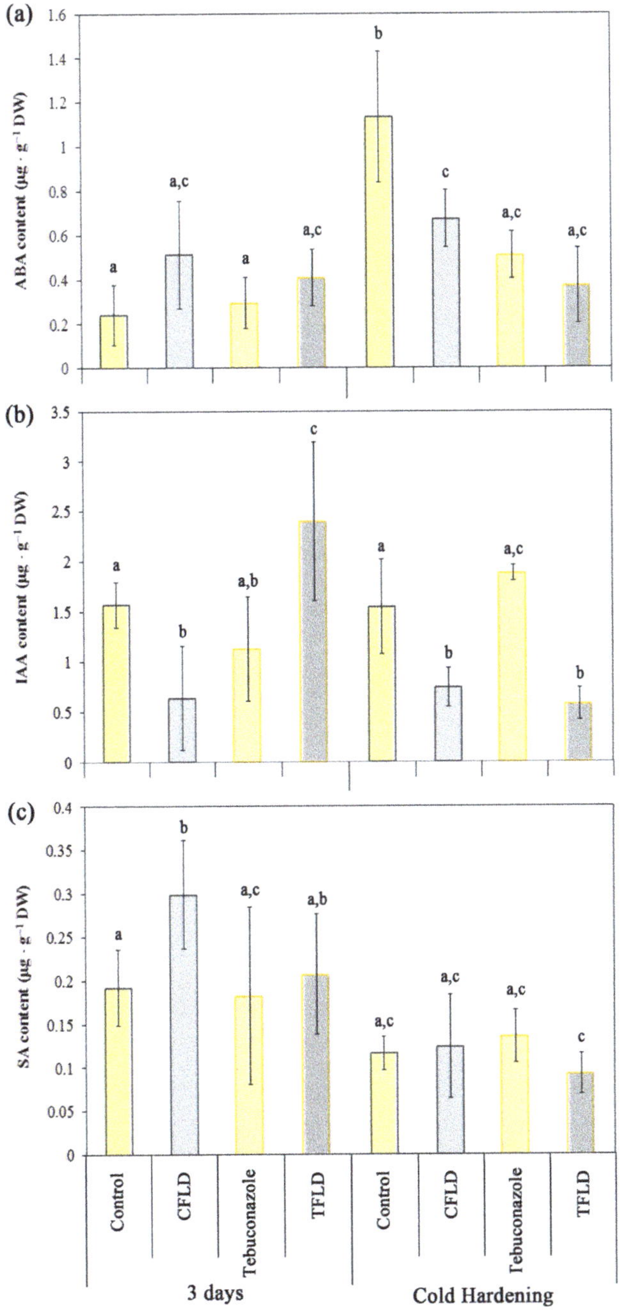

Figure 5. Influence of fluridone and tebuconazole on the abscisic acid (ABA) (**a**), indole-3-acetic acid (IAA) (**b**), and salicylic acid (SA) (**c**) in shoots of 3-day-old and cold-hardened (2 °C, 7 days) etiolated winter wheat seedlings. Data are presented as mean ± SD (n = 3). Different letters in the graph indicate the presence of statistically significant differences between the mean values of variants according to Fisher's LSD method ($p \leq 0.05$). Designations of variants coincide with designations in Figure 1.

Another important hormone involved in seed germination, growth regulation, and plant response to various stress factors is SA. Its basic levels vary in different plants [34]. In our study, SA contents in both control and hardened winter wheat seedlings were quite low (Figure 5c). Although there were no statistically significant changes in SA, the hardened seedlings showed a tendency towards a decrease in the level of this hormone. The effects of fluridone, tebuconazole, and their combination on SA content varied depending on the temperature conditions. In control seedlings, fluridone increased SA levels by approximately 1.5-fold, whereas in hardened seedlings, it had no effect on SA content (Figure 5c). Tebuconazole alone did not have a significant effect on SA content, either under control conditions or during low-temperature hardening. However, when combined with fluridone, it prevented an increase in SA levels in control seedlings and contributed to a decrease in SA in hardened seedlings (Figure 5c). Thus, the data obtained on the changes in SA levels in response to fluridone and tebuconazole treatments also indicate the complex effect of these substances on signaling cellular systems.

3. Discussion

Enhanced ABA biosynthesis and subsequent accumulation are among the general plant responses to stress [35]. Thus, during cold hardening, the plant hormonal balance shifts towards elevated concentrations of growth inhibitors, mainly ABA, which causes growth suppression [36]. The retardant effect of triazoles is also associated with their influence on the balance of phytohormones [8,10,12,13,23]. It is known that triazole compounds affect the isoprenoid pathway, which is responsible for the synthesis of essential plant hormones including GA, ABA, and cytokinins [37]. Triazoles affect GA biosynthesis by inhibiting the oxidation of ent-kaurene to ent-kaurenoic acid [8,9]. It can be assumed that the disruption of GA synthesis should lead to increased ABA production. Indeed, in a number of studies, it has been stated that triazole compounds enhanced cytokinin and ABA levels while reducing GA content [38].

The tebuconazole-containing seed dresser "Bunker" (60 g L^{-1} of tebuconazole) increased cold and frost resistance in etiolated cereal seedlings (winter wheat and rye, spring wheat) by inhibiting shoot growth, increasing sugar content, increasing unsaturation of fatty acids in membrane lipids, inducing the synthesis of stress proteins such as dehydrins, and reducing respiratory metabolism [27–29]. The growth-inhibiting effects of this tebuconazole-containing protectant persisted when winter wheat plants were grown under light conditions [39]. It remained to be determined to what extent these effects were specifically attributable to tebuconazole itself. It was unclear whether the identified effects were ABA dependent or ABA independent.

Being a 1, 2, 4-triazole derivative, tebuconazole presumably acts at the level of the cytochrome P450 monooxygenase enzyme, which catalyzes the synthesis of GA_{12} (a common precursor of other GBs) from ent-kaurene [23]. Inhibition of this enzyme leads to the suppression of GB synthesis and the switch of the metabolic isoprenoid pathway to ABA synthesis from mevalonic acid (a common precursor for GB and ABA synthesis) [9]. Suppressed GB synthesis and increased ABA contents result in the inhibition of the growth of plant axial organs [40]. ABA, the main stress hormone in plants, triggers the synthesis of stress proteins, including ABA-dependent dehydrins [1]. In addition, ABA stimulates the accumulation of water-soluble carbohydrates [41], which in turn can enhance dehydrin synthesis [42]. An increase in ABA can lead to an increase in the activity of desaturases, resulting in higher concentrations of polyunsaturated fatty acids [43]. ABA was shown to be involved in the stabilization of cell membranes in the cold [44]. Tebuconazole also suppresses the cytochrome respiratory pathway at the level of complex I of the mitochondrial respiratory chain; a decrease in respiration intensity saves substrates (water-soluble carbohydrates), which also favors accumulation [29]. Electron transport through the alternative pathway via alternative oxidase (AOX) is not inhibited. At low temperatures, it can be important for the prevention of oxidative stress in mitochondria.

Growth inhibition, sugar accumulation, increased fatty acid unsaturation in cell membrane lipids, and dehydrin synthesis all enhance plant survivability at low temperatures. For ABA-dependent pathways of increasing cold and frost resistance, ABA is involved in the process. To verify the assumptions made, it was necessary to test the dependence of the revealed effects of the tebuconazole-containing protectant on ABA synthesis. During cold acclimation of plants, ABA-independent pathways are also induced [31]. Blocking ABA synthesis with targeted inhibitors may help to reliably determine its role in triazole-mediated effects. Fluridone, a herbicide inhibiting endogenous ABA synthesis, was used in the present research. Fluridone was confirmed to be effective in inhibiting ABA synthesis in various plants including wheat [32,45,46].

The effect of "Bunker", as well as that of pure tebuconazole, was more pronounced in coleoptiles at an early seedling growth phase (on the 3rd day) and persisted during cold hardening of 3-day-old seedlings (Table 1). Inhibition of shoot growth (coleoptile and leaves) for seedlings from seeds treated with "Bunker" and tebuconazole was also more pronounced on day 3 of seedling growth and persisted during cold hardening, but diminished under control growth conditions (Table 1). Fluridone stimulated the growth of coleoptiles and shoots in seedlings from both untreated seeds and those treated by "Bunker" and tebuconazole. Thus, the ABA synthesis inhibitor partially reversed the growth-retarding effects of "Bunker" and tebuconazole. Similar data on wheat coleoptile growth stimulated by fluridone were obtained by other researchers [47]. When studying the effect of fluridone on the growth of coleoptiles of different types of cereals, they noted that fluridone treatment enhanced coleoptile growth by 14% in wheat and 67% in rice, compared to the control groups.

Notably, fluridone exhibited no stimulatory effect on the shoots of hardened seedlings grown from seeds not treated with "Bunker". Conversely, fluridone reduced the retardant effect of "Bunker" and tebuconazole. Based on previous reports of triazoles increasing ABA synthesis [8,10,12] and our experimental results showing changes in coleoptile and shoot length under fluridone treatment (Table 1), we conclude that the retarding effect of tebuconazole and tebuconazole-containing "Bunker" is largely independent of ABA under control conditions and is likely related to GB biosynthesis inhibition. However, during cold hardening in organs still growing (first leaf), ABA has a more noticeable effect on growth and, possibly, on resistance.

The effect of triazole retardants on leaf photosynthesis is well documented. Triazole derivatives, including tebuconazole, are known to increase carotenoid and chlorophyll contents [48]. The use of pure tebuconazole and the mixture of chlorocholine chloride with tebuconazole in winter rape during the spring growing season increased the content of photosynthetic pigments in leaves [25]. In turn, fluridone effectively inhibits the formation of carotenoids, which are precursors of ABA biosynthesis [32]. The data we obtained confirm that fluridone suppresses carotenoid synthesis in the shoots of seedlings grown from untreated seeds and prevents carotenoid formation caused by treatment with "Bunker" and tebuconazole (Figure 1). Inhibition of carotenoid synthesis in shoots by treatment with fluridone proves the appropriateness of its use to inhibit endogenous ABA synthesis.

One of the consequences of seedling growth inhibition after being subjected to triazole derivatives is a more economical consumption of sugars, which was demonstrated by Korsukova et al. [28] in winter wheat and rye by using the tebuconazole-containing protectant "Bunker". Tebuconazole treatment resulted in a rise in sugar yield in sugar beet under drought stress conditions [20]. Water-soluble carbohydrates fulfill several functions in plant cells during low-temperature hardening due to their osmoregulatory, antifreezing, cryoprotective, and antioxidant properties. In addition, water-soluble carbohydrates are the main substrates for respiration. ABA is evidenced to affect the metabolism of water-soluble carbohydrates and the activity of enzymes involved in water-soluble carbohydrate accumulation during hardening [41]. However, the results obtained with fluridone did not reveal the dependence of the elevated sugar contents on ABA synthesis in the shoots of winter wheat seedlings treated with the tebuconazole-containing protectant and tebuconazole (Figure 2). Thus, it can be assumed that the increase in the sugar content in cereals grown from seeds treated with "Bunker" and tebuconazole is not associated with ABA. Previous research demonstrated that "Bunker" effectively inhibited both phosphorylating and non-phosphorylating respiration in winter wheat mitochondria during malate oxidation [29]. This treatment also decreased the respiratory control coefficient by affecting the function of respiratory chain complex I. Inhibition of the substrate oxidation rate by mitochondria, induced by the tebuconazole-containing protectant and tebuconazole, can be one of the reasons for an increase in water-soluble carbohydrates in the shoots of unhardened and hardened cereal seedlings (Figure 2).

Reduced growth and increased sugar content are some of the factors that increase plant resistance to low temperatures [1,41]. The tebuconazole-containing protectant increases cold and frost resistance in cereal seedlings, causing the following physiological effects: reduced growth and accumulation of sugars, induction of dehydrin synthesis, and increased fatty acid unsaturation in membrane lipids, etc. [27,28]. In seedlings grown from "Bunker"-treated seeds, there was a decrease in the release of electrolytes and an increase in the survival rate compared to the control (Figure 3). The results obtained are consistent with the data on the ability of triazoles (tebuconazole and paclobutrazole) to reduce the release of electrolytes [12,26]. The survival rates of cold-hardened seedlings from seeds treated with "Bunker" and pure tebuconazole after freezing did not differ significantly (Figure 3b). However, seed plants treated with "Bunker" exhibited a reduced electrolyte yield, which was significantly different compared to the control (Figure 3a). Thus, it can be assumed that "Bunker" and pure tebuconazole have different effects on the permeability of cell membranes. This discrepancy could be attributed to the composition of Bunker, which contains additional components beyond the active ingredient (tebuconazole). Similarly, differences in composition might explain the varied effects of Bunker and pure tebuconazole on the respiratory activity of winter wheat mitochondria [29].

Shaki et al. [49] concluded that there is a positive and strong correlation between triazole treatment and the activity of antioxidant systems in plants under stress. Presumably, triazoles are able to induce stress-like symptoms to stimulate the antioxidant system. This may be the reason for the overall increase in H_2O_2 levels in plants [49]. However, our experiments only showed a trend towards an increase in H_2O_2 in seedlings from tebuconazole-treated seeds (Figure 4a). In contrast, fluridone led to an increase in H_2O_2 in shoots (Figure 4a). It should be noted that under the same treatment (TFLD), cold-hardened seedlings showed a decrease in the content of carotenoids (Figure 1b), playing a protective role against oxidative stress. The reduction in carotenoids combined with increased electrolyte release (Figure 3a) and decreased survival rates (Figure 3b) suggests that fluridone can act as a pro-oxidant. This finding aligns with previous research [33].

According to our results, the increase in frost resistance under the influence of the tebuconazole-containing protectant "Bunker" occurs with the participation of mechanisms unrelated to ABA, although ABA-dependent mechanisms certainly play an important role in the mechanisms of frost resistance development. We did not observe an ABA increase in control or in hardened winter wheat seedlings treated with tebuconazole (Figure 5a). On the contrary, in cold-hardened seedlings, ABA concentrations even decreased in response to tebuconazole treatment, whereas in control seedlings, this treatment had no visible effect (Figure 5a). Considering the reasons for such changes, it is important to remember that the endogenous hormone content is determined by the balance of its synthesis, decay, and transport, as well as conjugation with amino acids and sugars and conversion into inactive forms.

Based on our results, we can conclude that ABA synthesis was minimal in 3-day-old etiolated winter wheat seedlings during active growth (Figure 5a). This was indirectly supported by high levels of IAA, which acts as an ABA antagonist and functions downstream of ABA, particularly during seed germination [50]. It should be noted that high IAA levels are typical for seedlings grown in the dark [50]. Similar results were obtained by Kosova et al. working with two cultivars of spring wheat when, against the background of high IAA contents under control conditions, very low ABA concentrations were detected [2]. In our work, the absence of active ABA synthesis in control winter wheat seedlings was also evidenced by the fact that the level of this hormone did not change when seedlings were treated with fluridone, an inhibitor of ABA and carotenoid synthesis (Figure 5a) [32]. The fluridone concentration used in the work was sufficient, as indicated by a significant decrease in carotenoids during this treatment (Figure 1b). Moreover, the inhibitory effect of fluridone was so pronounced that it was not removed by the action of tebuconazole when they were applied together (Figure 1b). It can be assumed that some amount of ABA in etiolated winter wheat shoots was preserved, due to its conversion from an inactive form (e.g., from glucosyl ester) to an active one, or due to a decrease in its degradation rate [51]. Thus, it is logical that in the absence of ABA synthesis, the effect of tebuconazole on the content of this hormone was absent. At the same time, the effect of tebuconazole on the isoprenoid pathway was undisputed. This can be judged, in particular, by an increase in carotenoid content in the shoots of etiolated winter wheat seedlings when treated with tebuconazole. As carotenoids are a product of the isoprenoid pathway in plants and precursors of ABA, the increase in their content indicates the intensification of synthetic reactions of other branches of this process due to the blocking of GA synthesis [37,50]. Since ABA synthesis did not occur, carotenoids were not consumed, and their content in the tissues of seedlings increased (Figure 1b). It should also be noted that tebuconazole treatment caused inhibition of shoot growth of control seedlings (Table 1). Since no changes in IAA levels were observed in this case, it is likely that this inhibition was due to the effect of tebuconazole on other growth hormones, namely GA (Figure 5b).

ABA is known to be essential in plant responses to cold stress. A number of works have shown intensification of endogenous ABA biosynthesis under cold stress or improvement of cold and frost stress resistance in plants after application of exogenous ABA [52]. Indeed, in our work, we observed a rise in ABA levels in winter wheat seedlings after cold hardening (Figure 5a). Moreover, the results of our experiments with fluridone confirmed the important role of ABA in the formation of winter wheat frost resistance. Fluridone decreased the survival of seedlings when exposed to negative temperatures, which was accompanied by a decrease in ABA levels in shoots (Figure 5a). Contrary to expectations, tebuconazole did not increase ABA levels in hardened seedlings, while their survival rates under negative temperatures increased (Figure 5a). Moreover, tebuconazole prevented a fluridone-induced decrease in winter wheat seedling resistance to negative temperatures, but ABA contents were also reduced compared to ABA levels in hardened seedlings with no treatment (Figures 3b and 5a). The increase in frost resistance in this case was probably due to ABA-independent signal transduction, realized, as is known today, through DREB1/CBF [53,54]. The protective effect of tebuconazole in our experimental setting might be attributed to its capacity to impede GA synthesis. For instance, the GA1 mutant of Arabidopsis with impaired GA production demonstrated higher freezing tolerance than wild-type plants [55]. At least partially, this effect is connected with GA-regulated GATA transcription factors (TFs) GNC and GNL, promoting the CBF1, CBF2, COR15A, and COR15B responsible for the frost resistance [55,56].

Analysis of changes in the active IAA and ABA levels in the tissues of winter wheat seedlings treated with fluridone and tebuconazole allows us to assume that their effect on the hormonal system of plants was multifaceted, involving various signaling pathways. Thus, despite the impact of tebuconazole on the isoprenoid pathway in plants, our study did not reveal evidence supporting its direct effect on ABA synthesis. Evidence suggests that the ABA level can be reduced, rather than elevated, under the influence of triazole compounds. This finding aligns with those of Zheng et al., who documented a decline in the content of GA, ABA, and IAA in the inflorescences of *Astragalus sinicus* L. when using paclobutrazole, another representative of the triazole group [57]. A stable decrease in the ABA content in winter wheat treated with fluridone under low-temperature treatment was a consequence of its direct effect on the synthesis of this hormone, but a decrease in the level of free IAA was precisely a consequence of its indirect effect. In the work of Stetsenko et al. on the effect of fluridone on the hormone content in crystal grass under NaCl-salinity conditions, a decrease in free IAA and an increase in IAA conjugates were shown in both control and NaCl stress conditions [33].

One of the potential mechanisms of this influence involves changes in activity and/or in expression of genes of GH3 family enzymes, which regulate homeostasis of hormones including IAA, jasmonic acid (JA), and SA. Amidosynthetases belonging to this family catalyze reactions of the conjugation of hormones with amino acids, converting them from active free form to inactive storage and transport form [51]. At the same time, changes in enzyme activity and gene expression of amidosynthetases affect the content of the entire hormone complex rather than individual hormones alone. Thus, it was shown that OsGH3-13-activated rice enhanced drought tolerance accompanied by reduced endogenous free IAA and ABA synthesis [58]. In our study, the IAA content decreased in control winter wheat seedlings treated with tebuconazole and both in control and hardened seedlings treated with fluridone, although these compounds are not known to have a direct effect on auxin synthesis (Figure 5b). The alterations in IAA levels observed in both control and hardened seedlings could be attributed to the impact of fluridone and tebuconazole on GH3. Due to the modulation of different signaling pathways, enzyme activity can both increase and decrease (Figure 5b). It is interesting that GH3.5 can act as a modulator not only in auxin signaling, but also in the SA-regulated system [59]. Our study demonstrated simultaneous changes in IAA and SA contents in winter wheat control seedlings under fluridone treatment and in hardened seedlings under combined treatment (Figure 5a,b). It is worthy of further investigation to determine whether the GH3 family is indeed involved in hormonal homeostasis in the context of fluridone and triazole treatments.

The canonical TIR1/AFB-Aux/IAA signaling pathway is an important component of auxin signaling, which might be involved in the regulation of energy metabolism, as well as hormonal homeostasis and cross-talk between signaling pathways [60]. It has been established that auxin binding to the receptor proteins TIR1 (Transport Inhibitor Response 1) and AFB (Auxin Signaling F-Box) causes ubiquitination and proteasomal degradation of proteins that are repressors of Aux/IAA transcription. Due to the cleavage of these proteins, ARF transcription factors are released, activating the corresponding genes [61]. It has been shown that auxin-dependent degradation of Aux/IAA is carried out by enzymes of the E3 ligase family, called SCFTIR1/AFB1–5. In this case, the implementation of auxin signal perception during substrate recognition requires direct binding of auxin to the F-box protein [62]. The process of identifying genes under the control of ARF is still ongoing, but among the identified genes of great interest are the aforementioned GH3 family genes, which encode amidotransferase enzymes responsible for conjugating not only IAA, but also JA with amino acids [63]. The GH3 family can potentially link the auxin signal transduction pathway with the activity of mitochondrial energy-dissipating systems, since JA is known to be involved in the regulation of AOX gene expression [64,65]. The fact that the functional activity of mitochondria can be regulated by auxin signaling pathways is evidenced by the results of studies with Arabidopsis mutants (rao3/big, rao4/pin-formed1, and rao5/multidrug-resistance1/abcb19), when the impairment of polar auxin transport resulted in an increase in AOX gene expression [66]. It was also shown that treatment of plants with the mitochondrial respiratory chain complex III inhibitor antimycin A blocked auxin signaling and induced the expression of the AOX1a gene [66]. Thus, the disruption of auxin homeostasis can lead to serious disturbances in cellular metabolism under the influence of unfavorable environmental factors. Our results suggest that reduced levels of active IAA following fluridone treatment may contribute to the decreased survival of seedlings exposed to freezing temperatures (Figures 3b and 5b).

As mentioned above, SA has been shown to participate in plant defense responses to abiotic and biotic stress factors [67,68]. Several studies have stated that increased SA production was associated with reduced auxin biosynthesis and transport [39,40,69]. Interdependence between IAA and SA may be mediated by shared transcription factors and protein kinase CK2 regulating both SA and auxin biosynthetic pathways. Supporting this connection, we observed that when control winter wheat seedlings were treated with fluridone, IAA levels decreased while SA content increased (Figure 5b,c).

It is known that SA signaling, like other hormonal signaling pathways, is not a simple linear pathway. SA interacts with several compounds, including other hormones, polyamines, ROS, and several others [34]. Therefore, it has been stated that SA directly binds to and inhibits CAT2, thereby promoting H_2O_2 levels [69]. In our work, there were no changes in catalase activity when SA increased in control seedlings treated with fluridone (Figures 4d and 5b). However, an increase in H_2O_2 levels was accompanied by an increase in SA levels (Figure 4a). It can be supposed that, in this case, SA contributed the ROS increase by inducing the succinate–quinone reductase activity (SDH, complex II in electron transport chain) in mitochondria [70]. SDH catalyzes the oxidation of succinate to fumarate and the reduction in ubiquinone to ubiquinol, and is a significant source of mitochondrial ROS production [70,71]. So, it can be assumed that the indirect effect of SA on the active IAA content was realized via the rise in mitochondrial ROS production. This was followed by H_2O_2 sulfenylation of Tryptofan Synthetase subunit 1 (TSB1), which is responsible for IAA synthesis, resulting in the inhibition of its activity [34]. Although SA can enhance antioxidant enzymes activities [72], indeed, when recording an increase in SA, we observed a parallel increase in guaiacol peroxidase activity in control seedlings treated with fluridone (Figures 4b and 5c). However, these changes did not prevent ROS increase. In addition, fluridone significantly reduced the synthesis of carotenoids, which are also antioxidants. Consequently, one can propose that the negative influence of fluridone on the winter wheat frost resistance was attributed to its pro-oxidant action, which is facilitated through various pathways, probably including SA, antioxidant enzymes, non-enzymatic antioxidants, and mitochondria. The opposite effect of tebuconazole was largely associated with the activation of the ABA-independent signaling pathway during cold hardening.

This treatment, however, could not completely exclude the negative effect of fluridone, as indicated by an increase in ROS and a decrease in carotenoids and IAA. The generalized scheme of the influence of tebuconazole and fluridone on frost resistance in winter wheat and the involvement of sugars, hormones, hydrogen peroxide and LPO products (MDA), catalase, and peroxidases are shown in Figure 6.

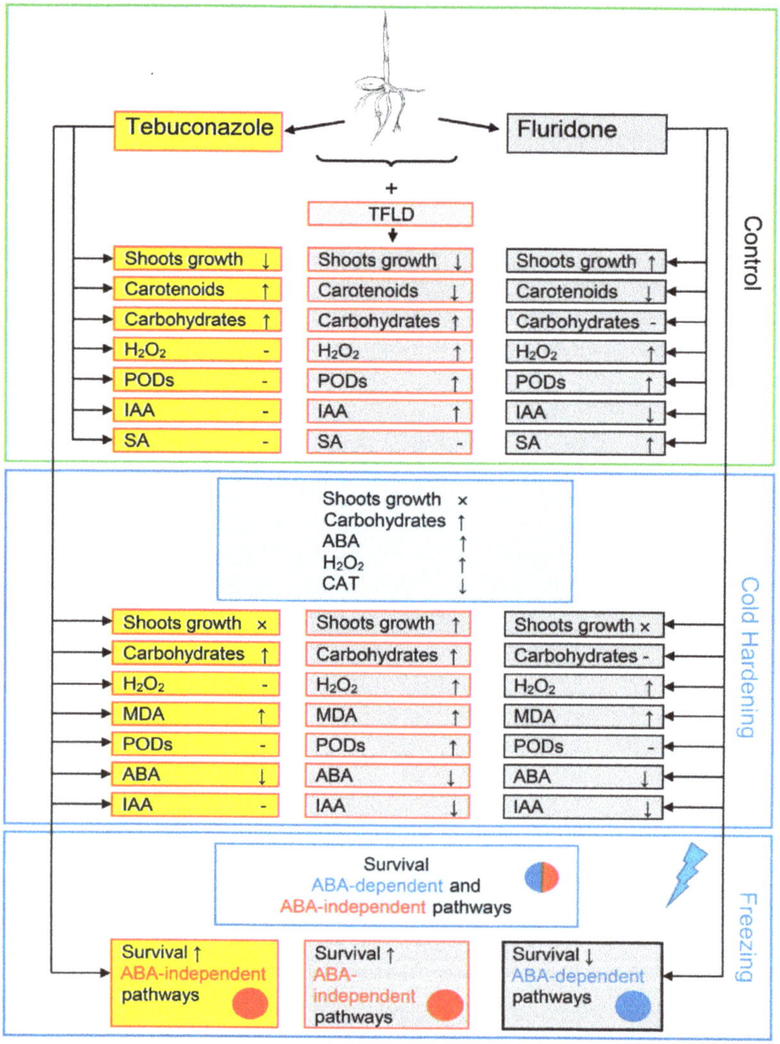

Figure 6. The generalized scheme of the influence of tebuconazole, fluridone, and its combined action (TFLD) on the increase in winter wheat frost resistance. Separate blocks show the influence of tebuconazole and FLD in optimal (Control, 24 °C), cold hardening (2 °C, 7 days), and freezing (−8 °C, 24 h) temperature conditions. ↑—parameter increase, ↓—parameter decrease, ×—growth stop, -—no parameter changes.

4. Materials and Methods

4.1. Plant Material, Fungicide and Inhibitor Treatments, and Growth Conditions

Etiolated winter wheat seedlings (*Triticum aestivum* L., 'Irkutskaya' variety) were used for the research. Winter wheat seeds were washed and disinfected with 0.1% potassium permanganate solution.

Some of the seeds were treated with an aqueous solution of the tebuconazole-based seed dresser (water-suspension concentrate) "Bunker" (1.5 μL g^{-1} of seeds) as detailed by Korsukova et al. [27]. The remaining seeds were treated with an aqueous solution of the tebuconazole (Tebuconazole PESTANAL®, Sigma-Aldrich, St. Louis, MO, USA) (water-suspension concentrate) (30 μg g^{-1} of seeds). For seed treatment, tebuconazole was dissolved in dimethyl sulfoxide (DMSO, Sigma-Aldrich, St. Louis, MO, USA).

Untreated seeds, seeds treated with Bunker, and those treated with tebuconazole were germinated on wet filter paper in water (FLD-) or FLD solution (FLD+) (5 mg L^{-1}) (Fluridon PESTANAL®, Sigma-Aldrich, St. Louis, MO, USA) in the dark at 24 °C in the incubator MIR-154 (Sanyo, Osaka, Japan) for up to 3, 5, and 10 days of age. FLD is a herbicide that slows down the synthesis of endogenous ABA by inhibiting the formation of carotenoids [45], and 0.1% DMSO solution was used for its preparation. The scheme of study of fluridone and tebuconazole influence on frost resistance in winter wheat seedlings is shown in Figure 7.

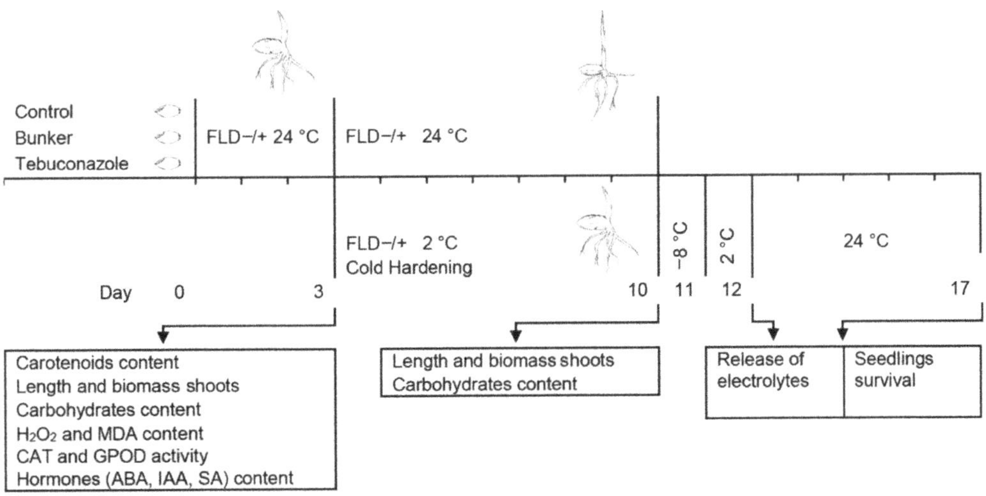

Figure 7. Scheme of study of effects of fluridone and tebuconazole on frost resistance in winter wheat seedlings. Control seeds (not treated at all), seeds treated with the tebuconazole-based protectant "Bunker" (1.5 μL·g^{-1} of seeds), and those treated with tebuconazole (30 μg·g^{-1} of seeds) were germinated in water or fluridone solution (FLD−/+) (5 mg·L^{-1}) for 3 days at 24 °C. Subsequently, they were kept at 2 °C for 7 days (Cold Hardening) or left to grow at 24 °C (Control). Cold-hardened seedlings were frozen at −8 °C for 24 h, and then thawed at 2 °C for 24 h, and after 5 days of recovery at 24 °C, their survival rate was determined. The arrows indicate sampling time for morpho-physiological and biochemical analyses.

4.2. Temperature Treatments and Determination of Frost Resistance of Seedlings

Three-day-old etiolated seedlings from untreated seeds, Bunker-treated seeds, and tebuconazole-treated seeds were cold hardened for 7 days at 2 °C in a MIR-154 incubator (Sanyo, Japan).

To evaluate the growth parameters of unhardened and hardened winter wheat seedlings, the coleoptile and shoot (first leaf) length, as well as fresh (FW) and dry (DW) shoot biomasses of 30 seedlings were measured. The degree of growth inhibition was determined as the ratio of the difference in coleoptile (or shoot) length of plants grown from untreated seeds and that of plants grown from treated seeds to the coleoptile length of plants grown from untreated seeds. Dry weight was measured after drying shoots of seedlings in an oven heated at 72 °C for 48 h.

To determine frost resistance, cold-hardened etiolated seedlings were frozen at −8 °C for 24 h in the incubator MIR-154 (Sanyo, Japan). The seedlings were then thawed at 2 °C for a day. After that, some of the seedlings were used to determine the release of electrolytes (V, %), and some of the seedlings were left to recover and grow in the dark at 24 °C for 5 days. The number of seedlings that survived after the freezing was expressed as a percentage of their total number. Evaluation of the release of electrolytes from wheat shoot tissues was carried out using the conductivity meter HI 8734 (Hanna Instruments Inc., Nusfalau, Romania) according to the formula $V = 100 \times (Lt/Lk)$, where Lt is the electrical conductivity of the sample after the freezing and subsequent incubation for 24 h at room temperature and Lk is the electrical conductivity of the same sample after a 15 min boil [73]. 0.1 g of shoots in 50 mL of distilled water was used for the analysis.

4.3. Determination of Carotenoids Content

Carotenoid contents in shoots (coleoptile and first leaf) were estimated according to Gavrilenko and Zhigalova [74]. The samples of shoots (50 mg) were homogenized with a mortar and pestle and extracted in 80% chilled acetone (1 mL) with the addition of $MgCO_3$. The homogenate was transferred into the test tubes (with the mortar and pestle being washed with 3 mL of acetone), left at 4 °C for 24 h, and then centrifuged at 7000× *g* for 10 min at 4 °C. The extract was poured into a 10 mL measuring tube. The pellet was mixed with a new portion of chilled 80% acetone (1 mL) and centrifuged again. The procedure was repeated several times until the pellet was completely discolored. Each time, the extracts were combined and the extract volume was adjusted to 7 mL with the solvent. Carotenoid content in the extract (C_{car}, mg·L^{-1}) was determined spectrophotometrically at a wavelength of 440 nm and calculated using the Wettstein's formula $C_{car} = 4695 \times D_{440} - 0268 \times C_{a+b}$. Pigment content was calculated using the formula $F = C \times V/(P \times 1000)$, where C is pigment concentration (mg·L^{-1}), V is volume of extract (mL), and P is plant fresh weight (g). Pigment content was expressed as mg·g^{-1} FW.

4.4. Determination of Water-Soluble Carbohydrate Content

The content of water-soluble carbohydrates in shoots was determined with 0.2% anthrone (Sigma-Aldrich, St. Louis, MO, USA) in concentrated sulfuric acid according, to Lyubushkina et al. [75], with some changes. The plant material was dried to a constant weight at 72 °C and ground in a mortar to powder. About 6–7 mg of dry ground material was transferred into test tubes, which were filled with boiling bidistilled water (10 mL). The tubes containing the extract were kept for a day in a thermostatic chamber at 24 °C. The supernatant was diluted 1:4 with bidistilled water, a 0.5 mL aliquot of the supernatant was taken, and double volume of anthrone solution in concentrated sulfuric acid was added (1 mL). After 15 min, the absorbance was measured at 620 nm. The content of water-soluble carbohydrates was calculated using a calibration curve based on sucrose, and expressed as a percentage (%) per g DW.

4.5. Determination of Hydrogen Peroxide and Lipid Peroxidation

Hydrogen peroxide content and lipid peroxidation were determined in extracts prepared from fresh powdered shoot tissue with 2.5 mL of 0.1% (*w*/*v*) trichloroacetic acid (TCA) per 0.5 g of tissue powder. The samples were previously frozen in liquid nitrogen and stored at −80 °C. The homogenate was centrifuged at 12,000× *g* for 15 min at 4 °C and supernatants were used for assays.

Hydrogen peroxide content was determined with xylenol orange [76]. The supernatant was diluted with 0.1% TCA in a ratio of 1:4 and a mixture of supernatant and reagent was prepared in a ratio of 1:1 (0.5 mL of each). The reagent composition was as follows: 0.5 mM $FeSO_4 \times (NH_4)_2SO_4 \times 6H_2O$, 0.5% H_2SO_4, 200 μM xylenol orange (AppliChem, Germany) and 200 mM sorbitol (Gerbu, Heidelberg, Germany). The solution was shaken, incubated at room conditions for 30 min, and absorbance of the reaction mixture was determined at 560 nm. The H_2O_2 content was expressed as μM of H_2O_2 per g FW based on the H_2O_2 calibration curve (ranged from 0.02 to 10 μM).

Lipid peroxidation was estimated using the thiobarbituric acid (TBA) method, based on production of the MDA [77]. To determine the TBA-reactive products content, 0.5 mL of supernatant and 1 mL of 0.5% (*w*/*v*) TBA (Dia-M, Moscow, Russia) in 20% TCA were mixed and incubated in a boiling water bath for 30 min and the reaction was stopped by cooling the samples in ice. The samples were then centrifuged for 5 min at 12,000× *g* (MiniSpin, Eppendorf, Hamburg, Germany) and absorbance of the supernatant at 532 and 600 nm was measured (SmartSpec Plus, BioRad, Hercules, CA, USA). The accumulated MDA was calculated using the extinction coefficient TBA 155 $mM^{-1}{\cdot}cm^{-1}$ after subtracting the nonspecific absorbance measured at 600 nm and expressed as nmols per gram of FW.

4.6. Determination of the Catalase and Guaiacol Peroxidase Activity

To measure the activity of catalase (CAT) and guaiacol peroxidase (GPOD) enzymes, fresh shoot samples (0.25 g) were frozen in liquid nitrogen and homogenized with 2.5 mL of 0.1 M K, Na-phosphate buffer (pH 7.8). After centrifugation at 15,000× *g* for 20 min, the resulting supernatant was used for the enzyme analyses. The supernatant was stored at 4 °C on ice in glass tubes.

The GPOD activity was determined by the increase in the optical density of the reaction medium due to guaiacol oxidation [78]. GPOD activity was measured in 0.1 M K, Na-phosphate buffer (pH 5.0), containing 10 mM hydrogen peroxide, 8 mM guaiacol (Sigma-Aldrich, St. Louis, MO, USA), and 0.05–0.1 mg protein at 470 nm for 3 min at 25 °C, using the extinction coefficient of 26.6 $mM^{-1}{\cdot}cm^{-1}$. GPOD activity was expressed as the rate of tetraguaiacol formation per 1 mg protein of the sample: $\mu mol{\cdot}min^{-1}{\cdot}mg^{-1}$ protein.

Catalase activity was measured polarographically, by the rate of oxygen evolution formed during the decomposition of H_2O_2 [79] using a Clark type oxygen electrode (Oxytherm system, Hansatech Inst., Pentney, UK). In a polarographic cell (volume 1.4 mL) at 25 °C with 0.1 M K, Na-phosphate buffer (pH 7.8) extract (0.1–0.2 mg protein) was added, then hydrogen peroxide to a final concentration of 0.003%. Catalase activity was expressed in nmoles of oxygen released per minute per 1 mg of protein: nmol $O_2{\cdot}min^{-1}{\cdot}mg^{-1}$ protein.

Protein concentration was determined using the Lowry method [80].

4.7. Determination of the Content of Endogenous Hormones

Hormone extraction and identification were carried out according to the method developed by Rudikovskii et al. [81]. The samples of shoots (1 g) were homogenized in liquid nitrogen. The internal standard was 4-methoxybenzoic acid. Extraction of phytohormones was carried out at 4 °C for 10 min in an ultrasonic bath, in 80% methanol with the addition of sodium diethylthiocarbamate as an antioxidant. Then, the samples were centrifuged (AllegraTM 64R centrifuge, Beckman Coulter, Brea, CA, USA) at 4 °C and 20,000× *g* for 20 min. The supernatant was removed and acidified with formic acid. The samples were purified using Sep-Pack C18 solid-phase extraction cartridges (Waters Corporation, Milfrod, MA, USA). Methanol was evaporated using an IR-1LT rotary evaporator (Labtex, Moscow, Russia) at 30 °C. The sample volume was adjusted to 5 mL with deionized water, acidified again with formic acid, and applied to an Oasis MAX cartridge (Waters Corporation, Milford, MA, USA). The cartridge was sequentially washed with 5% NH_4OH and 100% methanol. The sample was then washed with 2% formic acid dissolved in methanol. Then, the sample was evaporated to dryness using a rotary evaporator. Trimethylsilyl derivatives of phytohormones were obtained by heating the samples at 70 °C for 30 min with N,O-bis-trimethylsilylacetamide and hexamethyldisilazane in a SHC-40-02 SPU drying cabinet (Smolensk SKTB SPU, Smolensk, Russia). Indole-3-acetic, abscisic, and salicylic acid preparations (Sigma Aldrich, St. Louis, MO, USA) were used as hormone standards. Endogenous hormones were analyzed by gas chromatography using a 5973N/6890N MSD/DS chromatograph mass spectrometer (Agilent Technologies, Santa Clara, CA, USA). The analysis was performed using a quadrupole mass spectrometer with electron impact (EI) ionization at 70 eV in total ion current recording mode. HP-5MS capillary columns (30 m × 250 μm × 0.50 μm) were used to separate endogenous hormones (TMS derivatives of phytohormones). The stationary phase was phenyl-methyl-polyxylosan. The mobile phase was helium, gas flow rate 1 mL·min^{-1}. The operating temperatures were as follows: evaporator 250 °C, ion source 230 °C, detector 150 °C, and transfer line (connecting the chromatograph to the mass spectrometer) 280 °C. The scanning range was 41–450 a.m.u. The volume of the injected sample was 1 μL, the flow split was 5:1. Chromatography was performed in isocratic mode at 200 °C. Quantitative analysis of hormone content was performed using calibration curves constructed according to the corresponding standards. The NIST 05, Christie, and WILEY7 mass spectra libraries were used for the detection of plant hormones, as well as the comparison of the retention time with the retention times of standard compounds. The characteristic ions of TMS derivatives were: 4-metoxybenzoic acid (the internal standard)—267, 223, 193, 282, and 73; IAA—202, 319, 203, and 304; ABA—190, 183, 134, and 162; and SA—268, 267, 193, and 147.

4.8. Statistical Analyses

Statistical data processing was performed using SigmaPlot v. 14.0. In the figure captions, 'n' indicates the number of independent replicates. Data are presented as arithmetic mean (M) and standard deviation (±S.D.), or as median (Me) and interquartile range [25%; 75%]. Normality of distribution was checked using the Shapiro–Wilk test. Providing the distribution was normal, one-way analysis of variance and the procedure of multiple comparisons of means according to the Fisher LSD method were used to prove the presence of significant differences between the means. Providing the distribution was non-normal, the Kruskal–Wallis one-way analysis of variance on ranks and all pairwise multiple comparison procedures (Student–Newman–Keuls method) were used to prove the presence of significant differences. Differences between experimental data were considered statistically significant at $p \leq 0.05$.

5. Conclusions

Our research showed that fluridone, at concentrations that inhibit the synthesis of carotenoids and consequent ABA biosynthesis, partially reduced the growth-inhibiting effect of the tebuconazole-based "Bunker" seed dresser and tebuconazole in coleoptiles and the first leaf of unhardened seedlings. In addition, fluridone reduced the retardant effect of the protectant and tebuconazole on the first leaf in cold-hardened seedlings. Fluridone caused electrolyte leakage and reduced the survival rate of seedlings grown from untreated seeds to a greater extent than that of seedlings grown from "Bunker"-treated and tebuconazole-treated seeds. Thus, we can conclude that the survival of etiolated winter wheat seedlings under freezing conditions was ensured by gene activation through ABA-dependent and ABA-independent signaling pathways. Fluridone, which blocks ABA synthesis, weakened signal transduction through the ABA-dependent pathway, thereby reducing freezing tolerance. Additionally, fluridone's negative effect on the formation of cold and frost resistance mechanisms likely occurred through disruption of auxin homeostasis, specifically via SA and ROS, which blocked IAA synthesis and accelerated free IAA oxidation. A stable decrease in active IAA was observed with all fluridone applications and resulted in a decrease in signaling through the main auxin pathway TIR1/AFB-Aux/IAA, which controls the expression of many genes. Thus, fluridone can affect the homeostasis of other hormones, such as jasmonic acid, and through them, potentially, the energy metabolism of the plant cell, which is a key physiological process that determines the effectiveness of the stress response.

The protective effect of tebuconazole, both as a single treatment and in combination with fluridone, was mainly associated with the activation of ABA-independent signal transduction at low temperatures, which significantly increased the survival of winter wheat seedlings under freezing conditions. At the same time, partial compensation for the negative effect of fluridone on auxin homeostasis by tebuconazole, even under control conditions, was sufficient to stabilize plant metabolism, and prepared it for the action of stress environmental factors. The mechanisms of IAA metabolism and signaling remain to be fully elucidated and require further investigation. Such studies will not only provide new insights into how triazoles and fluridone affect plants, but will also enhance our understanding of auxin signaling pathways through TIR1/AFB-Aux/IAA and help identify new genes regulated by ARF.

Author Contributions: Conceptualization, N.V.D. and O.I.G.; methodology, A.V.K., I.V.L., N.S.Z., E.V.B., E.A.P., T.P.P., K.A.K., N.V.D., L.V.D. and O.I.G.; validation, A.V.K., I.V.L., E.V.B., N.V.D., L.V.D. and O.I.G.; formal analysis, A.V.K., I.V.L., E.V.B., N.V.D. and O.I.G.; investigation, A.V.K., I.V.L., N.S.Z., E.V.B., E.A.P., T.P.P., K.A.K. and O.I.G.; resources, A.V.K. and E.V.B.; data curation, O.I.G.; writing—original draft preparation, A.V.K., I.V.L. and O.I.G.; writing—review and editing, A.V.K., I.V.L., T.P.P., N.V.D., L.V.D. and O.I.G.; visualization, A.V.K., I.V.L., N.V.D. and O.I.G.; supervision, O.I.G.; project administration, N.V.D. and O.I.G.; funding acquisition, N.V.D. and O.I.G. All authors have read and agreed to the published version of the manuscript.

Funding: This research was funded by the federal program (Project State Registration No—122041100049-0) funded by the Ministry of Science and Higher Education of the Russian Federation for the Siberian Institute of Plant Physiology and Biochemistry of the Siberian Branch of RAS within the framework of the basic project "Study of genetic control, molecular and physiological-biochemical pathways of formation and regulation of plant resistance in rapidly changing environmental conditions".

Data Availability Statement: The original contributions presented in the study are included in the article. Further inquiries can be directed to the corresponding author.

Acknowledgments: The authors are grateful to Natalya A. Sokolova for the assistance in analyzing the content of endogenous hormones and to anonymous referees for helpful discussions. Collections of the Bioresource Center of Collective Use and Equipment at the Bioanalitika Center of Collective Use of the Siberian Institute of Plant Physiology and Biochemistry SB RAS (Irkutsk, Russia) were used.

Conflicts of Interest: The authors declare no conflicts of interest.

References

1. Theocharis, A.; Clément, C.; Barka, E.A. Physiological and molecular changes in plants grown at low temperatures. *Planta* **2012**, *235*, 1091–1105. [CrossRef] [PubMed]
2. Kosová, K.; Prášil, I.T.; Vítámvás, P.; Dobrev, P.; Motyka, V.; Floková, K.; Novák, O.; Turečková, V.; Rolčik, J.; Pešek, B.; et al. Complex phytohormone responses during the cold acclimation of two wheat cultivars differing in cold tolerance, winter Samanta and spring Sandra. *J. Plant Physiol.* **2012**, *169*, 567–576. [CrossRef] [PubMed]
3. Kosová, K.; Klíma, M.; Prášil, I.T.; Vítámvás, P. COR/LEA proteins as indicators of frost tolerance in Triticeae: A comparison of controlled versus field conditions. *Plants* **2021**, *10*, 789. [CrossRef] [PubMed]
4. Ejaz, I.; Pu, X.; Naseer, M.A.; Bohoussou, Y.N.D.; Liu, Y.; Farooq, M.; Zhang, J.; Zhang, Y.; Wang, Z.; Sun, Z. Cold and drought stresses in wheat: A global meta-analysis of 21st century. *J. Plant Growth Regul.* **2023**, *42*, 5379–5395. [CrossRef]
5. Hameed, A.; Farooq, T. Chapter 7—Triazole-based plant growth-regulating agents: A recent update. In *Advances in Triazole Chemistry*; Elsevier: Amsterdam, The Netherlands, 2021; pp. 169–185. [CrossRef]
6. Hof, H. Critical annotations to the use of azole antifungals for plant protection. *Antimicrob. Agents Chemother.* **2001**, *45*, 2987–2990. [CrossRef]
7. Popov, S.Y.; Dorozhkina, L.A.; Kalinin, V.A. *Fundamentals of Chemical Plant Protection*, 1st ed.; Art-Lion: Moscow, Russia, 2003; pp. 102–107. (In Russian)
8. Chizhova, S.I.; Pavlova, V.V.; Prusakova, L.D. Abscisic acid content and growth of spring barley plants treated with triazoles. *Russ. J. Plant Physiol.* **2005**, *52*, 93–98. [CrossRef]
9. Vettakkorumakankav, N.N.; Falk, D.; Saxena, P.; Fletcher, R.A. A crucial role for gibberellins in stress protection of plants. *Plant Cell Physiol.* **1999**, *40*, 542–548. [CrossRef]
10. Jiang, K.; Asami, T. Chemical regulators of plant hormones and their applications in basic research and agriculture. *Biosci. Biotechnol. Biochem.* **2018**, *82*, 1265–1300. [CrossRef]
11. Saito, S.; Okamoto, M.; Shinoda, S.; Kushiro, T.; Koshiba, T.; Kamiya, Y.; Hirai, N.; Todoroki, Y.; Sakata, K.; Nambara, E.; et al. A plant growth retardant, uniconazole, is a potent inhibitor of ABA catabolism in *Arabidopsis*. *Biosci. Biotechnol. Biochem.* **2006**, *70*, 1731–1739. [CrossRef]
12. Soumya, P.R.; Kumar, P.; Pal, M. Paclobutrazol: A novel plant growth regulator and multi-stress ameliorant. *Indian J. Plant Physiol.* **2017**, *22*, 267–278. [CrossRef]
13. Tafazoli, E.; Bey, C. Changes in endogenous abscisic acid and cold hardiness in *Actinidia* treated with triazole growth retardants. *J. Plant Growth Regul.* **1993**, *12*, 79–83. [CrossRef]
14. Zhang, C.; Wang, Q.; Zhang, B.; Zhang, F.; Liu, P.; Zhou, S.; Liu, X. Hormonal and enzymatic responses of maize seedlings to chilling stress as affected by triazoles seed treatments. *Plant Physiol. Biochem.* **2020**, *148*, 220–227. [CrossRef] [PubMed]
15. Li, J.; Geng, R.; Kong, X.; Li, L.; Zhang, Z.; Liu, R. Transcriptomic and physiological properties reveal the tolerance mechanism to difenoconazole toxicity in wheat (*Triticum aestivum* L.). *Ecotoxicol. Environ. Saf.* **2023**, *255*, 114787. [CrossRef] [PubMed]
16. Mohsin, S.M.; Hasanuzzaman, M.; Bhuyan, M.H.M.B.; Parvin, K.; Fujita, M. Exogenous tebuconazole and trifloxystrobin regulates reactive oxygen species metabolism toward mitigating salt-induced damages in cucumber seedling. *Plants* **2019**, *8*, 428. [CrossRef]
17. Mohsin, S.M.; Hasanuzzaman, M.; Nahar, K.; Hossain, S.; Bhuyan, B.; Parvin, K.; Fujita, M. Tebuconazole and trifloxystrobin regulate the physiology, antioxidant defense and methylglyoxal detoxification systems in conferring salt stress tolerance in *Triticum aestivum* L. *Physiol. Mol. Biol. Plants* **2020**, *26*, 1139–1154. [CrossRef]
18. Mohsin, S.M.; Hasanuzzaman, M.; Parvin, K.; Hossain, S.; Fujita, M. Protective role of tebuconazole and trifloxystrobin in wheat (*Triticum aestivum* L.) under cadmium stress via enhancement of antioxidant defense and glyoxalase systems. *Physiol. Mol. Biol. Plants* **2021**, *27*, 1043–1057. [CrossRef]
19. Morrison, M.J.; Andrews, C.J. Variable increases in cold hardiness induced in winter rape by plant growth regulators. *J. Plant Growth Regul.* **1992**, *11*, 113–117. [CrossRef]
20. Namjoyan, S.; Rajabi, A.; Sorooshzadeh, A.; AghaAlikhani, M. The potential of tebuconazole for mitigating oxidative stress caused by limited irrigation and improving sugar yield and root quality traits in sugar beet. *Plant Physiol. Biochem.* **2021**, *162*, 547–555. [CrossRef]
21. Pobezhimova, T.P.; Korsukova, A.V.; Dorofeev, N.V.; Grabelnych, O.I. Physiological effects of triazole fungicides in plants. *Proc. Univ. Appl. Chem. Biotechnol.* **2019**, *9*, 461–476. (In Russian) [CrossRef]
22. Upadhyaya, A.; Davis, T.D.; Walser, R.H.; Galbraith, A.B.; Sankhla, N. Uniconazole-induced alleviation of low-temperature damage in relation to antioxidant activity. *Hort. Sci.* **1989**, *24*, 955–957. [CrossRef]

23. Rademacher, W. Plant growth regulators: Backgrounds and uses in plant production. *J. Plant Growth Regul.* **2015**, *34*, 845–872. [CrossRef]
24. Ijaz, M.; Mahmood, K.; Honermeier, B. Interactive role of fungicides and plant growth regulator (Trinexapac) on seed yield and oil quality of winter rapeseed. *Agronomy* **2015**, *5*, 435–446. [CrossRef]
25. Matysiak, K.; Kaczmarek, S. Effect of chlorocholine chloride and triazoles—Tebuconazole and flusilazole on winter oilseed rape (*Brassica napus* var. *oleifera* L.) in response to the application term and sowing density. *J. Plant Prot. Res.* **2013**, *53*, 79–88. [CrossRef]
26. Zhang, Y.-J.; Zhang, X.; Chen, C.-J.; Zhou, M.-G.; Wang, H.-C. Effects of fungicides JS399-19, azoxystrobin, tebuconazloe, and carbendazim on the physiological and biochemical indices and grain yield of winter wheat. *Pestic. Biochem. Physiol.* **2010**, *98*, 151–157. [CrossRef]
27. Korsukova, A.V.; Borovik, O.A.; Grabelnych, O.I.; Voinikov, V.K. The tebuconazole-based protectant of seeds "Bunker" induces the synthesis of dehydrins during cold hardening and increases the frost resistance of wheat seedlings. *J. Stress Physiol. Biochem.* **2015**, *11*, 118–127. (In Russian)
28. Korsukova, A.V.; Grabelnych, O.I.; Borovik, O.A.; Pobezhimova, T.P.; Voinikov, V.K.; Dorofeev, N.V. The influence of the treatment of seeds by tebuconazole on the carbohydrates content and frost resistance of winter wheat and winter rye. *Agric. Chem.* **2016**, *7*, 52–58. (In Russian)
29. Pobezhimova, T.P.; Korsukova, A.V.; Borovik, O.A.; Zabanova, N.S.; Dorofeev, N.V.; Grabelnych, O.I.; Voinikov, V.K. The influence of tebuconazole and tebuconazole-based protectant "Bunker" on the functioning of winter wheat mitochondria. *Biochem. (Mosc.) Suppl. Ser. A Membr. Cell Biol.* **2020**, *37*, 224–234. (In Russian) [CrossRef]
30. Lazukin, A.; Pinchuk, M.; Korsukova, A.; Nikiforov, A.; Romanov, G.; Stepanova, O.; Grabelnych, O. Comparison of presowing wheat treatments by low-temperature plasma, electric field, cold hardening, and action of tebuconazole-based protectant. *Appl. Sci.* **2022**, *12*, 6447. [CrossRef]
31. Gusta, L.V.; Trischuk, R.; Weiser, C.J. Plant cold acclimation: The role of abscisic acid. *J. Plant Growth Regul.* **2005**, *24*, 308–318. [CrossRef]
32. Gamble, P.E.; Mullet, J.E. Inhibition of carotenoid accumulation and abscisic acid biosynthesis in fluridone-treated dark-grown barley. *Eur. J. Biochem.* **1986**, *160*, 117–121. [CrossRef]
33. Stetsenko, L.A.; Vedenicheva, N.P.; Likhnevsky, R.V.; Kuznetsov, V.V. Influence of abscisic acid and fluridone on the content of phytohormones and polyamines and the level of oxidative stress in plants of *Mesembryanthemum crystallinum* L. under salinity. *Biol. Bull.* **2015**, *42*, 98–107. [CrossRef]
34. Janda, T.; Szalai, G.; Pál, M. Salicylic acid signalling in plants. *Int. J. Mol. Sci.* **2020**, *21*, 2655. [CrossRef] [PubMed]
35. Rock, C.D.; Sakata, Y.; Quatrano, R.S. Stress signaling I: The role of abscisic acid (ABA). In *Abiotic Stress Adaptation in Plants*, 1st ed.; Pareek, A., Sopory, S.K., Bohnert, H.J., Govindjee, G., Eds.; Springer: Dordrecht, The Netherlands, 2010; pp. 33–73. [CrossRef]
36. Xue-Xuan, X.; Hong-Bo, S.; Yuan-Yuan, M.; Gang, X.; Jun-Na, S.; Dong-Gang, G.; Cheng-Jiang, R. Biotechnological implications from abscisic acid (ABA) roles in cold stress and leaf senescence as an important signal for improving plant sustainable survival under abiotic-stressed conditions. *Crit. Rev. Biotechnol.* **2010**, *30*, 222–230. [CrossRef]
37. Deepthi, V.P. Triazoles as plant growth regulators and stress protectants in fruit crops: A review. *Int. J. Agric. Sci. Vet. Med.* **2016**, *4*, 53–64.
38. Ahmad, I.; Kamran, M.; Ali, S.; Cai, T.; Bilegjargal, B.; Liu, T.; Han, Q. Seed filling in maize and hormones crosstalk regulated by exogenous application of uniconazole in semiarid regions. *Environ. Sci. Pollut. Res.* **2018**, *25*, 33225–33229. [CrossRef]
39. Grabelnych, O.I.; Polyakova, E.A.; Korsukova, A.V.; Zabanova, N.S.; Berezhnaya, E.V.; Lyubushkina, I.V.; Fedotova, O.A.; Stepanov, A.V.; Pobezhimova, T.P.; Dorofeev, N.V. Differently directional effects of tebuconazole-based protectant of seeds "Bunker" on the growth of winter wheat shoots and roots. *Bull. Irkutsk. State Univ. Ser. Biol. Ecol.* **2020**, *34*, 3–19. [CrossRef]
40. Finkelstein, R.R.; Gampala, S.S.L.; Rock, C.D. Abscisic acid signaling in seeds and seedlings. *Plant Cell.* **2002**, *14*, 15–45. [CrossRef]
41. Liu, L.; Cang, J.; Yu, J.; Wang, X.; Huang, R.; Wang, J.; Lu, B. Effects of exogenous abscisic acid on carbohydrate metabolism and the expression levels of correlative key enzymes in winter wheat under low temperature. *Biosci. Biotechnol. Biochem.* **2013**, *77*, 516–525. [CrossRef]
42. Borovik, O.A.; Pomortsev, A.V.; Korsukova, A.V.; Polyakova, E.A.; Fomina, E.A.; Zabanova, N.S.; Grabelnych, O.I. Effect of cold acclimation and deacclimation on the content of soluble carbohydrates and dehydrins in the leaves of winter wheat. *J. Stress Physiol. Biochem.* **2019**, *15*, 62–67.
43. Bakht, J.; Bano, A.; Dominy, P. The role of abscisic acid and low temperature in chickpea (*Cicer arietinum*) cold tolerance. II. Effects on plasma membrane structure and function. *J. Exp. Bot.* **2006**, *57*, 3707–3715. [CrossRef]
44. Chen, W.P.; Li, P.H. Membrane stabilization by abscisic acid under cold aids proline in alleviating chilling injury in maize (*Zea mays* L.) cultured cells. *Plant Cell Environ.* **2002**, *25*, 955–962. [CrossRef]

45. Allagulova, C.R.; Maslennikova, D.R.; Avalbaev, A.M.; Fedorova, K.A.; Yuldashev, R.A.; Shakirova, F.M. Influence of 24-epibrassinolide on growth of wheat plants and the content of dehydrins under cadmium stress. *Russ. J. Plant Physiol.* **2015**, *62*, 465–471. [CrossRef]
46. Kondhare, K.R.; Hedden, P.; Kettlewell, P.S.; Farrell, A.D.; Monaghan, J.M. Use of the hormone-biosynthesis inhibitors fluridone and paclobutrazol to determine the effects of altered abscisic acid and gibberellin levels on pre-maturity a-amylase formation in wheat grains. *J. Cereal. Sci.* **2014**, *60*, 210–216. [CrossRef]
47. Hoffmann-Benning, S.; Kende, H. On the role of abscisic acid and gibberellin in the regulation of growth in rice. *Plant Physiol.* **1992**, *99*, 1156–1161. [CrossRef] [PubMed]
48. Yang, D.; Wang, N.; Yan, X.; Shi, J.; Zhang, M.; Wang, Z.; Yuan, H. Microencapsulation of seed-coating tebuconazole and its effects on physiology and biochemistry of maize seedlings. *Colloids Surf. B Biointerfaces* **2014**, *114*, 241–246. [CrossRef]
49. Shaki, F.; Rezayian, M.; Ebrahimzadeh, H.; Niknam, V. Role of triazolic compounds in underlying mechanisms of plant stress tolerance: A Review. *Iran. J. Plant Physiol.* **2022**, *12*, 3943–3954.
50. Emenecker, R.J.; Strader, L.C. Auxin-abscisic acid interactions in plant growth and development. *Biomolecules* **2020**, *10*, 281. [CrossRef]
51. Du, H.; Wu, N.; Fu, J.; Wang, S.; Li, X.; Xiao, J.; Xiong, L. A GH3 family member, OsGH3-2, modulates auxin and abscisic acid levels and differentially affects drought and cold tolerance in rice. *J. Exp. Bot.* **2012**, *63*, 6467–6480. [CrossRef]
52. Eremina, M.; Rozhon, W.; Poppenberger, B. Hormonal control of cold stress responses in plants. *Cell Mol. Life Sci.* **2016**, *73*, 797–810. [CrossRef]
53. Mishra, S.; Kumar, S.; Saha, B.; Awasthi, J.; Dey, M.; Panda, S.K.; Sahooet, L. Crosstalk between salt, drought, and cold stress in plants: Toward genetic engineering for stress tolerance. In *Abiotic Stress in Plants*; Wiley: Hoboken, NJ, USA, 2016; pp. 55–86. [CrossRef]
54. Kim, J.-S.; Kidokoro, S.; Yamaguchi-Shinozaki, K.; Shinozaki, K. Regulatory networks in plant responses to drought and cold stress. *Plant Physiol.* **2024**, *195*, 170–189. [CrossRef]
55. Richter, R.; Bastakis, E.; Schwechheimer, C. Cross-repressive interactions between SOC1 and the GATAs GNC and GNL/CGA1 in the control of greening, cold tolerance, and flowering time in Arabidopsis. *Plant Physiol.* **2013**, *162*, 1992–2004. [CrossRef] [PubMed]
56. Richter, R.; Behringer, C.; Müller, I.K.; Schwechheimer, C. The GATA-type transcription factors GNC and GNL/CGA1 repress gibberellin signaling downstream from DELLA proteins and PHYTOCHROME-INTERACTING FACTORS. *Genes Dev.* **2010**, *24*, 2093–2104. [CrossRef] [PubMed]
57. Zheng, C.; Liu, C.; Ren, W.; Li, B.; Lü, Y.; Pan, Z.; Cao, W. Flower and pod development, grain-setting characteristics and grain yield in Chinese milk vetch (*Astragalus sinicus* L.) in response to pre-anthesis foliar application of paclobutrazol. *PLoS ONE* **2021**, *16*, e0245554-1. [CrossRef]
58. Zhang, S.-W.; Li, C.-H.; Cao, J.; Zhang, Y.-C.; Zhang, S.-Q.; Xia, Y.-F.; Sun, D.-Y.; Sun, Y. Altered Architecture and Enhanced Drought Tolerance in Rice via the Down-Regulation of Indole-3-Acetic Acid by TLD1/OsGH3.13 Activation. *Plant Physiol.* **2009**, *151*, 1889–1901. [CrossRef]
59. Zhang, Z.; Li, Q.; Li, Z.; Staswick, P.E.; Wang, M.; Zhu, Y.; He, Z. Dual regulation role of GH3.5 in salicylic acid and auxin signaling during Arabidopsis-Pseudomonas syringae interaction. *Plant Physiol.* **2007**, *145*, 450–464. [CrossRef]
60. Du, W.; Lu, Y.; Li, Q.; Luo, S.; Shen, S.; Li, N.; Chen, X. TIR1/AFB proteins: Active players in abiotic and biotic stress signaling. *Sec. Plant Abiotic Stress* **2022**, *13*, 1083409. [CrossRef]
61. Calderón Villalobos, L.I.A.; Lee, S.; De Oliveira, C.; Ivetac, A.; Brandt, W.; Armitage, L.; Sheard, L.B.; Tan, X.; Parry, G.; Mao, H.; et al. A combinatorial TIR1/AFB-Aux/IAA co-receptor system for differential sensing of auxin. *Nat. Chem. Biol.* **2012**, *8*, 477–485. [CrossRef]
62. Dharmasiri, N.; Dharmasiri, S.; Estelle, M. The F-box protein TIR1 is an auxin receptor. *Nature* **2005**, *435*, 441–445. [CrossRef]
63. Wojtaczka, P.; Ciarkowska, A.; Starzynska, E.; Ostrowski, M. The GH3 amidosynthetases family and their role in metabolic crosstalk modulation of plant signaling compounds. *Phytochemistry* **2022**, *194*, 113039. [CrossRef]
64. Ng, S.; De Clercq, I.; Van Aken, O.; Law, S.R.; Ivanova, A.; Willems, P.; Giraud, E.; Van Breusegem, F.; Whelan, J. Anterograde and retrograde regulation of nuclear genes encoding mitochondrial proteins during growth, development, and stress. *Mol. Plant* **2014**, *7*, 1075–1093. [CrossRef]
65. Van Aken, O. Mitochondrial redox systems as central hubs in plant metabolism and signaling. *Plant Physiol.* **2021**, *186*, 36–52. [CrossRef] [PubMed]
66. Ivanova, A.; Law, S.R.; Narsai, R.; Duncan, O.; Lee, J.H.; Zhang, B.; Van Aken, O.; Radomiljac, J.D.; Van der Merwe, M.; Yi, K.; et al. Functional antagonistic relationship between auxin and mitochondrial retrograde signaling regulates *Alternative Oxidase1a* expression in Arabidopsis. *Plant Physiol.* **2014**, *165*, 1233–1254. [CrossRef] [PubMed]

67. Ali, S.; Baloch, A.M. Overview of sustainable plant growth and differentiation and the role of hormones in controlling growth and development of plants under various stresses. *Recent Pat. Food Nutr. Agric.* **2020**, *11*, 105–114. [CrossRef] [PubMed]
68. Zhang, Y.; Fu, X.; Feng, Y.; Zhang, X.; Bi, H.; Ai, X. Abscisic acid mediates salicylic acid induced chilling tolerance of grafted cucumber by activating H_2O_2 biosynthesis and accumulation. *Int. J. Mol. Sci.* **2022**, *23*, 16057. [CrossRef] [PubMed]
69. Zhong, Q.; Hu, H.; Fan, B.; Zhu, C.; Chen, Z. Biosynthesis and roles of salicylic acid in balancing stress response and growth in plants. *Int. J. Mol. Sci.* **2021**, *22*, 11672. [CrossRef]
70. Jardim-Messender, D.; Margis-Pinheiro, M.; Sachetto-Martins, G. Salicylic acid and adenine nucleotides regulate the electron transport system and ROS production in plant mitochondria. *Biochem. Biophys. Acta* **2022**, *1863*, 148559. [CrossRef]
71. Quinlan, C.L.; Orr, A.L.; Perevoshchikova, I.V.; Treberg, J.R.; Ackrell, B.A.; Brand, M.D. Mitochondrial complex II can generate reactive oxygen species at high rates in both the forward and reverse reactions. *J. Biol. Chem.* **2012**, *287*, 27255–27264. [CrossRef]
72. Song, W.; Shao, H.; Zheng, A.; Zhao, L.; Xu, Y. Advances in roles of salicylic acid in plant tolerance responses to biotic and abiotic stresses. *Plants* **2023**, *12*, 3475. [CrossRef]
73. Borovik, O.A.; Grabelnych, O.I. Mitochondrial alternative cyanide-resistant oxidase is involved in an increase of heat stress tolerance in spring wheat. *J. Plant Physiol.* **2018**, *231*, 310–317. [CrossRef]
74. Gavrilenko, V.F.; Zhigalova, T.V. *Large Workshop on Photosynthesis*; Academa: Moscow, Russia, 2003; pp. 54–55. (In Russian)
75. Lyubushkina, I.V.; Polyakova, M.S.; Pomortsev, A.V.; Korsukova, A.V.; Zabanova, N.S.; Pobezhimova, T.P.; Grabel'nykh, O.I.; Voinikov, V.K. Efficiency of androgenesis in vitro in the culture of isolated winter wheat anthers: 1. The influence of the duration of low-temperature pretreatment of donor plants on respiration, carbohydrate content, and ROS levels in flowers from different parts of the spike. *Russian J. Plant Physiol.* **2023**, *70*, 213. [CrossRef]
76. Bellincampi, D.; Dipierro, N.; Salvi, G.; Cervone, F.; De Lorenzo, G. Extracellular H_2O_2 induced by oligogalacturonides is not involved in the inhibition of the auxin-regulated rolB gene expression in tobacco leaf explants. *J. Plant Physiol.* **2000**, *122*, 1379–1385. [CrossRef] [PubMed]
77. Heath, R.L.; Packer, L. Photoperoxidation in isolated chloroplasts: I. Kinetics and stoichiometry of fatty acid peroxidation. *Arch. Biochem. Biophys.* **1968**, *125*, 189–198. [CrossRef] [PubMed]
78. Garmash, E.V.; Velegzhaninov, I.O.; Grabelnych, O.I.; Borovik, O.A.; Silina, E.V.; Voinikov, V.K.; Golovko, T.K. Expression profiles of genes for mitochondrial respiratory energy-dissipating systems and antioxidant enzymes in wheat leaves during de-etiolation. *J. Plant Physiol.* **2017**, *215*, 110–121. [CrossRef] [PubMed]
79. Fedoseeva, I.V.; Gamburg, K.Z.; Varakina, N.N.; Rusaleva, T.M.; Tauson, E.L.; Stupnikova, I.V.; Borovskii, G.B.; Stepanov, A.V.; Davydenko, E.A.; Rikhvanov, E.G.; et al. The effect of sodium azide and 2,4-dinitrophenol on the development of thermotolerance and induction of Hsp101 in cultured *Arabidopsis thaliana* cells. *Russ. J. Plant Physiol.* **2008**, *55*, 225–231. [CrossRef]
80. Lowry, O.H.; Rosebrough, N.J.; Farr, A.L.; Randall, R.J. Protein measurement with the Folin phenol reagent. *J. Biol. Chem.* **1951**, *193*, 265–275. [CrossRef]
81. Rudikovskii, A.V.; Stolbikova, A.V.; Rudikovskaya, E.G.; Dudareva, L.V. Role of phytogormones in the formation of dwarf and tall Siberian crabapple (*Malus baccata* L. Borkh.). *Zemdirb.-Agric.* **2019**, *106*, 167–172. [CrossRef]

MDPI AG
Grosspeteranlage 5
4052 Basel
Switzerland
Tel.: +41 61 683 77 34

Plants Editorial Office
E-mail: plants@mdpi.com
www.mdpi.com/journal/plants